REANIMATING PLACES

For Anne Buttimer

Reanimating Places

A Geography of Rhythms

Edited by
TOM MELS
University of Kalmar, Sweden

Routledge
Taylor & Francis Group

LONDON AND NEW YORK

First published 2004 by Ashgate Publishing

2 Park Square, Milton Park, Abingdon, Oxon OX14 4RN
711 Third Avenue, New York, NY 10017, USA

Routledge is an imprint of the Taylor & Francis Group, an informa business

First issued in paperback 2016

British Library Cataloguing in Publication Data
Mels, Tom
    Reanimating places : a geography of rhythms. -
    (Re-materialising cultural geography)
    1. Geographical perception 2. Landscape assessment
    I. Title
    304.2'3

Library of Congress Cataloging-in-Publication Data
Reanimating places : a geography of rhythms / edited by Tom Mels.
        p. cm. -- (Re-materialising cultural geography)
    Includes index.

1. Geographical perception. 2. Landscape assessment. I. Mels, Tom. II. Series.

    G71.5.R42 2004
    304.2'3--dc22

                                                                2004004818

    ISBN 978-0-7546-4187-2 (hbk)
    ISBN 978-1-138-27533-1 (pbk)

Transfered to Digital Printing in 2013

Contents

List of Figures and Tables   *vii*
List of Contributors   *ix*
Preface   *xiii*
Acknowledgements   *xv*

PART I: INTRODUCTION

1   Lineages of a Geography of Rhythms
    Tom Mels   3

PART II: REANIMATING PLACE AND DISPLACEMENT

2   Sense of Place: Its Relationship to Self and Time
    Yi-Fu Tuan   45

3   The Stranger's Lifeworld: The Chinese Diaspora
    and Immigrant Entrepreneurs in Canada
    David Ley   57

4   Softly Heaves the Glassy Sea: Nature's Rhythms in an Era of
    Displacement
    Edmunds Bunkse   71

5   Rhythms and Identity in Boyne Valley Landscapes
    Gerry O'Reilly   87

PART III: REANIMATING URBAN LIFEWORLDS

6   Temporality and the Rhythms of Sustainable Landscapes
    Edward Relph   111

7   Grasping the Dynamism of Urban Place: Contributions from
    the Work of Christopher Alexander, Bill Hillier, and Daniel Kemmis
    David Seamon   123

8   Rent Rhythm in the Flamenco of Urban Change
    Eric Clark   147

9   Time-Space Rhythms and Everyday Urban Life
    Ann-Cathrine Åquist   161

PART IV: REANIMATING EMBODIED LANDSCAPES

10 Place and Identity: The Life of
    Marie de l'Incarnation (1599-1672)
    Anne Godlewska   175

11 Placing the Holy
    Gunnar Olsson   201

12 *Masquing* and Dancing the Theater of State: The 'Invention' of Britain
    as a Landscaped Body Politic
    Kenneth R. Olwig   215

13 Heritage Landscapes, Geographical Imaginations and Material Cultures:
    Tracing Ulster's Past
    Nuala C. Johnson   227

PART V: REANIMATING GEOGRAPHIES

14 Place-Making and Time
    Robert David Sack   243

Index   255

List of Figures and Tables

Figures

5.1 The Boyne Valley and catchment area   88
5.2 Newgrange, main ceremonial burial chamber   94
5.3 Newgrange, entrance stone and light box of main chamber   95
5.4 View of Mellifont Abbey   99
5.5 Trim town skyscape   100
6.1 The future becomes the past   116
6.2 The closure of temporality: a farm on the edge of Toronto   117
7.1 San Francisco waterfront site to be developed   130
7.2 Photograph of a wooden model of the San Francisco waterfront project after 89 projects   131
7.3 The San Francisco waterfront site after completion of five projects   132
7.4 The French village of Gassin   134
7.5 Gassin's convex map   135
7.6 Gassin's axial map   136
7.7 Gassin's 'deformed wheel'   138
8.1 Lilla Trädgårdsgatan 1961   149
8.2 Lilla Trädgårdsgatan 1987   150
8.3 Modified conceptual outline of a development cycle, given increasing potential land rent (PLR)   153
8.4 Building value (BV) capitalized land rent (CLR), and potential land rent (PLR) of Stralsund 33   154

Tables

7.1 Christopher Alexander's 'Seven Rules for Urban Design'   128
8.1 Four abstract concepts of land rent distinguished in terms of time and (intensity and type of) use, and two concrete categories of cash flow distinguished in terms of time   158
9.1 Monday time-diaries of parents in a Swedish household with three young children   169

List of Contributors

Ann-Cathrine Åquist is an Associate Professor of Human Geography at the University of Örebro, and a member of the Center for Housing and Urban Research, Sweden. She is the author of *Time-geography and Social Theory* (in Swedish, Lund University Press, 1992). Her research and teaching interests feature environmental issues, planning and everyday life, as well as the history of geography.

Edmunds Bunkse is a Professor of Geography at the University of Delaware, USA, and is also an Adjunct Professor at the University of Latvia in Riga. He has written extensively on humanistic themes and landscape, including a book (in Latvian) on the Berkeley School of landscape geography (1998). In the spring of 2004 Johns Hopkins University Press published his *Geography and the Art of Life* – an existential or lifeworld geography. Bunkse holds an honorary doctorate (Dr. *honoris causa*) from the University of Latvia and is a Foreign Member of the Latvian Academy of Sciences.

Eric Clark is a Professor of Geography at Lund University, Sweden, and executive committee Member, ISIS (International Small Island Studies Association). He is the author of *The Rent Gap and Urban Change* (Lund University Press, 1987), co-editor of *Identity Dynamics and the Construction of Boundaries* (Nordic Academic Press, 2002), and editor of the journal *Geografiska Annaler B: Human Geography*.

Anne Godlewska is a Professor of Geography at Queen's University at Kingston, Canada. She has published numerous articles in French and English. Her most recent book is *Geography Unbound: French Geographical Science from Cassini to Humboldt* (University of Chicago Press, 1999). Godlewska co-edited *Geography and Empire* (Blackwell, 1994).

Nuala C. Johnson is a Lecturer in Geography at the School of Geography, University of Belfast, Northern Ireland. She has published a range of journal articles on national identity, the heritage industry, and social memory. Johnson is the author of *Ireland, the Great War and the Geography of Remembrance* (Cambridge University Press, 2003) and she is co-editor of *A Companion to Cultural Geography* (Blackwell, 2003).

David Ley is Canada Research Chair in Geography at the University of British Columbia, Vancouver, Canada, and Co-Director of the Vancouver Centre of the Metropolis Project, a national and international network for research on race, ethnicity and immigration. His books include *A Social Geography of the City* (Harper and Row, 1983), *The New Middle Class and the Remaking of the Central City* (Clarendon, 1996), and *Neighbourhood Organizations and the Welfare State* (with Shlomo Hasson, The University of Toronto Press, 1994).

Tom Mels is a Lecturer in Human Geography at Kalmar University, Sweden. His teaching and research focus on ideologies of nature and landscape, environmental justice, and critical geographies of modernity. He is the author of *Wild Landscapes: the Cultural Nature of Swedish National Parks* (Lund University Press), and a forthcoming history of Swedish geography (co-authored with Anne Buttimer).

Gunnar Olsson is a Professor of Geography, Emeritus, at Uppsala University, Sweden. Among his books are *Distance and Human Interaction* (Regional Science Research Institute, 1965), *Birds in Eggs/Eggs in Bird* (Pion, 1980), *Lines of Power/Limits of Language* (University of Minnesota Press, 1991), and a collection of interventions on Swedish ideology entitled *Antipasti* (Korpen, 1990). Olsson has edited or co-edited several books, including *Philosophy in Geography* (Reidel, 1979), *A Search for Common Ground* (Pion, 1980), and *Poste restante* (Nordplan, 1996).

Kenneth Robert Olwig is a Professor of Landscape Theory at the Swedish University of Agricultural Sciences in Alnarp, Sweden. He is the author of *Nature's Ideological Landscape* (Allen and Unwin, 1984) and *Landscape, Nature, and the Body Politic* (University of Wisconsin Press, 2002). Most recently, Olwig has co-edited *Nordscapes: Landscapes and Regional Identity on Europe's Northern Edge* (University of Minnesota Press, forthcoming).

Gerry O'Reilly is a Lecturer in Geography at St. Patrick's College, Dublin City University, Ireland. His teaching and research interests include sustainable development, political and economic geography. O'Reilly authored *Ceuta and the Spanish Sovereign Territories: Spanish and Moroccan Claims* (IBRU Press, 1994) and was involved in *Sustainable Landscapes and Lifeways* (Cork University Press 2001, edited by Anne Buttimer).

Edward Relph is a Professor of Geography at the University of Toronto, Canada. He is the author of several books and reports, including *Place and Placelessness* (Pion 1976), *Rational Landscapes and Humanistic Geography* (Croom Helm, 1981), *The Modern Urban Landscape* (Johns Hopkins University Press, 1987), and *The Toronto Guide* (Centre for Urban and Community Studies, University of Toronto, 2003).

Robert David Sack is Clarence J. Glacken and John Bascom Professor of Geography and Integrated Studies at the University of Wisconsin-Madison, USA. Sack is the author of *Conceptions of Space in Social Thought* (University of Minnesota Press, 1980), *Human Territoriality* (Cambridge University Press, 1986), *Place, Modernity, and the Consumer's World* (Johns Hopkins University Press, 1992), and *Homo Geographicus* (Johns Hopkins University Press, 1997). He recently edited *Progress: Geographical Essays* (Johns Hopkins University Press, 2002) and his new book is entitled *A Geographical Guide to the Real and the Good* (Routledge, 2003).

David Seamon is a Professor of Architecture at Kansas State University, USA. His teaching and research emphasize a phenomenological approach to place, architecture, environmental experience, and environmental design as place making. He has authored *A Geography of the Lifeworld* (St. Martin's Press, 1979) and edited or co-edited

Dwelling, Place and Environment (Columbia University Press, 1989), *Dwelling, Seeing, and Designing* (State University of New York Press, 1993), and *Goethe's Way of Science* (State University of New York Press, 1998). Seamon is editor of the *Environmental and Architectural Phenomenology Newsletter.*

Yi-Fu Tuan is the John K. Wright and Vilas Professor of Geography, Emeritus, at the University of Wisconsin-Madison, USA, and the author of numerous articles and books, including *Topophilia* (Prentice-Hall, 1974), *Space and Place* (University of Minnesota Press, 1977), *Morality and Imagination* (University of Wisconsin Press, 1989), *Escapism* (Johns Hopkins University Press, 1998), and *Who am I?* (University of Wisconsin Press, 1999).

Preface

The Finnish philosopher Georg Henrik von Wright once remarked that during the nineteenth century, science gradually came to be recognized as a 'force of production.' It thereby constantly risked being streamlined and reified in a way which strengthened prevailing orthodoxies. Adorno called this 'the academic industry,' in which scholars were disciplined to practice repetition and duplication rather than deviation. One might of course wonder whether this image of mass-production still holds sway in the post-positivistic knowledge industry. Are we not all geared towards newness? Are not the old knowledge-economies of scale replaced by those of scope? Perhaps so, but academic 'turns' also reveal that what was once familiar can be eclipsed at another time. I feel that in the late modern academic industry, the exchange value of an idea tends to increase no longer simply by closeness to the familiar, but perhaps more often by means of repetition (return) cloaked as novelty. Furthermore, there is a certain colonial quality to the very announcement of newness. Declarations of independence at the self-proclaimed research 'frontiers,' after all, tend to hold the Phoenix promise of illuminated brightness as well as the Faustian torments of imperial trenching. One challenge therefore remains to tease out just how much of the present 'newness' (e.g. the corporeal, contextual, cultural, spatial, temporal turns) is reified, packaged, and sold as such by industrial demands.

Where in the current moment in the rhythm of geography could the present volume be located? While connecting to various contemporary 'turns,' *Reanimating Places* also openly acknowledges the resilience of earlier, unresolved themes and worries sensed in the undogmatic pulse of humanism. Throughout the volume, there is a recurring recognition of the importance of lived, everyday, thinking, feeling, contextually embedded, and struggling human subjects. This recognition reflects a now rather widespread impatience with the reified conceptualizations of time, space, people and lifeworlds in mainstream science. Against such reification, we advance contextual and non-dualistic notions of time-space rhythms. As will become clear, while the authors of the book diverge in their understanding and explanatory seizure of how places and landscapes can be reanimated by invoking time-space rhythms, our common point of departure was Anne Buttimer's innovative work on this theme. I say point of departure because I continue to think of the notions of time-space rhythms and reanimation as open passages rather than an accomplished journey.

Each chapter in the book is devoted in quite different ways to accounts of time-space rhythms that accentuate their material, discursive, and lived plurality. While affording fairly extensive insights into a variety of time-space rhythms, the present collection of essays does certainly not pretend to be the complete word on this vast topic. On the contrary, the outcome remains open and incomplete. As a glance on the contents of this book will readily reveal, many of the chapters are written by authors known for their existentialist and phenomenological predisposition, although Marxian and feminist angles also shine through in some of the chapters. My initial and perhaps somewhat optimistic intention, the marks of which can be traced in the introductory chapter, was to include a wider variety of critical perspectives on rhythms in the volume. Notwithstanding the volume's ultimately more limited scope, I hope that the result will be read in the spirit of Anne's own oeuvre: as an invitation to further dialogical adventures in the reanimation of places within human geography and beyond. In this dialogical gesture, I imagine, lies one of the core values of the book. Equally important, it is in that dialogical sense we dedicate the book to Anne and her lifelong adventures in humanistic geography.

As the initiator of this volume, I would like to extend gratitude to the authors, not only for their solid dedication to completing their essays under demanding academic circumstances of time-space compression, but also for the numerous encouraging exchanges with their now somewhat less inexperienced editor. I also greatly appreciate the support and editorial advice of Valerie Rose at Ashgate and the anonymous reviewers who provided comments on an earlier draft of the volume. My own work on the project emanated from a period of leave at the Department of Geography, University College Dublin. I would like to thank Anne Buttimer, Gerald Mills, Willy Nolan, Gerry O'Reilly, and Anngret Simms for their unceasing generosity and hospitality. At home, Torsten Hägerstrand and Gunnar Olsson fired my imagination about material and symbolic time-space rhythms in a double dialogue that forced me to carve out my own version. Finally, the conventions of the academic industry remind me to acknowledge the Swedish Foundation for International Cooperation in Research and Higher Education (STINT) for a post-doctoral scholarship during my year in Dublin, and the Swedish Research Council for supporting my ongoing research, part of which is reflected in this book.

Tom Mels, Winter 2003

Acknowledgements

The authors and publisher would like to thank the following for their kind permission to reproduce illustrations: Christopher Alexander table 7.1, figures 7.1, 7.2, 7.3; Cambridge University Press figures 7.4, 7.5, 7.6, 7.7.

PART I
INTRODUCTION

Chapter 1

Lineages of a Geography of Rhythms

Tom Mels

Prologue

Repetitions, movements, cycles, intervals, serenity. 'Space melts like sand running through one's fingers. Time bears it away and leaves me only shapeless shreds' (Perec 1997, 91). The fragmentation of space by memory. I still remember that bedroom window overlooking and overhearing Mellifont Avenue in Dun Laoghaire at rush hour. Its dense blend of maritime traffic and subterranean rumbling trains, scaffolds of gentrification, teenagers balancing on skateboards, Catholics on their way to church at Easter, and cars shifting gears. I remember suburbanites hurrying home from work through salty whipping rains, ignoring political campaigners during the general elections, and friends socializing at the pub just around the corner. I remember wondering what the politico-economic rhythm of the 'Celtic Tiger' might mean for those homeless addicts gathering near the pier. I remember the sculptural immobility of the Irish Sea at night-time. Seemingly motionless, but sandy and alive at dawn. Soon I got familiar with these spectacles and their multifarious visual, haptic, symphonic, natural, social, real-and-imagined rhythms around our temporary new home. At first sight, there was nothing hidden about the place and its habitual ebb and flow, its religious, economic, political, circadian (approximately 24-hour) and circaceptan (seven-day) cycles. Yet these rather trivial rhythms seemed impossible to register. How could I weave 'a net of words' sensitive enough to catch 'the rhythm of that sound-dance that bends and moves without ever destroying the penumbrae between external and internal, subject and object, body and soul?' (Olsson 1978, 114). No decimal notation of time, no geometrical commandment, no camera or tape-recorder could easily articulate the experience of this lifeworld. Obviously, seen as an object, cut off from the spectator behind the window, I could conjure up some admirable theoretical descriptions, and measures designed to enclose and master beings or objects, but not the enveloping open grounds of existence. Moreover, for all their noises, smells, openly visible qualities, I always suspected that these spectacles concealed as much as they revealed of the principles, practices, and powers at work in my rhythmic environment: those who had abandoned the site, and those waiting to enter it. The rhythmicity of everyday activities and occurrences may very well exist unnoticed or experienced without being understood. Whoever wants to define rhythms with dogmatic epistemology easily misses its object.

Whether or not we recognize the rhythmicity of the world, and whatever theoretical conclusions we draw from its complexity, human beings have always been rhythm-makers as much as place-makers. Past and present arhythmic conflicts and polyrhythmic dialogues are equally children of their historical time and place. In this pedestrian geography of rhythms, the theater set experienced from my window was not solely an immediate here-and-now, visual thing projected on my mind's eye. Woven into the cosmological time of nature's waves and cycles at Dun Laoghaire's coast were the different qualities and ordering powers associated with subjective experiences of duration and space, including the timeframes of workplace, school, dwelling, and homelessness. Indeed, the social construction of time-space rhythms comes forward as a braiding of domains, including human interaction with rhythms of the natural environment, the world of socialization, technological infrastructure and equipment, symbolic communication and the production of meaning.

Standard English language dictionaries suggest that rhythm is a rather complex term. The concept originates from the Greek *rhythmos*, meaning 'measure, rhythm, measured motion,' a word akin to the verb *rhein*, or 'to flow.' Rhythm occurs in a wide range of circumstances, and can be measured in various different ways. It 'is applicable to sound in poetry and music and also to any recurrent sound, movement, arrangement, or condition in virtually any sphere. Sometimes the word connotes little more than regular alternation,' while in other cases 'it suggests subtlety and variation in recurrence' or 'a recurrence pattern too varied to be easily grasped' (Merriam-Webster 1993: rhythm). In physical geography and geology, rhythm is sometimes used to describe 'regularity in the way something is repeated in space' (OED 1989: rhythm). While for obvious reasons rhythm is essential to musicians, poets, and dancers, how could rhythms be of any interest to human geographers? And if so, as this collection of essays concedes, how would we grasp those rhythms?

The Dun Laoghaire account suggests that even a spur-of-the-moment glance through a window shows how deeply various rhythms structure everyday life and experience of place (Lefebvre 1996; Parkes and Thrift 1979; Zerubavel 1981). Everyday life, our day-to-day events and actions, may indeed be seen as a spatial as much as a temporal term, permeated as it is by a multiplicity of rhythms. In a passage from her acclaimed essay, aptly entitled 'Grasping the Dynamism of Lifeword,' Anne Buttimer offered a more theoretical contribution to the understanding of time-space rhythms:

> To record behavior in an isometric grid representing space and time is only an opening onto the horizons of lived space and time. Neither geodesic space nor clock/calendar time is appropriate for the measurement of experience. The notion of rhythm may offer a beginning step toward such a measure. Lifeworld experience could be described as the orchestration of various time-space rhythms: those of physiological and cultural dimensions of life, those of different work styles, and those of our physical and functional environments. On a macrolevel one is dealing with the synchronization of movements of various scales, taking a sounding, as it were, at the particular point where our own experience has prodded us to explore (Buttimer 1976, 289).

While one may, as I will below, quarrel about which way of representing time and space is appropriate for various aspects of everyday life experience, this remains a profound insight. Buttimer's passage questions at once the appropriateness of positivist method and brings out the material and mental complexity of experience. The idea of time-space rhythms reveals a complexly contested world of sensory experiences and representations of time-space. It prepares a central place for the understanding of spatiotemporal patterns, routines, repetitions, scales, and flows. The importance of assessing these rhythms lies in its ability to decipher the dynamic nature of lifeworlds – those busy and often unquestioned junctions connecting worlds of objective affairs and subjective experiences – and thereby of social space. For subjects and things are necessarily embedded in particular places, and it is only in those places that the rhythmic dialogues between nature and society, symbol and substance, body and culture are staged.

Buttimer's passage alludes to a rather expansive and scalarly complex notion of lifeworld as an orchestration of time-space rhythms. The influence of Edmund Husserl's phenomenology can be felt in her emphasis on how the deepest levels of human experience are as spatial as they are temporal and need to be explored as spatiotemporal configurations. Husserl also highlighted the extraordinary importance of the physical and lived body in his exploration of lifeworld (Zahavi 2003). In turn, Buttimer's idea of rhythms can be linked to the body and place, discourse, social space, and dwelling (all of which are, of course, interconnected).

First, rhythms are bound up with place and corporeality. An examination of individual bodies, with all their specificities of shapes, abilities, and sex/gender, is enough to unearth that they encompass a great physiological whole of rhythmic spatio-temporal orchestration (Adam 1990, 73-74; Bachelard 1994, 2000; Buttimer 1987; Longhurst 2001; Merleau-Ponty 2002; Rodaway 1994; Sartre 1949; Whitehead 1929). In a piece published some years after Buttimer's essay, Yi-Fu Tuan connects the measurement and experience of distance in time and space with bodily movement (cf. Tuan, this volume). Enshrined in this description of conscious and unconscious corporeal rhythms is a common inclination to temporalize space:

> The rhythms we are conscious of are the longer cycles of bodily needs for food and rest; and in general, life is a succession of stresses and strains and their resolution. Repetitious or recurrent time is a common measure of distance. How far is it from one village to another? If the distance is short it can be given as so many paces. Walking is a rhythmic process the individual units of which can be added up. Giving distance as *x* paces is not an abstraction for the paces are intimately felt both as the pendulum-like swing of steps and as the sum of cumulative muscular efforts. Great distances are judged by rhythms in nature and particularly by the alternations of day and night. It is so many 'sleeps,' days, or moons from one place to another. The daily cycle has pre-eminence in distance estimations because it matches the human biological rhythm of wakefulness and sleep, and because the alternation of day and night is a highly visible drama. Nature's linear dimensions, other than the human body itself, such as the height of trees or the breadth of a stream, have rarely been used to designate distance: nature's periodicities rather than lengths provide the yardstick (Tuan 1978, 13-14).

We walk, talk, eat, sleep, work, think, and communicate in a rhythmic manner while 'our consciousness has a profoundly temporal gear with accumulated memories reaching backwards in time and imaginations reaching into the future. Impressions, imaginations and feelings follow upon each other in an unceasing series of *before* and *after*. Therefore, it is certainly not far-fetched to understand our interior as a movement in time, even when we remain completely motionless in space' (Hägerstrand 1991, 141). Corporeality is essential to place and lifeworld, because we experience objects, their place and our own place *with* our lived-living body. Place, in this respect, 'arises *within the withness* essential to the body's primitive prehensions and repetitions of its environing world. Just as we are always with a body, so, being bodily, we are always within a place as well. Thanks to our body, we are in that place and part of it' (Casey 1997, 214).

Second, there is more to rhythms (and place) than the body or individual experience. It extends into social space (see below). As Buttimer points out, corporeal rhythmic nodes are connected to daily labor, vocational meaning, and cultural context, i.e. the composite existential position or 'horizon' of knowledgeable and capable individuals. For those working nightshift, dinner might be routinely served at the break of dawn, noon might be bedtime, and evening the time for commuting to the factory. Individuals tend to build up habitual and pre-discursive movements or time-space routines, fusing and interacting with other individuals' routines in city streets, buildings or elsewhere as what David Seamon studied as a phenomenology of 'place ballets' (Seamon 1979, 56; cf. Clark, this volume). Not only are time-space rhythms of the body, meaning, and voluntary or enforced behavior connected, they are also thoroughly embedded in the cultural context of social space (Buttimer 1968, 142; 1969, 419; 1974, 24-25), and hence orchestrated by imperatives and constraints of situated knowledges and power inequalities (Foucault 1981; Pickles 1987; Pile and Thrift 1995). By implication, 'the whole of (social) space proceeds from the body, even though it so metamorphoses the body that it may forget it altogether – even though it may separate itself so radically from the body as to kill it … The analysis of rhythms must serve the necessary and inevitable restoration of the total body' (Lefebvre 1991b, 405).

Third, rhythms are connected to particular discourses, geographical imaginations and modes of representation. The metamorphosis of the body through the deliberate creation of a public time-space is mediated through differences of gender, class, age, or vocation, from within overarching discourses such as those of nationality, ethnicity, religion, or tradition. These discourses are patterned by partial value propositions and more or less powerful modes of symbolic communication (Buttimer 1982). In artistic composition as much as in geographical discourse, rhythm is often orchestrated through the regular recurrence of metaphorical style. Repetitive rhythms involve such things as geographical association or dissociation and temporal imaginations of past, present, and future. For instance, nationalists usually treat their nation as an organic home, a garden with avowedly natural boundaries and roots in the past, thereby repetitively normalizing their own discourse. As consciously or unconsciously repeated, overt or covert modes of structuring and reproducing the world, symbols, maps, and metaphors also tend to encourage particular dominant rhythms and routine ways of organizing human spaces and actions, while excluding, controlling, or masking the rhythms of others (Buttimer 1978; 1980; 1989; 1993; Buttimer et al. 1999; cf. Godlewska 1995).

Fourth, rhythms are implicated in the issue of dwelling. Dwelling is not bound to a particular scale, but traverses scales from the domestic household to the globe. Place-bound social practices, coded gestures, metaphorical styles, technological applications, and experiences are at the same time constitutive of rhythms that operate over a variety of spatial and temporal scales. In a time when overcoming the ancient separation and dissociation between the 'local' lifeworld and 'global' totality has earned recognition in academic discourse and politics, this seems indeed more important than ever (Buttimer 1990). Ultimately, the theme of rhythm will bring us from the local scale to a more global liberation cry (*cri de coeur*) of humanity, a *Poesis* of emancipatory reflection on the relationships between humanity and the earth. For Anne Buttimer, this involves the challenge of developing sustainable modes of dwelling. To dwell is to reach beyond mere inhabiting and organizing space. It means to consider the earth as an abode, 'to see one's life as anchored in human history and directed toward the future, to build a home which is the everyday symbol of a dialogue with one's ecological and social milieu' (Buttimer 1976, 277; cf. Buttimer 2001, 10; Bachelard 1994, on home as a metaphor for humanness).

Unsurprisingly, dwelling is fraught with difficulties. While dwelling is centered on 'home,' care and affection, the hold of contemporary home-creation on the desire for domination remains strong. The environmental dialogue to which Buttimer alludes often ends up in calculating monologues, naturalizing themselves, but deaf to the social and environmental havoc they produce (Evernden 1985). Purified 'natural' rhythms have long been transposed into moral regimes commending 'normality' and what is considered 'healthy' or 'sane.' Sometimes this leads to nostalgic yearnings for a 'return' to nature and natural rhythms (Rifkin 1987; Young 1988). Yet Buttimer's emphasis on dialogue, dynamism, and orchestration suggests that these 'natural' rhythms cannot be divorced from 'social' ones. There is no such thing as a pre-given set of rhythms, independent of history and systems of valued reference (cf. Buttimer 1974; Lash et al. 1998). This is not to reduce natural rhythms to social meaning, to deny that nocturnal sleep follows diurnal exhaustion, that ebb follows flow, or that life processes of growth are followed by decay. It is rather to question the existence of a prehistoric 'natural' rhythm to which we can somehow return. Human existence has always involved a plurality of encounters with material rhythms of this kind, always a reworking with shifting social roles and meanings. Rhythms are never stumbled upon independently of social context or of the organizing practices and technological devises that speed them up or slow them down, reverse them or magnify their current.

It seems to me that if authentic dwelling places cannot be revived from the past or from nature (however defined), rhythms and dwelling cease to be self-evident and become susceptible to manipulation and distortion as well as to dialogue. This is why dwelling can never really be a convenient verb or concrete noun, but remains a doubt. The open-endedness is deeply embedded in the etymology of the word, for the Old English *dwellan* means to linger as well as to go astray (Casey 1993, 114).

Introduction: Geographical Rhythms and the Rhythms of Geography

My focus on Buttimer's work is deliberate, not least because her original humanistic geography offers a convenient and sophisticated vantage point from which to tease out some of the theoretical and practical complexities involved in a rhythmic view of the world. Thus far, I have invoked rhythm as an essentially conceptual tool conveying the intrinsic dynamism of the world, i.e. what might be called geographical rhythms. What I have been less occupied with is where and how Buttimer's idea of a geography of rhythms may be connected to the rhythms of geography, i.e. to recurring intellectual concerns in the discipline's discourse.[1]

Obviously, Buttimer's work has generally been associated with the recurring pulse of humanism in geography. Indeed, as I have attempted to show, Buttimer's notion of time-space rhythms cannot be divorced from her extensive engagement with this humanistic pulse. Buttimer (1993, 48) understands the recurring liberation cry of humanism in geography and other disciplines within 'the cyclical movement in intellectual life' as a form of resistance against the widespread abstraction of human beings in science:

> From whatever ideological stance it has emerged, the case for humanism has usually been made with the conviction that there must be more to human geography than the *dance macabre* of materialistically motivated robots which, in the opinion of many, was staged by the post-World War II 'scientific' reformation (Buttimer 1993, 47).

Humanistic geography may therefore be characterized as a form of *reanimation* against the reified world of positivist science and technocracy. This reanimation of people and places is also reflected in the substantive studies of the individual chapters of the present book, which all in their own particular ways connect to and express this 'humanistic spirit' in the rhythm of geography.

My argument is organized as follows. I will first try to locate Buttimer's effort in the wider context of the humanistic revival in geography during the 1970s and its resistance against the reified world of abstract space. I argue that thinking about time-space rhythms can be seen as a part of this revival, which also featured a recovery of landscape, nature, social space, and place. Secondly, I will review a series of disciplinary conjunctions, including Buttimer's dialogue with time-geography and subsequent efforts to develop a stronger sense of structural context. I will argue that this structural dimension is essential (but not enough) for an understanding of human subjects and subjectivity. Therefore, I will move on to structurationists' work, and finally to Marxian dialectic and in particular Henri Lefebvre's remarkable and unorthodox rhythmanalysis. My – for obvious reasons – limited and biased account is then followed by an introduction to the individual chapters of the volume and a very brief conclusion.

Clearly, my synopsis covers an array of currents in critical geography, which have often been portrayed in antagonistic or even mutually exclusive terms. This is particularly true (but certainly not uniquely so) of Marxist and humanistic positions in geography. Whilst unwilling to deny fundamental disagreements in political and philosophical outlook, my working assumption is that there are important and potentially synergistic lessons to be drawn from both currents for the development of a progressive and heterogeneous, relational view of time-space rhythms. From my personal vantage point, keeping a glowing tension between lifeworld and structure, matter and meaning, freedom and determination, remains a vital Phoenix promise. Differences in approaches to subjectivity, place, time, and other concepts, indicate something of the complexity involved in the dynamism of the world and the reanimation of places in geography. For that reason, I want to use this opportunity to employ Buttimer's work as a critical, post-positivistic prism through which some dimensions of the extended discourse about time-space rhythms can be reflected.

Topological Thinking and the Eclipse of Lifeworld

As my summary exposition has suggested, Buttimer's geography of rhythms expresses a concern with understanding the multiple articulations of individual and collective; the subjective and intersubjective; nature and society; body and world; and the spaces of experience, memory, symbol, and action (Buttimer 1983, 1986, 1998; cf. Rose 1993, 48). To appreciate the originality of what Buttimer was trying to do, it is useful to recall that the notions of time and space and time-space rhythms in geography were undertheorized. Granted, there were earlier attempts in philosophy to question the absolute notion of time and space by mobilizing lived experience, social or relational notions of time-space, landscape and place. Even so, despite such critical attempts, positivism continued to paralyze geographical thinking with Euclidean space and clock time. Moreover, this theoretical underdevelopment (and the various dualisms coming along with it) remains a concern shared by many contemporary human geographers and social scientists in general (cf. Massey 1994; Wallerstein 1998).

Without for a moment denying fundamental differences in theoretical outlook, I like to think of this concern in terms of a number of approaches to counteract *reification*. As Adorno and Horkheimer appositely wrote, 'all reification is forgetting' (1997, 256). Reification, in its most simple form, occurs when a living entity or transitory state of affairs – be it places, practices, people, life, thought, or objects – is understood or treated as if it were an immutable and isolated Thing. They are no longer recognized as the products of past, present or future human activity in a constant relational process of becoming. Reification circumscribes the widespread tendency in private or political life to ground existence, seeking 'to attain the rocklike and unshakable solidity of a thing' (Barrett 1962, 246). Paradoxically, as long as it is conscious and alive, existence will never be a thing and hence we will have to live with its unavoidable insecurity (Sartre 1956). What many readings of reification suggest is that it is instrumental for the expansion of capitalism and positivist science: it describes the propensity to control the world through formal and quantifiable measures (be it coins, clocks, laws, or rules). Reification is thus tightly linked to particular ways of acting in and organizing the world. In this respect, the environmentalist Neil Evernden gives a useful characterization of the process of reification of a place, landscape, or person:

> The frightening thing about being turned into a thing is that the person who does this to you need no longer treat you as a person, and so does not have the obligation towards you that you would normally expect. You are deprived of his [*sic*] kinship, and are therefore open to his manipulation. In treating you as an object he frees himself to do with you what he wills; all restrictions are lifted (Evernden 1985, 88).

Evernden goes on to argue that this has resulted in 'our near-total reliance on the gaze of objectivity,' denying for instance the caring relational worldview associated with notions of dwelling or environmental movements (ibid. 101).

In science, reification typically attains the form of what Adorno called 'identity thinking' or, more appropriate for my current purposes, 'topological thinking' (Adorno 1996). This may be thought of as a process by which abstract concepts are reified as completely covering the world of concrete phenomena. Although Adorno traces topological thinking to the distant past, the scientific revolution offers a clear example, with the victory of analytical atomism, the reification of nature as object, and a manipulative relationship to objects of study. All of these were pivotal to the humanist principle of free enquiry (von Wright 1986, ch. 5).

Let us now return to Buttimer's humanistic geography. She thought of rhythms as a conceptual tool for grasping the dynamism of lived experience, place and social space, against the mute thingness of container space, machine time, and the *dance macabre* of thingified robots. Her very invocation of lifeworld experience questioned and shook the plane geometry of abstract time-space, and awakened its deadened externality by way of the body, discourse and power, dwelling, place, and lived social space. Hence, I take Buttimer's concept of rhythms to be quite literally an expression of reanimation. Reanimation means restoring to life or infusing new life or spirit into peoples, things, and places. It conveys an intransigience towards all forms of reification. From this reading, the invocation of rhythms signals an effervescent undercurrent that comes to the surface in different times and theories as a heightened awareness of the dynamism of lifeworld. In a general sense, I think it may be useful to see the case against reification as a vital premise for any critical humanistic geography – against the prospect of a society of interchangeable puppets in an environment with no values but exchange value, and against the prospect of a geography enthralled by instrumental rationality.

Humanism and the Reanimation of Lifeworld

Buttimer's maneuver against positivist geography's mathematical abstractions was, of course, informed by Husserl's rather than Adorno's critique of how modern science 'forced lifeworld onto a Procrustean bed of alien concepts.' Husserl traced this topological penchant for abstracting nature 'to the ancient art of measurement as it first emerged in the practical activity of surveying land. In such surveying, the identification and tracing of certain basic shapes and their subsequent normalization led to the creation of a plane geometry of ideal shapes such as we find paradigmatically in Euclidean geometry' (Casey 1997, 221).

The reified surfaces of Euclidean geometry reduced concrete lifeworlds and places to a mere location in abstract space. This translated into an expansionist obsession with the *absolute* and, more particularly, the *infinite* among theologians, natural scientists, and philosophers during the sixteenth and seventeenth centuries:

> The colonizing tendency of Christianity is echoed in the attempts of Galilean, Cartesian, and Newtonian physics to appropriate whole realms formerly consigned to alchemy and 'natural philosophy,' not to mention local custom and history. In both instances, the power of place, uncontested in the ancient world (and still potently present in medieval times), was to put into abeyance – indeed often literally abolished, and with as much relentless force as that with which native peoples were subjected to Christian indoctrination. By the end of the eighteenth century, the idea of universal space came to be regarded as obtaining not just for the external world and for God but also for the mind of the knowing subject (Casey 1997, 77-78).

Space, in this context, was seen as the core figure of topological thinking: it referred to an abstract phenomenon of quantitative measuring, the construction of universalizing frameworks, abstraction and radical essentialism. An important explanation for this reification, the spatial primacy, is that it has long held the promise of unlimited intellectual or material command. Infinite 'space is not only measurable and predictable (hence mathematizable) but altogether "passable" … [O]ne imagines oneself cleaving the air of infinite space freely and without hindrance' (Casey 1997, 338).

This topological thinking condensed in the concept of (abstract) space had fundamental consequences for understandings of place and time (and subsequently for the development of concepts in geography), with far-reaching consequences for lifeworld experience. It also evoked an undercurrent of resistance. Thus, twentieth-century humanism protested against the increasing repression of reflexivity in modern society. Following this, in humanistic geography, humanism revolted against its own earlier scientistic triumphs: against the instrumental rationality of its own making (Relph 1981, 17). This was not merely a philosophical protest, for the reawakened verve of humanism in geography may best be explained from the volatile societal context of the late 1960s:

> For all its munificence, the overarching growth ethic, highlighted by an explosive technology both on the planet and in extraterrestrial space increasingly appeared as the bearer of monumental self-destruction. The convergence of science and technology, once the Promethean harbinger of utopian society, began to emerge more as a central villain in the exhaustion and despoliation of man's [*sic*] own environment. The linking of scientific rationality and politics, once the hallmark of and enlightened democracy, moreover, began to emerge as the chief mechanism for a stronger, if more subtle and therefore less penetrable despotism (Ley and Samuels 1978, 1).

This reopened to debate the issues of (among many other things) values, the politics of science, 'rational' planning, and the widespread suppression of subjectivity in modern society. Humanistic geography proposed the core notions of the 'body-subject' to describe in a non-dualistic way the reciprocity of body and world, as well as the 'intersubjectivity' or the generally taken-for-granted sociocultural heritage through which we learn to behave in certain ways. Closely related to this humanistic theme was the retrieval of *place, social space, nature,* and *landscape* within the widespread disciplinary dominance of (abstract) space and spatial measures in the wake of the quantitative revolution (Entrikin 1976; Glacken 1967; Ley 1977; Meinig 1979; Olwig 1984; Relph 1976; Tuan 1977). Importantly, this connected to and expanded earlier challenges against the paradigm of abstract time-space, funneled through the undercurrents of relational and platial thought. Needless to say, it would also influence Buttimer's later notion of rhythms.

Place

The ascendancy of abstract space as uniform extension reduced place (or, for that matter, lifeworld) to a simple node to be located with mathematical precision in a grid-system. The tendency was 'to take space as paradigmatically a concept of physics, to treat place as secondary to it, and to view any other concept of place or space as purely a psychological phenomenon' (Malpas 1998, 25). As the historian Aaron Gurevich shows in his monumental study of medieval worldviews (Gurjewitsch 1978), 'relational' concepts of time, space and the human body were still prevailing in the pre-modern medieval world, but subsequently receded to the background in the early modern period of 'scientism.' Yet, even this recession would not bring a total triumph of Newtonian space. Relational and social notions of space would resurface, and with them place. Already in his own era, Newton's absolute notion of time-space was challenged by the rationalist Gottfried Wilhelm Leibniz, who thought of time-space as a social construct (albeit firmly placed in the materiality of the world) out of the relations between bodies and things. In the phenomenological tradition of the twentieth century, a relational position was developed in the work of Edmund Husserl and Maurice Merleau-Ponty. The latter thought of space as constitutive of social life, the body-subject, and lived experience, and ultimately retrieved the importance of *place* (Merleau-Ponty 2002, 284; cf. Priest 1998). As Casey (1997, 238) claims, 'Merleau-Ponty culminates a late modern effort to reclaim the particularity of place from the universality of space by recourse to bodily empowerment.'[2]

Immersed in the critical (humanist and Marxist) turns in geography of the 1970s, the philosophical questioning of conventional notions of time and space expressed a wider social critique of modernity, class society, and technocracy. To borrow from the philosopher Jeff Malpas, one could say that this exemplifies the 'recalcitrance' of place against the dominance of space: place tends to constantly ascend from 'the underworld of the modern cultural and philosophical unconsciousness' (Casey 1997, 337). Time and again, place revolts against the 'emphasis on quantitative calculation over qualitative "judgment," on the formal process over concrete practice, on the global and international over the local and proximate' (Malpas 2003, 2343). For philosophers of place like Bachelard, Malpas, Casey, and others, then, this heightened concern with the more idiosyncratic nature of place is a general reaction inherent to the apparent hegemony of a reified spatial logic of the modern West.

Social Space

Abstract space not only subsumed place, but also the relational notion of social space. During the 1960s, Anne Buttimer unveiled how the concept of *l'espace social* was initially developed in a post-war disciplinary borderland between French sociology and geography. The geographer Maximilien Sorre believed that the distribution of social forms, the so-called *substrat social*, 'should incorporate both the physical and social environments, and for this twofold substrate he used Durkheim's term "social space," qualifying the original meaning to include the physical environment' (Buttimer 1969, 419). From these continental theoretical and empirical explorations of the concept of social space, Buttimer concluded that it captured a relational worldview, allowing us to see the connections between subjective dimensions of social significations, attitudes, perceptions, behaviors and traditions, and the objective spatial environment on various geographies of scale (Buttimer 1987, 309-316). Social space was thus considered to be constitutive of the entire spatial and platial *genre de vie* or a whole pattern of daily reproduced and transformed life. This brought out the dialectic of societal, symbolic, technological, and other artifactual and ideational forces and the landscapes carved out by such modes of life (Buttimer 1971, 170-195). Explorations of social space subsequently put a strong emphasis on 'how social reference systems, collective memories, and customary forms of interaction are expressed in the ways groups assign a common meaning to space' (Buttimer 1972, 307). Like philosophies of place, this represented an important reanimation of space, infusing it with new life.

Experienced Time

The idea of time would also be submitted to a process of abstract spatialization and reanimation. Malpas describes the spatialization of time in the following manner: 'If space is understood as necessarily conjoined with time, time itself is understood in a way that assimilates it to space – as another dimension of the so-called "block universe" in which location can be plotted according to the axis of time as well as the axes of space' (Malpas 2003, 2344. Note that spatial dominance will not merely manifest itself in the apparent disappearance of time, but rather in modifications in the way time is treated).

Lefebvre suggests in a similar way that abstract space 'relegates time to an abstraction of its own – except for labour time which produces thing and surplus value. Time might thus be expected to be quickly reduced to constraints placed on the employment of space: to distances, pathways, itineraries, or modes of transportation' (Lefebvre 1991b, 393). In other words, time is first abstracted and then permanented within the rationality of a now fetishized space (ibid. 21). This touches the core of the Marxian humanist Georg Lukács' theory of reification, in which reification (and the gauche 'false consciousness') is conceived as the dominance of space over time: 'Thus time sheds its qualitative, variable, flowing nature; it freezes into an exactly delimited, quantifiable continuum filled with quantifiable "things" … in short, it becomes space' (Lukács 1971, 90).[3] The assembly line offers an apt example of the 'spatialization' of time, with profound consequences such as the fragmentation of labor and the alienation of the human subject.

These critical voices connect to earlier interventions against absolute, and in some cases spatialized, time. Emile Durkheim famously pointed at the difference between the single measure of time and the time-scales produced in the rhythm of social life, especially through the repetition of rites and ceremonies. One may also think of Bergson's intuitively knowable duration; Freud's psychic temporalities; Heidegger's temporal wholeness of human existence; Merleau-Ponty's dependence of time upon the perceiving body-subject; or Husserl's lectures on internal time-consciousness. Each shifted attention from what was perceived as the rather impersonal historical records – as described by the great historicists Darwin, Hegel, Marx, and Spencer – to the personal past and experience of time (Casey 1997; Kern 1983). This allowed for a view of time as changeable and animated by the social, subjective, or natural processes producing it.

Landscape and Nature

In geography, the dominance of abstract space over place and lived or experienced time has equivalents in the erasure and reanimation of several more concepts from the disciplinary vernacular. Because of its referential promiscuity, *landscape* has often been deemed unscientific and replaced by ostensibly more straightforward, solid, and controllable spatial concepts such as region or area. In a similar vein, *nature* with all its daunting ideological, ethical, political, and aesthetic weight has frequently been omitted from 'scientific' geography. Yet, as Kenneth Olwig claimed, no matter how elegant our discursive maneuvers to avoid those values may be, their 'ghostly traces' tend to live on. Despite efforts to erase such seemingly unscientific concepts from geography, their load of values 'remained in the discipline as silent "vestiges" or even as entombed spirits that threatened to rise from the grave no matter how well sealed with intellectual stones' (Olwig 1996, 86). This is the revenge of subjectivity or ideology on those who believe they can completely transcend it by way of topological thinking. Arguably, the return to these 'recalcitrant' concepts in humanistic geography was no mere incident, but at least in part an open declaration of the unavoidability of values in geography and the mistaken strategy to avoid them through reified Newtonian concepts and topological thinking (cf. Buttimer 1974).

Reasserting Rhythms

The notion of rhythms, finally, occurred in the midst of these recalcitrant terms. While all this may seem obvious after decades of post-positivism, the details of how to study (the reification of) time-space rhythms is not. Yet after this detour, I think we are at least in a somewhat better position to recontextualize some reanimating efforts in geography. Simplified, where humanistic geographers emphasized the complexity of human lifeworld experience, Marxian scholars have been more inclined to study structural forces and their colonization of lifeworld. Both, I argue, have something pertinent to say about social and lived rhythms – especially when disciplinary boundaries were transcended. Let us briefly look at some of the recent scholarly conjunctions in the rhythm of geography, which sought to remember the work, people and scaffolds of time-space rhythms behind the scene of thing-like appearances, dislocations, and mutually exclusive dichotomies.[4]

Before we go on, one final caveat: it would be a mistake to think that the dominance of abstract time-space can be properly dealt with through the uncritical reassertion of landscape, nature, social time and space, or place. At least not before cultivating an awareness of other widespread fallacies: the exaggerations of historicism (reification of time), insular notions of place and landscape, or romanticized views of nature. For instance, modern social theory typically privileges time over space, assuming 'either the existence of some pre-existing spatial order within which the temporal processes operate, or that spatial barriers have been so reduced as to render space a contingent rather than fundamental aspect to human action' (Harvey 1990, 205). The reassertion of place and nature, in turn, may rely on narrow, romanticized, or even xenophobic notions of Being, and on taken-for-granted notions of normality and naturalness (cf. W.J.T. Mitchell 2000; Said 2000). Buttimer mentions the insider's trap of home place: 'one lives in places and may be so immersed in the particulars of everyday life and action that he or she may see no point in questioning the taken-for-granted or in seeing home in its wider spatial or social context' (1980, 172). The vital point is that in all these cases, the risk of reification is evident. Critical awareness of such reification unavoidably alerts one to think of the distortions that occur when, for instance, space is reduced to time or time to space. It is exactly this reification of spatial or temporal logic that rhythm disrupts when it re-asserts itself as a heightened concern with the more idiosyncratic character of time-space, place, nature, and landscape. The question, then, is not of pitting time against space, or place and landscape against space, but to develop modes of giving appropriate recognition to the lived and social, and unremitting temporal, platial, and spatial nature of carving out human existence.

First Conjunction: Humanism and Time-Space Traces

As I have suggested, Buttimer's writings on rhythms could be located in the wider philosophical deconstruction of Newtonian absolute time and 'container' views of geometrical space, i.e. of time and space as entities on their own which are independent of their constituents. But Buttimer's whole endeavor with time-space rhythms was just as much a synergetic result of the somewhat counter-intuitive conjunction of her humanistic geography and Torsten Hägerstrand's time-geography (Buttimer 1987). I say counter-intuitive, because while Hägerstrand's dynamic geography rested firmly on material everyday realism, Buttimer's was at that time deeply inspired by existentialism and phenomenology, with their focus on subjective meaning and conscious experience in everyday contexts. Now this was precisely what made their dialogue interesting, for in exceedingly simplified terms, it demarcated a conjunction between an essentially lived and relational and a more positivistic ontology of space.

Why time-geography? Buttimer claimed that while notions such as place and body-subject offered important insights, phenomenologists like Husserl and Merleau-Ponty left some important geographical issues unexamined: 'Although they refer to "world" as an already constituted intentional structure, they have not yet explicitly recognized the dynamics of processes already operative which set the rhythms of time and space for everyday life situations. Geographers resorted to a similar kind of exaggeration when trying to counter charges of environmental determinism' (Buttimer 1976, 286; cf. Seamon, this volume). For Buttimer, societies, everyday life, lifeworlds, place, and nature are intrinsically spatio-temporal, and they could not be comprehended without reference to material and imaginative processes. Time-space, that is, remained necessarily a dynamic medium *and* outcome of social life. Buttimer believed that this was where the promise of time-geography, with its focus on time, space, people, and finitude came in. Time-geography's ontology was inherently dynamic. After all, it was Hägerstrand who assumed 'that people and things are processes, and that the essence of any geographic now (a landscape in its fullness, if you like) is not best understood in terms of its stable individuality but in terms of its double face of graveyard and cradle of creation' (Hägerstrand 1983, 239).

The Matter of Time-Geography and the Meaning of Time-Space

In important respects, time-geography may indeed be said to be more closely akin to the dynamic flow of musical notation (representing wave motion, rather than immobility) than to the reified Euclidean geometry of traditional maps. One of the basic premises of time-geography was that people's lives should be understood as an incessant flow or path through time-space, subject to various constraints. Individuals repeatedly couple and uncouple their paths with other people's paths, institutions, technologies, and physical surroundings. More specifically, through the performance of particular intentional, goal-oriented tasks making up personal or institutional projects, individual paths become tied to societal context, space to place.

Even so, phenomenology and time-geography could not be easily wedded. While Hägerstrand and his research associates at the University of Lund sought to transcend the map by tracing people's paths and projects through time-space, Buttimer was wary of the physical realist and logical positivist world-view from which much of this endeavor emanated. Was not this a clear example of the 'spatialization' of time? The extended critique of time-geography that followed became almost as extensive, and certainly as well-known, as the perspective itself. For all its emphasis on practice and everyday life, time-geography remained a non-verbal 'corporeal ontology' attempting to 'draw the lines of the tangible' (Gren 1994, 87). This estranged world of bodies in motion contrasted with humanistic geography's phenomenological, existential, and hermeneutic concern with distinctively human aspects of value, meaning, subjective goals and experience. As the Swedish geographer Martin Gren has argued, although time-geography was founded on a corporeal ontology of the tangible, it has failed to tease out the full consequences of embodied (in part unreflective) practical action. While moving against reification and thereby lending a clear sense of process to the world, time-geography tended to reify the external observer's view of paths in time-space, rendering people of flesh and blood as if they were bone-dry things.

Despite its efforts to plunge geographical thought back into the stream of life, time-geography reproduced an antinomy between being and knowing. If time is not an abstract intellectual construct, but a constituting part of our bodily immersion in the world, time-geography rarely made efforts to articulate plural notions of experienced time. Time-geography was also accused for forgetting that the rhythms of social life, with the repetition of ceremonies and rites, bring into being quite different time-scales in different cultural formations (Gosden 1994; cf. Durkheim 1965). In a similar way, many felt that space except as an idealized abstract container tends to fall into oblivion in time-geography, which thereby also failed to fully understand place.[5]

Notwithstanding some obvious exaggerations in this critique, what was seen as the disembodied and placeless conception of time-space worried Buttimer and inspired her to develop a more complete geography of rhythms, gesturing at a whole way of life or an inherently dynamic *genre de vie*. 'Focus on a particular genre de vie may yield some insight into the conflict of time-space rhythms which an individual may experience, but to assess the implications of their juxtaposition in place is more difficult. Each genre de vie, analytically speaking, could be considered as a world unto itself, but existentially they interweave and jointly shape the common time-space horizons' (Buttimer 1976, 290). This obviously left very little space for any reified notions of Being-in-itself, Newtonian time, or Euclidian geometry.

Second Conjunction: Structuration and the Limits to Time-Space Dynamism

The second critical disciplinary conjunction, from the late 1970s onwards, involved a new interest in social theory. In hindsight, the dialogue between Buttimer's dynamic humanistic geography and Hägerstrand's matter-realistic time-geography revitalized a critique of the rather widespread tendency in social science to separate space and time, to prioritize either space or time, or to ignore lived space, time, and place. It also made clear that any understanding of purposeful individuals could not be separated from constraints in time-space or the wider structures of social space and its many taken-for-granted practices. Yet, concerning this last issue, it was felt that the encounter between existential phenomenology and time-geography left some relevant matters suspended. At least, this was what structuration theories claimed.

Structurationists argued that earlier attempts in humanistic social science – while seeking to transcend Weberian voluntarism and Durkheimian determinism – failed to explain how social structures (or 'systems') were both the outcome and medium of the agency that constitute them. As the critical realist philosopher Roy Bhaskar claimed in a review of sociological work, these earlier attempts did not develop any profound understanding of social change in time-space, or 'rhythmic' (Bhaskar 1978, 1993; Collier 1994, 137-151). This proved to be an essential theoretical issue in subsequent developments in geography too, in which focus came to lie upon enlivening, through various ideas of structuration, not only individuals but also what was perceived as the reification of society in geographical thought. According to their critics, time-geography and humanism both failed to fully arrive at what Derek Gregory (1994, 89) called the 'explanatory-diagnostic moment' in critical theory: it never became entirely clear as to how constraints were socially produced, why certain projects and orchestrations became dominant, or why and how abstract time-space and its prevailing representations imposed themselves upon lived time-space. If abstract time and space were inappropriate measures for understanding the complexity of human experience, how and why had they become so powerful? Society's real-but-hidden structural properties needed to be explained and dialectically related to agency (Gregory and Urry 1985). Therefore, sociologists and geographers sought increasingly to accommodate theories of knowledgeable and capable human beings (the humanistic insight) and pre-existing, social structures of signification, domination, and legitimation that are involved in the reproduction of action (the structuralist insight). Some of these issues gained some incipient interest in the first two volumes of the *Timing Space and Spacing Time* series, a collection involving several representatives of the Lund group of time-geography (Carlstein et al. 1978; cf. Parkes and Thrift 1980). Just as imperative as the extension of time-geography, however, was the way structurationists borrowed insights from humanism and Marxism.

Clearly, humanistic geography often emphasized experience, the human subject, and subjectivity (or 'agency'). But while indeed some humanistic geographers moved close to idealism and voluntarism, many actively engaged with structural or systematic powers (cf. Ley, this volume). Buttimer's geography, towards which some of the critics turned, was and is particularly rich of terms with structure-agency connotations: body-subject, *genre de vie*, lifeworld, milieu, practice, horizon, intersubjectivity, the taken-for-granted, routines, social space, and so forth. And her heuristic framework (1982; 1993) for examining the rhythm of geography has a steady structural backbone and does have its explanatory-diagnostic moments.

Nonetheless, it was obvious that the Hägerstrand-Buttimer encounter was followed by a heightened attention to structural powers. In geography it partly inspired the need for a more fully developed exploration of what Habermas (1987) fittingly saw as the colonization of the lifeworld by an increasingly instrumental rationality of economic, scientific and political systems (Weber's *Zweckrationalität*). While Buttimer's notions of rhythms and her reactivation of *genre de vie* were presented as bridges between time-geography and humanistic geography, they would subsequently be cast in the language of structuration. In a remarkable essay, Derek Gregory argued that *genre de vie*, with its emphasis on materiality of everyday life and its conception of contingency as well as structure, could be a useful theoretical starting point to connect the erstwhile mutually hostile worlds of humanistic and Marxist geography (Gregory 1981). In practice, however, structuration theory would move the explanatory moments towards an engagement with politico-economic structures of, in particular, capitalism and the state.

For structuration theory to become a powerful analytical tool, geographers felt they needed develop a fuller understanding of time-space geography. Although theories of social reproduction and structuration (notably those of Anthony Giddens, Pierre Bourdieu, and Roy Bhaskar) were sensitive to the ever-present time-space constitution of society and practices in everyday life, it remained by and large unclear as to precisely *how* social reproduction and transformation was expressed in specific time-space locations. This would, again, be possible through the use of time-geography. As Allan Pred acknowledged, the fusion of structuration theory and time-geography indirectly owed much to the Vidalian tradition 'with its emphasis on local practical life and its conceptualization of *genre de vie* as a creative adoption to natural environment based upon the traditional attitudes, values, ideas, beliefs, and psychology of a population' (Pred 1984, 280). The structurationist geographers' ambition also echoed Buttimer's project in its efforts to examine 'exactly how the functioning and reproduction of particular cultural, economic, and political institutions in time and space are continuously bound up with the temporally and spatially specific actions, knowledge build-up, and biographies of particular individuals' (Pred 1984, 281). Human agency was seen as an ongoing performance in time-space, moving through particular sites of socialization, that is: through places structured by institutions, authority, regulations, routines, and rules bound up with particular temporalities (Thrift 1983).

What structuration theories described as social reproduction and transformation, was, albeit in a quite different phenomenological language, at the core of Buttimer's notion of rhythm: 'everyday behavior demonstrates a quest for order, predictability, and routine, as well as a quest for adventure and change ... This tension between stability and change within rhythms of different scales, expressed by the body's relationship to its world, may be seen as prototype of the relationship between places and space, home and range in the human experience of world' (Buttimer 1976, 285). But beyond these striking similarities with Buttimer's geography of rhythms, the structurationist worldview obviously emanated more from the power, contradiction and conflict imbued presumptions, explanatory-diagnostic and political commitments of Western Marxism, than from phenomenology and existentialism (Pred 1981, 5). Notions of social power relations and socio-spatial reproduction moved further to the cote of geographical theory, telling more about how time-geographies were produced and policed, and why particular social routines and projects became hegemonic.

While much theoretical language was cultivated to explain the relation between system and lifeworld, in the more substantive geographies that followed it always proved far easier to theorize the structure-agency dialectic and to describe structural determination than to empirically portray the transformative practices of human subjects. In many cases, this gap between the philosophical confirmation of the living human subject and its substantive dissolution in abstract elaborations of capitalist structures also haunted much Marxist work. Even so, the effort to bring humanism and Marxism into a dynamic dialogue has deep roots in the examination of cultural politics and economy in Western Marxism (Cosgrove 1984; Jackson 1989; D. Mitchell 2000). This brings us to a third conjunction, inspired mainly by Marxian philosophers and geographers from the 1970s onward. Most basically, they argued that modern rhythms are distinctively capitalist rhythms: fabricated and framed from within the logics of the capitalist social formation. They shared Marx's ambition of 'not only describing but evoking and enacting the desperate pace and frantic rhythm that capitalism imparts to every facet of modern life' (Berman 1988, 91). They envisaged in a radical register how abstract notions of time-space were constructed as modes of social control with commercial empires in mind. I will argue that this offered important insights into the colonization of lifeworld and the horizons of lived space and time.

Third Conjunction: Capitalism and the Dialectic of Being and Becoming

The mediations between societal pressures and individual life course in space and place as analyzed by structurationists delineate one important aspect of a more complete and inherently disruptive spatio-temporal process. This was obvious for Karl Marx, who personally witnessed and engaged himself in the dramatic transformations of life and space propelled by capitalism in the nineteenth century (Giddens 1990; Hobsbawm 1975; Jameson 2002; Kern 1983; Liedman 1997). As one of his most reknown devotees in geography famously argued, this was a profound period of 'time-space compression.' David Harvey suggests that time-space compression is a highly complex and in part contradictory restructuring of the experience of time-space that involves acceleration, or what Marx (1973, 539) observed as 'the annihilation of space by time,' but also a host of other changes. It circumscribes a discontinuous and uneven trend, principally understood as a 'speed-up and acceleration in the pace of economic processes and, hence in social life,' which in turn connects to 'shifts in systems of representation, cultural forms and philosophical sentiment' (Harvey 1990, 230 and 239). Put differently, capitalist modernization entailed 'the perpetual disruption of temporal and spatial rhythms, and modernism takes as one of its missions the production of new meanings for space and time in a world of ephemerality and fragmentation' (Harvey 1990, 216).

Key insights into this complexity have been offered by cultural Marxist thinkers like Raymond Williams (Jackson 1989). As the latter asserts, modernism and capitalist modernity cannot be divorced from the metropolis and 'metropolitan perceptions.' These are drenched by a global social and spatio-temporal unevenness. Geographically, the 'effective metropolis ... is now the modern transmitting metropolis of the technologically advanced economies' (Williams 1989, 38). Simultaneously, something like a systematic 'epistemological separation' of colony from metropolis and colonial labor from imperial prosperity structures modernism, resulting in a condition in which 'the truth of metropolitan experience is not visible in the daily life of the metropolis itself; it lies outside the immediate space of Europe, in the colonies. The existential realia of the metropolis are thus severed from the cognitive map that would alone lend them coherence and reestablish relationships of meaning and of its production' (Jameson 2003, 700). Added to this spatial (imagined and material) unevenness and the systematic occultation of that unevenness, Williams turns attention to the cultural openness and complexity of what he dubs the 'miscellaneity' of the metropolis. This gathers a seemingly contradictory set of real-and-imaginative practices:

> Although Modernism can be clearly identified as a distinctive movement, in its deliberate distance from and challenge to more traditional forms of art and thought, it is also strongly characterized by its internal diversity of methods and emphasis: a restless and often directly competitive sequence of innovations and experiments, always more immediately recognized by what they are breaking from than by what, in any simple way, they are breaking towards. Even the range of basic cultural positions within Modernism stretches from an eager embrace of modernity, either in its new technical and mechanical forms or the equally significant attachments to ideas of social and political revolution, to conscious options for past or exotic cultures as sources or at least as fragments *against* the modern world, from the Futurist affirmation of the city to Elliot's pessimistic recoil (Williams 1989, 43).

Harvey captures these diverging reactions to the anxieties engendered by time-space compression as a tension between the privileging of Being and Becoming. 'Becoming' embraced the collapse of space through time (e.g. by technologies of transportation and communication) as a proceeding of the Enlightenment project and sought to impose a unified, rational organization of space and time (conceptualized as abstract time-space). This 'external' spatio-temporal rationality, in turn, was expected to secure welfare, individual liberty and 'internal,' private senses of time and place. In contrast to this universalistic spatial trend, 'Being' celebrated the particularity and identity of *place*. It reinforced the permanence of place amid the growing abstraction of a rapidly transforming time-space. The invention of tradition, cultural heritage preservation, and philosophies of Being of the late nineteenth century, all reflect this trend.

Although it may seem as if the Being-Becoming distinction enthrones a place-space or time-space dualism, Harvey empathically denies this and warns that these two positions should not be considered in dichotomous terms, but rather 'as two currents of sensibility that flowed side by side, often within the same person, even when one or other sensibility became dominant in a particular place and time' (Harvey 1990, 275). Indeed, as Williams's passage already argued, the archetypical modernist works of art oscillated between the mixed emotions of joy and despair, and polarities of affirmation and nostalgia – 'their inner dynamism will reproduce and express the inward rhythms by which modern capitalism moves and lives' (Berman 1988, 102). Furthermore, the interrelationships between Being and Becoming are pertinent to Harvey's relational theory of time, space, nature, and place.[6] Rephrasing this matter, Harvey (inspired by Leibniz and Whitehead) employs the notions of permanence and process to articulate this theory:

A 'permanence' arises as a system of 'extensive connection' out of processes. Entities achieve relative stabilities in their bounding, and their internal ordering of processes creating space, for a time. Such permanences come to occupy a piece of space in an exclusive way (for a time) and thereby define place – their place – (for a time). The process of place formation is a process of carving out 'permanences' from the flow of processes creating spaces. But the 'permanences' – no matter how solid they may seem – are not eternal: they are always subject to time as 'perpetual perishing.' They are contingent on the processes that create, sustain and dissolve them (Harvey 1996, 261).

Importantly, Harvey (and before him Berman, Williams, and others) problematizes monolithic readings of capitalist modernity as subjugated by Becoming and, as a consequence, spatialized time. Essential to Harvey's diagnostic trait of acceleration is that novel notions of time and space – emerging from new tools and machinery, modes of social regulation, scientific representation, technologies and institutions of modern art, changing environmental practices, etc. – circulate through radically uneven geographies of power and control. The ordering of this unevenness can be identified at all spatial scales, from the local to the global, diffusing at very different paces through a multitude of cultural, social, economic and political power structures (cf. Harvey 1996, 242-247; Jameson 1991, 366-367).

Arguably, in our current age the articulations between postmodern culture and the infrastructure of late capitalist globalization have fashioned novel power-geometries (Massey 1993, 1999). Frederic Jameson, in an absorbing recent essay, has characterized this as an 'existential uneven development' involving a new mode of abstraction. This novel abstraction is less about the 'old-fashioned value of firms and factories or of their products and their marketability' and more about 'the less palpable abstractions of the image or the logo, which operate with something of the autonomy of the values of present-day finance capital. It is a distinction between an object and its expression and an object whose expression has in fact virtually become another object in its own right' (Jameson 2003, 703). Moreover, while the modern era was one in which time (that is, abstract, spatialized time) was habitually acknowledged as the cultural and existential dominant, in the post-modern era juxtaposition and the spatial prevail (Foucault 1986). Even so, in contra-distinction to the easier popular dictums about space as the dominant of the postmodern, Jameson argues against such dualism or 'great transformation' (Jameson 1991, 156), that this new value abstraction has profoundly temporal consequences for everyday life:

> For the dynamics of the stock market need to be disentangled from the older cyclical rhythms of capitalism in general: boom and bust, accumulation of inventory, liquidation, and so forth, a process with which everyone is familiar and that imprints a kind of generational rhythm on individual life. This process, which also creates the impression of a political alternation between Left and Right, between dynamism and conservatism or reaction, is of course to be sharply distinguished from the far longer cycles of the so-called Kondratiev waves, fifty- or sixty-year periods that are as it were the systole and diastole of the system's fundamental contradiction (and that are, by virtue of their very dimensions, less apparent to those biological individuals we also are). From both these temporal cycles, then is to be distinguished the newer process of the consumption of investments as such, the anxious daily consultation of the listings, deliberations with or without your broker, selling off, taking a gamble on something as yet untested ... The narrowing and the urgency of the time frame need to be underscored here and the way in which a novel and

the more universal microtemporality accompanies and as it were condenses the rhythms of quarterly 'profit taking' (and is itself intensified in periods of crisis and uncertainty). The futures of the stock market ... come to be deeply intertwined with the way we live our individual and collective futures generally ... By the same token, the new rhythms are transmitted to cultural production in the form of the narratives we consume and the stories we tell ourselves, about our history fully as much as about our individual experience. It is scarcely surprising that the historical past has diminished accordingly (Jameson 2003, 706-704).

The new rhythms are by no means only temporal for it was exactly the power of a new *spatial* dominant that defined the distinction between monopoly capitalism and the experience of globalization under late capitalism. Instead of the modern occultation of colony and metropolis, the latter is characterized by instant information transfers, amounting to an epistemological transparency (going hand in hand with a global Americanization as well as its opposition). This new transparency is not, however, to be confused with a return to a somehow realistic rendering of the object or to representational languages. It is rather, Jameson asserts, a suppression of history, time and temporality, 'a realism of image or spectacle society,' and in that sense 'a symptom of the very system it represents in the first place' (Jameson 2003, 701). This particular realism, as the French situationist Guy Debord clarified in the 1960s, is characterized by the supplantation of everyday time by commodified 'spectacular time,' which presents itself as a 'pseudo-cyclical time.' The latter echoes 'the natural vestiges of cyclical time' such as weekly work and rest or day and night. Yet, it 'is in fact merely the consumable disguise of the time-as-commodity of the production system, and it exhibits the essential traits of that time: homogeneous and exchangeable units, and the suppression of any qualitative dimension' (Debord 1995, 110-111).

Now where in all this deep structural impact and reified realities is the complex human subject? Where in this *dance macabre* geography of materialistically motivated robots in the spectacle society and capitalist choreography? Clearly, such questions reactualize the humanist's *cri de coeur* in which not everything touched by and touching the human subject can be relegated to the rhythms of capital. But neither can we afford to ignore the latter's reifying tendencies and powers, for they will manipulate everyday experience, lifeworld, and the horizons of lived space and time. It is this tension between resistance and adoption, between creativity and the taken-for-granted, between subjective expression and subjection to authoritative control, that remains an undeniable and intricate power in the dynamism of lifeworld (cf. Godlewska, this volume). This brings me to my final conjunction, in which rhythms are re-worked and contextualized from a more humanistic angle, while leaving no doubt as to how lifeworld experience is subjected to colonization.

Fourth Conjunction: Henri Lefebvre's Rhythmanalysis

The accounts of modernity and postmodernity demonstrate a longstanding interest in the operation of time-space rhythms among Marxian scholars. It is not difficult to find in Jameson's or Debord's insights on spatial and temporal transparency and opacity, the spectacle society, and the rhythms of modernity and postmodernity, a clear affiliation with the enigmatic French philosopher Henri Lefebvre. As David Harvey also acknowledges, his own insights into the significance of time, space, and money in the production, reproduction and transformation of social power owe much to Lefebvre. Lefebvre played a key role in the development of a humanistic Marxism, sharing with existentialism a concern for the person in everyday experience as the prime source of meaning. He has suitably been described as 'an important "conducting wire" of motivating ideas and sentiments from group to group and generation to generation' (Shields 1999, 4). Lefebvre also developed an intriguing theory of time-space rhythms, which has only recently started to attract some attention in English-language geography (cf. Clark, this volume).[7] In contrast to surprisingly persistent accounts in social theory and philosophy on the stasis, immobility, and unrepresentability of space, Lefebvre understands space as a abundantly open dimension of circulation, possibility, and dynamic change (Massey 1994). Simultaneously, he thereby overcomes the widespread tendency in theory to tear apart time and space and treat them in isolation.

To understand Lefebvre's version of time-space rhythms, it is useful to revisit his general argument against reification (for this remains a theoretical cornerstone of his assessment of everyday life and his reanimation of the production of space).

As Lefebvre recognizes, one important common feature of landscapes of exchange value is that they crystallize out of process, and while this embodies a multiplicity of temporal and geographical forces and scales, they also attain a thing-like appearance or permanence in the landscape (Harvey 1996, 55). Now while the notion of time-space rhythms forces us to think in terms of change and process, permanences easily fall victim to the typical double amnesia that is reification. First, the projection of the commodity structure (via exchange, property, aestheticization) and other values into the landscape is forgotten, and then the morphology and representation of landscape cloak themselves in triviality and an appearance of immutable naturality. Rhythm disappears into the mist-enveloped reified realm of fixed things. While 'space contains and disseminates social relationships … it tends to become the absolute Thing' (Lefebvre, 1991b, 82-83). Similar reifications tend to invade daily life. Lefebvre advisedly remarks that

> we must not overlook the constancies and (relative) stabilities of everyday life, but nor must we fetishize them. The everyday per se does not have the viscous consistency which is often attributed to it, nor the evascent, fluid mobility which is a corollary of that representation. Stabilities are definitive in appearance only, and we must avoid confusing them with 'the real,' the 'existential,' 'being,' 'substance' or 'human nature.' Once we penetrate a stability, its limits become apparent on all sides. We may think it is compact and durable. The critical thought inherent in knowledge dissolves this illusion (Lefebvre 2002, 255).

To avoid this kind of reification, Lefebvre proposes we need 'to get back from the object (product or work) to the activity that produced and/or created it' (ibid. 113; cf. Mels 2002; Mitchell 1996, 30). Lefebvre shares Buttimer's concern with social space, the rhythmic understanding of place – the space where everyday life is situated – and also laments the present hegemony of quantitative clock time and abstract (geodesic) space. Capitalist modernity for Lefebvre was not merely the annihilation of space by time (speeding up of transportation and circulation), but the eradication of lived time by abstract commodified space. 'Since time can apparently be assessed in terms of money, ... since it can be bought and sold just like any object ("time is money"), little wonder that it disappears after the fashion of an object' (1991b, 96). In his monumental second volume of the *Critique of Everyday Life*, Lefebvre clarified:

> There exist social time or social time scales which are distinct from biological, psychological and physical time scales. There is a social space which is distinct from geometric, biological, geographic and economic space. Everyday space differs from geometric space in that it has four dimensions, which are in a two-by-two opposition: 'right/left – high/low.' Similarly, everyday time has four dimensions which differ from dimensions as mathematicians and physicists would define them, namely the accomplished, the foreseen, the uncertain and the unforeseeable (or again: the past, the present, the short-term future and the long-term future). The more deeply we analyse them, the more subtle and the more differentiated these ideas become (Lefebvre 2002, 231).

From this image of time-space, Lefebvre then proposed a 'rhythmology' or, borrowing from Gaston Bachelard, a 'rhythmanalysis.'[8] The latter thought of rhythmanalysis as a theoretical act of 'descending from the great rhythms forced upon us by the universe to the finer rhythms that play upon man's [*sic*] most exquisite sensibilities,' thereby 'reconciling the disturbed psyche' (Bachelard 1994, 65). For Lefebvre, rhythmanalysis was a critique of reification as well as a project of reanimating social space and place (Lefebvre 1992). Taking rhythm further into a Marxian vocabulary of a suppressed subject, commodification, and the production of space, he stated that any comprehension of social space must begin 'with the spatio-temporal rhythms of nature [including, as we have seen, the body] as transformed by social practice' (Lefebvre 1991b, 117. Note that these social practices themselves can also be seen as rhythms with a particular dynamic geography. Cf. Lefebvre 1996). This requires a dialectical way of thinking, in which an unmistakable hostility against reification and positivism is communicated:

> A rhythm invests places, but is not itself a place; it is not a thing, nor an aggregation of things, nor yet a simple flow ... Every rhythm possesses and occupies a spatio-temporal reality which is known by our science and mastered so far as its physical aspect (wave motion) is concerned, but which is misapprehended from the point of view of living beings, organisms, bodies and social practice. Yet social practice is made up of rhythms – daily, monthly, yearly, and so on. That these rhythms have become more complicated than natural rhythms is highly probable. A powerful unsettling factor in this regard is the practico-social dominance of linear over cyclical repetition – that is to say, the dominance of one aspect of rhythms over another (Lefebvre 1991b, 206).

The similarities between Buttimer's and Lefebvre's assessment of abstract time-space are quite evident. Both bring out what might be called epistemological breaks or ruptures between different time-spaces.[9] Yet where Buttimer argued that notions of abstract time and space were inappropriate for a phenomenological understanding of lived experience, Lefebvre thumped the same abstractions from a more overtly politicized critique of everyday life and space – a critique geared toward their full transformation. For him, the widespread ideal and reality of linear repetition was a one-dimensional and narrow mode of repetition, prioritizing the here and now and repressing subjectivity and Becoming. Abstraction and linear repetition were part of a vast system of calculation in which labor, time, and space are brought under uniform measures. Lefebvre asserted that, inappropriate as they are, these abstractions and uniformities could be sensed in daily life and in various other realms of reality. This conclusion is confirmed in the historical process of capitalist modernization as sketched earlier in this introduction.

Lefebvre's diagnosis is as devastating as it is eloquent. While modernity initially set out to invade the space and time of daily life in order to convert it into 'a rationally exploited colonial province of society' during the twentieth century, this colonization had been completed and everyday life has turned into 'a province of organization; the time-space of voluntary programmed self-regulation' (Lefebvre 1984, 72). Everyday places become the terrain that is envisioned, constrained, structured, and appropriated by the space of the commodity and state bureaucracy; they become a product of modernity, the abstract space of political centralization, exchange value, and the market. This administration of society by an impersonal bureaucracy, the internalization of social control in daily life and its concurrent 'normalization' of whole populations parallels Michel Foucault's (1977) analysis of the disciplinary society. Lefebvre understood the diffusion of self-repression as the hallmark of what he called – interestingly contrasting to the currently hegemonic geopolitical lexicon – a 'terrorist society,' in which the source of oppression and violence is partly internalized and hence no longer easily identifiable (1984, ch. 4).

While linear repetition, and its constituent abstracted subjects, spaces, and times, was but one of the many rhythms identifiable through rhythmanalysis, it arguably remains the cultural dominant measure of late capitalism. Perhaps this scenario may seem pessimistic and inescapable, but, Lefebvre observed, within this cultural dominant lie the traces of utopian surplus, available through conscious intervention by the critically minded. Beyond this scenario lied the avant-garde, defying the uniformation of value through money-price and preserving the subjectivity embodied in space-time and works of art. For through its resistance to the commercial flatland of value, the avant-garde frees into the bargain the possibility of a renewal of value (Adorno 1997). Notwithstanding Lefebvre's murky world of bureaucratic domination and surveillance, there are moments of *juissance*, freedom and resistance, notably located at the level of the human body and differentiated space (Lefebvre 1976, 89; 1991b). Everyday life, despite its overwhelming falsity, is also the only domain of a vestigial authenticity, untouched as it were by modernity; a truly lived space of difference, use-value, love, desire, play, and dreams. It is the location of dis-alienation, promising 'a revolutionary plan to recreate a style, resurrect the Festival and gather together culture's scattered fragments for a transfiguration of everyday life' (Lefebvre 1984, 38; cf. 1976; 1991a; 1991b).

In Buttimer's rhythmanalysis, a not altogether dissimilar awareness of the limits of positivist, modern time-space and its power to invade everyday experience, lends voice to a humanistic liberation cry for centeredness and dwelling:

> To many people, for whom *reaching* appears to be more important than *home making*, places may be simply points on a topological surface of access. Eventually, the architectural monotony or discordance within any one physical place could be the result of coincidence

rather than design: each component is part of a spatial linkage and the center of each lies at a national or multinational headquarters ... The skyscrapers, airports, freeways, and other stereotypical components of modern landscapes are they not the sacred symbols of a civilization which has deified reach and derided home? ... Or the gaping wounds of mining and industrial landscapes are they not the refuse of a civilization intoxicated with Promethean hubris? (Buttimer 1980, 174).

Buttimer's unease 'is not because of aesthetics alone or sensory overload; rather it derives from a nausea about values which make machines, commodities, movement and salesmanship more important than human encounters or letting nature have some breathing place' (ibid. 178). To reanimate places is to register the close harmony, dissonances, and transpositions in the polyphony of places and tempos beyond reified existence. It reminds us how social rhythms are in fact *lived* and cannot be separated from the complex everyday experience of human beings in their diverse social and material environments. It is such complexity, framed in various ways from different philosophical vantage points that will be encountered throughout this volume.

Prospectus

The essays in this collection gather a diverse group of geographers who wanted to reanimate places with the help of the notion of a geography of rhythms. We not only think that the theme is pertinent to an understanding of landscape, place, social space, and nature, but also consider this collection an appropriate way of celebrating Professor Anne Buttimer's unique contribution to the rhythm and reanimation of our discipline. Although the chapters in this volume approach some important issues involved in the reanimation of places – ranging from symbolic orchestrations of landscape and treatments of rhythms in urban and holy places, to deeply personal memories of particular natural rhythms – I trust it is clear that the niceties of this vast terrain cannot be covered in one set of essays.

What may be less immediately clear, but none the less important, is that the authors held different opinions about reanimation and in particular about how to communicate time-space rhythms. Some authors referred explicitly to geographical rhythms and rhythmanalysis. Others were more inclined to lend a voice to the humanistic 'liberation cry' through accounts of reanimated landscapes or places. These differences translated quite well into the double contents of a geography of rhythms towards which the volume moved. Analogous to the very meaning of the word, rhythm gave conceptual guidance to the reanimation of places but was no unbending authority. No effort was made to force one particular vantage point from which to approach rhythms on the authors.

For all its disparity, at least four common and interrelated elements may be identified in the individual chapters. The first theme that runs though these chapters is a critical, conceptual-practical awareness of the diverse and often contradictory uses of place, time and space, society and nature. In this respect, the authors construe rhythm and reanimation as a general term for a heightened attentiveness to the heterogeneity of time-space relationships and their dynamism. This is then used for unpacking ordinary ways of thinking about time and space and acting in time-space. Second, and as a consequence, all authors connect rhythms and reanimation to the expansive issue of values – be it in terms of exchange values and use values, subjective experience, or collective values of social and environmental justice. Needless to say, these are never air-tight mental preserves for bookish intellectuals, and the essays explore their rendition into practices of inclusion and exclusion, mobility and rest, transformation and reproduction in built and imagined geographies. The third theme that appears especially striking throughout the volume concerns the making and transformation of persons and social groups through different time-space rhythms. This can take the form of, for instance, mobility and memories of childhood seasons or segregating rhythms of political economy. Fourth, the chapters convey a leaning toward methodological pluralism. Different modes of enquiry bring different rhythms to the surface and different rhythms require a variety of research procedures. I will revisit this issue at the end of this introduction.

While each essay stands on its own and can be read separately, the remainder of this introduction will make clear that the collection does follow a particular sequence. For reasons of clarity, the rest of the volume is organized into three major thematic parts, with theoretically informed case-studies, and an afterword. Although each of the core parts centers around distinct reanimations of places, this classification into sections is intended as an open heuristic device rather than an ontological statement about rhythms or reanimation. For instance, questions of urban dynamism (Part III) easily convert back into problems of self-identity and displacement (Part II), or forward into politics and embodiment (Part IV). In line with this fluidity, the book ends with a theoretical chapter by Robert Sack (Part V). Without explicit reference to any particular case-study, Sack's chapter not only envelops most of the issues discussed in the volume but also communicates a personal view about place-making in the dynamic world of rhythms.

The thematic focal point of *Part II* is rhythms of place, nature, and identity. Although each of the four chapters deals with dissimilar cases, they share an interest in the implications of memory and mobility, attachment and detachment in the reanimation of places. Yi-Fu Tuan's opening essay introduces wide-ranging and pertinent ideas concerning the reanimated connection between self, place, and time. The chapter maintains that the relationships woven into this trio as a sense of place are of vital importance in trying to understand what it means to be human. Tuan locates a sense of place in the rich rhythmic of change and stability characterizing both human lives and places. Put generally, place enables humans to slow down time, to make a pause. While certainly not inert, place offers us a means to overcome the seemingly relentless flow of time. This is clearly the case in the artistic 'cousins' of geographical place, i.e. in senses of place as articulated and augmented in poems, novels, stories, movies, paintings, or photographs. When we revisit a place and note its permanencies, time can in a way even be reversed. Given that the shaping of a sense of place tends to take time, strong and multi-layered senses of place are most obviously attached to home: the place of our everyday social, emotional, and bodily life. Even so, Tuan claims, a sense of place is seldom limited to the local scale and the relative stability of home. Brief encounters with some novel place can sometimes provoke immediate attachment. Higher geographical scales of neighborhood, country or even planet Earth can also arouse senses of place.

A recurring theme throughout Tuan's essay is how such rhythms of rest and mobility affect people's sense of home and identity. On different occasions, Anne Buttimer has also addressed this theme. In her *Grasping* article, she asked 'Could the gestalt or coherent pattern of one's life-space not emerge from mobility as a kind of topological surface punctuated by specific anchoring points?' Tuan notes that in contemporary mobile society, people often wander far from home without breaking the emotional bond between sense of place and self. On the positive side, such spatial mobility often stimulates self-discovery. Experiences of strangeness or surprising familiarity of new places will often help people disclose previously unfamiliar features of themselves and can even arouse intense feelings of love at first sight. The question is also reflected upon in the second and third essay. David Ley fashions his account from a case study of transnational entrepreneurs circulating within the trans-Pacific field of the Chinese diaspora. He argues that this new configuration is a variation upon a much older theme in social phenomenology, the problematic status of the stranger. In his chapter, Ley examines recent immigrant entrepreneurs to Canada with continuing transnational ties to Hong Kong and Taiwan from the perspective of the stranger, whose taken-for-granted world is disoriented by the rhythms and routines of a new place. Not only do established recipes fail to facilitate entrepreneurial practice, but national commitments are also equivocal. While media and official statistics laud the achievements of these new entrepreneurs, in-depth interviews show a much more problematic experience of limited business success, with tendencies to cultural enclosure either by recreating the old land in an ethnic enclave or by repatriation to a more familiar economic regime in East Asia. Ley shows that dual or multiple affiliations in two or more nation states are raising with new urgency the questions of place, identity and citizenship in an increasingly transnational age.

In the essay that follows, this transnational age with its condition of fragmented cultures and spaces is further explored through the prism of nature's rhythms. As Edmunds Bunkse argues, questions of place and identity, and feelings of at-homeness and displacement are deeply imbued with, and mediated by, experiences of nature's rhythms. Bunkse's chapter offers a series of profoundly personal narratives about how those rhythms and embodied memories fuel senses of spatial and temporal displacement as well as closeness and familiarity. In Bunkse's topoanalysis, all of nature's rhythms – from childhood observations of sound and movement in a stormy forest along a Baltic shoreline, to the adult's awareness of changing seasons in America's urban Eastern Seaboard – are laden with personal and cultural associations. A vital thread in Bunkse's argument is that these experiences and feelings are necessarily entangled in the temporal shifts of past and present lifeworlds. Biographical mediation and conscious or unconscious perception and remembrance of nature's rhythms are simultaneously molded through cultural filters of poetry and prose. Deprived of his homeland at an early age, Bunkse writes, 'I have created another country, both through the vicissitudes of contemporary life and by choice. It is an extended country that reaches from Narvik beyond the Arctic circle, to Oaxaca near the tropics, from British Columbia to Maine, a collage country of bits and pieces of home, made up largely of northern rhythms that first revealed themselves in childhood.' While this story is largely autobiographical, it also represents a generic human tendency to select motifs from past experiences and platial contexts to shape the present and thereby the future. This also connects lived rhythms to modes of dwelling.

In the final chapter of Part II, Gerry O'Reilly takes us to a fascinating part of Ireland, in what perhaps can be described as an existential phenomenology of landscape: an effort to reconstruct the Boyne Valley in the experience of geography students and occupants in the light of historical process. The Valley is home to thriving agricultural communities, people employed in tourism, mining, industry, and daily commuters. Contemporary dwellings nestle, and daily rhythms interact, amongst Neolithic chambers, Celtic earthworks and monastic settlements, estate parklands and symbolic battle locations. Like pentimento in a painting, O'Reilly points out, reappearances from the past protrude on the landscapes of the Boyne Valley. Historical events documented in stone and manuscripts enhance legends and myths passed on to the cultural rhythms of each generation. Just as power brokers in the past shaped the landscape, the state is making its imprint by buying symbolic areas in the creation of Irish heritage. Somewhat generalized, students tend to interpret sustainability in terms of perceived cultural continuity, nature and seasonality, and millennia of human activity mirrored in landscapes. On O'Reilly's field trip, the student's sense of place, historical landscape continuity and identity come to the fore. As the next two essays make clear, awareness of valued, lived and practiced rhythms such as those discussed throughout O'Reilly's and in the other chapters of Part II, are of vital importance in the development of sustainable human environments.

Part III is mainly concerned with various ways in which rhythms relate to everyday life, planning and practice in urban environments. Each of the chapters is concerned with reanimating the making of urban rhythms, be it in terms of fixed capital, architecture, environmental planning, or the everyday rhythms of gendered practices and routines. In chapter 6, Ted Relph argues that the dominant values of a society are expressed in that society's rhythmic relationship to landscapes and built places. A society, which aims for sustainability, will therefore require sustainable time-space rhythms. In mainstream approaches to sustainable development, these insights remain underestimated. The temporal pattern of the built environment of Markham, a town near Toronto, offers an apt illustration. At several locations in Markham, fragments from different eras are juxtaposed in a confused, discordant, disruptive, and arhythmic way. Markham is not unique in that the twentieth century has seen a proliferation of such 'time edges,' and with it increasingly complex problems of environmental degradation. This mirrors currently dominant uses and perceptions, treating time and landscape as commodities that are destined to be instantly parceled, priced, and marketed. Simply simply adding more sophisticated forms of education, research, technology, and international co-operation, Relph maintains, will not solve these problems. If it is true that cultural attitudes to time are built into the landscape, an important key to sustainability will lie in understanding the intricate relations between experiences of past, present, and future. The phenomenological concept of temporality, revolving around an authentic sense of care for the world, offers such an understanding. Short-term commodification and its resulting time-edged fragmentation of landscape need to be replaced by places, where sustainable lifeways have a chance to unfold. Meaningful rhythmic coexistence of elements from various periods 'as though in a visual equivalent of a musical phrase that linked different melodies' arguably helps reduce future needs for ingenuity.

Relph's phenomenological argument about sustainable rhythms is confirmed in the next chapter. Here, David Seamon notes that a recurring problematic in humanistic geography and place-making is to understand and practically respond to the lived rhythms of physical and human worlds. Recent philosophies of urban planning and design have offered fruitful new modes of mediating on various geographies of scale the physical body and lived rhythms of cities. By consciously starting from the multi-layered rhythm of places, three urban thinkers – Christopher Alexander, Daniel Kemmis, and Bill Hillier – show remarkable new ways of facilitating platial senses of belonging. The work of these thinkers trades on organic metaphor, claiming that a promising rehumanization of urban places would evolve if cities were perceived throughout as complex wholes. If decision-making and design are devoted in every step to the communal creation of wholeness, they can make rhythms in urban worlds socially and physically more identifiable, coherent, and meaningful. The focal point in Alexander's vision is the process of physical design, and notably how new developments can harmonically resonate with existing urban environments. While this offers a comprehensive starting-point, Seamon argues that any creation of wholeness requires special attention to the mutual interplay between city dwellers and politicians in the political process. This institutional rhythm is elaborated upon in Kemmis' writings. Added to this, Seamon argues that Alexander tends to overemphasize particular design parts, which gets in the way of a more integrated understanding of organic wholeness. This theoretical lacuna is provided for in Hillier's holistic attention to topological qualities of places, movement, and the integrative or segregative effects of pathways. His 'space syntax' amounts to a startling vision of the rhythms of place, by pulling together private and public, journey and dwelling, resident and visitor, interior and exterior, part and whole. Seamon shows that, taken together, these three theorists complement earlier phenomenological understandings of how the dialogue between human and physical worlds can contribute to lived place and more humane modes of dwelling.

Notwithstanding Seamon's call for the rehumanization of urban places and attention to lived rhythms, urban rhythms often turn out to be structured by conflicts and calculation. These landscapes of calculation, as Relph suggested in the previous chapter, have often been highly detrimental to the aims and hopes of sustainable development. In chapter 8, Eric Clark discloses how the political economy of land rent and rent gaps contributes to, and regulates the production of a restless urban space in Malmö, Sweden. Buildings and other architectural permanences of the city could be regarded in terms of what Goethe called 'petrified music,' directed by various social and economic forces. Clark argues that the history of a city block in Malmö may be described metaphorically as a long slow flamenco dance, dense with overlaying punctuating rhythms. The main body of the chapter analyzes how two of these rhythms relate in ways constitutive of each other. One is the interplay of two forms of differential land rent, central to both Marx's and Thünen's understandings of spatial dynamics. The other is the much slower beat of rent gaps, central to the effort of gentrification theory to grasp conflicting processes of urban change and redevelopment. In the history of thought on land rent, these two theories have remained unconnected. The distinctive contribution of Clark's analysis is to show how they relate in a rhythmic way. Both of these animating powers regulate the spatial flow of fixed capital in the urban landscape, at different temporal scales and in different but connected rhythms. Finally, tentative thoughts are raised regarding why the choreography generated by the interplay of differential rents and rent gaps becomes less conflicting and tumultuous in some places than in others. Clark suggests that a more comprehensive understanding of the vast rhythmic world of urban space requires extending the geography of rents with insights into its contextual interplay with other social rhythms. Such an extended geography of rhythms will raise urgent questions about values beyond the economic measure of rents and pecuniary calculation in general.

It is to values in a more social and corporeal sense that the final chapter of Part III turns. Ann-Cathrine Åquist asserts that everyday life and unpaid housework have been highly undertheorized in economic geography. In Richard Florida's widely read recent book on the rise of the so-called 'creative class,' creativity is the central theme, but everyday life is also taken into account. While Florida argues that the increasing importance of creativity is transforming leisure, community and everyday life in general, his description of the everyday life of the creative class is overwhelmingly dominated by work. The borders between work and leisure are blurred, as the members of this class seem to mix work and play. They spend long hours at their workplaces and afterwards they hang out with their colleagues at sports arenas or hip urban nightspots. Florida's theory, Åquist claims, is interesting and thought provoking, and indeed seems to match several social and economic trends. At the same time, the image of everyday life that Florida sketches is structured by time-space rhythms that emerge only from working life and the specific characteristics of creativity. As Åquist makes clear, it ignores other structuring factors in everyday life such as the time-space rhythms resulting from biological dimensions of existence and responsibility for the well-being of children or others depending on care. Florida appears to be imagining a new and decisive force in economic development that requires women and men to live bachelor lives filled with a mixture of paid work and play, with no commitments to anyone other than colleagues and without consideration of the far more complex world of rhythms affecting human bodily existence. What Åquist and the other authors in Part III make explicit is that many fundamental values of a society come to life in different ways in the practical rhythms of place and landscape. The ideological and corporeal dimensions of embodiment are further examined in the next part.

Part IV gathers four essays on political landscapes and places of corporeality. Nature, as we shall see, recurs mainly in an ideological disguise as the naturalization of political embodiment. Throughout, the relationships between different politics of scale and the personal body or body politic are of vital concern. The rhythm of displacement, belonging, and identity on various geographical scales is a power-filled theme in Anne Godlewska's essay on the seventeenth-century Ursuline mystic, Marie de l'Incarnation. Godlewska locates place at the heart of the complex human struggle for identity. Marie de l'Incarnation carved out a deeply religious identity through conscious manipulation of her bodily existence. In her world, the female body was strictly controlled by means of the religious rejection of the physical body as well as the gendered constitution of society at large. As Godlewska shows, Marie de l'Incarnation imposed a radical ascetic regime on her own body, which enabled her to turn rigorously gendered space of social control into a *topos* of considerable freedom. By actively shaping this corporeal place, she not only transformed her own identity, but simultaneously immersed complicity and resistance into a series of 'nested' places, from the convent and the Church to the town of Quebec, French North America, the spaces of French Catholicism and the eternal realm of God. Flows of information carried by an extensive social network of spiritual writings, personal correspondence, indigenous seminarists, family, and association with the Jesuit fathers in Quebec and France, were pivotal in animating these nested places beyond the confined rhythm of the convent.

'The criteria of rationality and truth in every culture,' Buttimer once noted, 'have always been derived from its foundational myths' (1993, 3). Godlewska's story of the Ursuline nun in her religious and worldly context brings out the importance of the body in reanimating places as sites of social power. It also illustrates how the Ursuline order was structured by orders of the Church and its foundational stories. Gunnar Olsson's chapter approaches ordering social power from the perspective of an ancient cosmology, the Babylonian creation epic *Enuma Elish*. A fascinating example of cartographic reason, this tale unearths astonishingly pertinent rhetorical techniques of social regulation. The metered rhythm of *Enuma Elish* was intended for recitation, rhetorically turning chaos into cosmos and thereby, Olsson argues, concerned with the creation of people and a body politic rather than with the production of things or a religion. In recounting the epic, Olsson guides us from a nameless void-in-itself, positioned in the coordinates of above and below, to the power struggles of the gods, and the eventual act of creation. The world is named, projected, triangulated, and assembled by the 'lord of lords' Marduk, from the corpse of the defeated goddess Tiamat. Throughout, this story of deconstruction, power, and construction discloses an animating series of ontological transformations, in which the power-geared conversion of invisible ideologies into the visible morphology of a Babylonian street pattern leads the way. Yet, as Olsson points out, the most authoritative *topos* of power is always located in a tautological, nameless, and empty control room. Even embodied power hides itself beyond countless names and duplicities.

The idea of social power as an embodied animation of places is explored in the next two chapters in a consideration of landscape and the geographical imaginations of body politic and heritage. Kenneth Olwig and Nuala Johnson turn explicitly to concepts of landscape as embodying 'the people' or their past in Britain and Ulster. Olwig's essay argues that Britain in important ways was constructed by the architects of the Stuart court through the veritable transubstantiation of landscape from the phenomena of place into a scenic illusion in which Britain is envisioned as a bodily entity. The *masque* – a rhythmic form of courtly spectacle, in which masked performers danced and acted, developing into a form of drama in which landscape scenery and music played a key part. Olwig shows how the masque came to be employed as an ideological vehicle in the slippery terrain of British political representation. By designing the landscape as a scenic backdrop in the masques, the Stuart court transposed the ancient idea of landscape as a platial world of customary practice and law, into a spatial discourse in which the monarch was seen as the masculine head of a natural body politic. Thus conceived, the geographical body of Britain could also be imagined as a living political organism, with loyal subjects incorporated in the royal body. As Olwig points out, this politico-geographical discourse and its theatrical enactment was not only gendered and racialized, but also demanded the submission of concrete, cyclical rhythms to abstract, linear ones. The platial world of landscape customarily obliged the monarch to make regular cyclical journeys to conserve a corporate bond with the populace. This was congruent with the ideal of landscape as a country or land shaped through the customary rhythms of dwelling in a particular place. In contrast, the novel scenic discourse was rooted in the geometric reason of single point perspective emanating from the Stuart court and natural law. It enthroned the surveilant royal head as the enlightened focal point of linear vision to commend from an elevated *topos* of power, the geographical body. Underlying much of Olwig's account is the principal mechanism of how landscape as scenery masks the differences between different lands by unifying them within an avowedly natural, abstract space fashioned through natural laws embodied by a natural leader or state. This theme in the conflict between court and country in Britain paralleled in important ways contemporary struggles over political representation and landscape on the European continent.

In the final chapter of Part IV, Nuala Johnson traces the reanimated past through the material and symbolic heritage landscape of the Ulster Folk Museum. Heritage landscapes bring out specific questions about identity and authenticity in reanimating places. The communication of a shared inheritance to the general public tends to provoke tensions between the popular demands of a heritage industry and the academic standards of professional historians. According to traditional views of these tensions, the heritage and tourism industries are guilty of simplification and distortion, and the promotion of nationalistic narratives of places and their past – in sharp contrast to the rigorously neutral realism of genuine historians. However, Johnson claims that according to postmodern questionings of the categorial split between authenticity and illusion, dichotomous interpretations of this kind are no longer tenable. Instead, she points out that the place-based activities of both heritage and history, have their own particular ways of passing on meaning and conveying novel ideas about rural place and the past. A case study of the Ulster Folk Museum (later called the Ulster Folk and Transport Museum) in Northern Ireland also problematizes the dualism. The early years of the museum, in which the cultural geographer Emyr Estyn Evans played a key role, drew attention to the intimate relationship between the representation of the past for popular audiences and the geographical imagination of a 'folk' museum. In the 1950s and 1960s, Evans vision was to preserve representative pieces of material culture highlighting a common regional identity. The preservation effort would avoid romantization of the past and remain politically neutral by emphasizing living tradition and religiously uncontentious material culture. In spite of this, in the volatile political climate of the time, immediate controversy arose about the location of the museum and the selection of exhibits. It is conceivable, as Johnson suggests, that this might have reflected an underestimation of the symbolic constitution of culture. Eventually, however, the museum was located at Cultra Manor outside Belfast and developed with considerable sensitivity for regional diversity, the correct location of elements in the museum landscape, and living culture. In a very real sense, the museum succeeded in materializing and reanimating the imagination of material history as articulated in the discipline of cultural geography. Johnson concludes that the current academic fascination with temporal and geographical authenticity can be seen as a part of a much longer history of imagining the past, and reconstructing and reanimating it for present popular audiences.

The above chapters suggest an intimate contribution of the lapse of time to how individuals experience lifeworld, how groups of people make sense of and plan their environment, or how politico-economic rhythms mold places. In addressing the dynamic existential relationships between time and the making of place, as well as urging for variety and complexity in place-making, Robert Sack's concluding essay in *Part V* tangents core issues of all the earlier chapters. Sack offers a more theoretically geared view of reanimating places in an account of how place and the process of place-making create experiences of time. Place, Sack explains, is made by people through their intentional attempts to bound, regulate, and control an area of space. The idea of place-making tells us how personal, social, economic, and other 'layers' of space are in fact interrelated. Correspondingly, instead of separating layers of time, place-making ties, bends, weaves, and alters its qualities. Importantly, place-making attaches spatial causality at once to physical interactions and boundaries, and to the interaction between appearance and ideology. Knowledgeable and capable human beings weave together the realms or threads constituting reality – nature, meaning, and social relations – in particular ways. The structure and dynamics of place can then be imagined as a loom, at which people bring together the threads of reality. The creative process of weaving transforms places into tools, enabling people to materialize their projects while frustrating others. Everyday place-making is not simply expressed in permanent structures, but remains a process in which the rhythms of nature, meaning, and social relations combine with the dynamism of the loom. At the end of his essay, Sack returns to the issue of dwelling. He considers a series of common existential reactions to place-making, including romantic yearnings for a natural state beyond place-making, totalitarian attempts to ossify place in eternal permanencies, and postmodern celebrations of a volatile and destabilized world. In contrast to these extremes, Sack's existential position revolves around the ideal of reanimating places in a way which increases variety, complexity, and human awareness of the world.

Regrasping Rhythms

This introduction has been fairly long and dense. Therefore, I will keep my conclusion concise. My main concerns have been to locate the idea of a geography of rhythms within the rhythm of geography. I have highlighted the important contribution of Anne Buttimer to the articulation of a geography of rhythms and tried to expand upon some of its core propositions with the help of a range of sources. My central approach was to read the concern with rhythms as a recalcitrant undercurrent in geography, bringing out lived notions of body-subject, lifeworld, place, nature, landscape, and social space. This concern can be understood more generally as a maneuver against reification or topological thinking – against the *dance macabre* of abstract time-space. To grasp the nature of this dance, one needs to identify where it comes from, how it is experienced by different individuals and groups, what its consequences are, and so forth. I argued that a substantial understanding could be distilled from a critical analysis of the alliance of instrumental rationality (scientism) and capitalist society. While humanistic geography has been at pains to criticize the former, Marxist geography has contributed with a critique of the latter. This motivated a movement in this essay from Buttimer's work to critical geographies with a Marxian bent, and in particular Henri Lefebvre's rhythmanalysis. Some readers will no doubt have principled objections against this particular route, and prefer to leave humanistic geography and Marxian geography in the more traditional permanence of mutual exclusivity. Although there is no reason to deny fundamental differences, or to find in all cases and on all levels of enquiry some 'third way' solution, I also think that such bifurcated views tend to efface commonalities brought out by the ideas of rhythms and rhythmanalysis. It seems to me, then, that both currents offer profound insights into the complexity of a relational, lived time-space. Marxian rhythmanalysis has importantly highlighted the functional relationships between time-space rhythms and capitalist modernity. But there are limits to this perspective. There is far more to everyday human experiences than the colonization of lifeworld, as my personal experience of home in Dun Laoghaire with which I started this essay may exemplify. Humanistic rhythmanalysis has rightly secured the intimate complexity of lifeworld experience of capable and social human beings. Taken together, these currents offer a composite relational picture of time-space dynamism on a variety of scales. Finally, the very tension between the rhythms of lifeworld and system reveals that they combine irregularly and in a variety of shifting ways. The precise nature of rhythms, therefore, remains a contingent question, which cannot be fully grasped by topological thinking of any kind.

If new time-space logics and mutual dependencies of politico-economic, technological, personal and intellectual development replace older ones and produce novel sites of social struggle with their own particular consequences, received procedures of research and antiquated cognitive maps may consequently need elaboration or redrawing. Even the limited quantity of time-space rhythms discussed in this collection of essays suggests a plurality of ways in which rhythms can be validly grasped: from their implication in social geographies of different human groups and individuals in daily life, political economies of landscape change, existential phenomenologies of past and present places, to biographical narratives of personal fate and the interpretation of various forms of representation. Which procedure is enrolled depends to some degree on the nature of the rhythms studied, but also on our purposes and expectations. If there is no all-embracing logic producing and transforming all imaginable rhythms, then no single procedure can be expected to explain or understand them. From this point of view, academic work should remain open for a variety of methods rather than encourage a narrowness that would at best suit a limited amount of research areas. In addition to these very practical considerations, if we recognize it or not, the question of method cannot possibly be isolated from broader terrain of social theory and philosophy of science. There remains a definite overlap between philosophical modes of understanding and explanation (including the use of concepts and abstraction) and procedural modes of research design and analysis. In that sense, the chapters in this book do not offer any definite method in the narrow sense, but rather render various possible ways of reanimating places and exploring particular rhythms.

This is certainly not to say that just any method is acceptable, but rather that a recognition of different methodological and philosophical angles will be required to gain a proper understanding of past, present, and possible rhythms. The methodological, epistemological, and ontological construction of these rhythms can be consequential for everyday life and our shared environment. The latter hints at the wider cultural and political resonance of academic work and hence how it may help articulate a liberating sense of a 'politics of rhythms.' For me, this politics would be one, which recognizes a multiplicity of objective and subjective rhythms of time-space and the part of human practices, hopes and fears in their creation. It would be a politics, which confronts the socially and environmentally destructive and unjust dynamics of modern capitalism. It would be one, which explores how rhythms are socially differentiated and contested along the lines of gender, sexual preference, class, and ethnicity. Finally, it would be a politics keen to understand how particular rhythms have become taken-for-granted or reified and thereby aid various kinds of prejudice. Though the contributors may not necessarily subscribe to my personal outlook, these themes do appear in various ways and degrees throughout the volume. They also call for further work on rhythms and the reanimation of places.

As Anne Buttimer once noted: 'For each facet of humanness – rationality or irrationality, faith, emotion, artistic genius or political prowess – there is a geography. For each geographical interpretation of the earth there are implicit assumptions about the nature of humanness. The common ground is vast indeed' (Buttimer 1990, 1-2). The essays in this volume hope to elucidate that for each material or metaphorical location on this common ground, there is an animated rhythm, too.

Notes

1   I believe that this notion of the rhythms of geography connects rather adequately to Buttimer's path-breaking reading of the history of geographical thought and practice in societal context, as used in *Geography and the Human Spirit* (1993). Simplified, this presents the history of geography in different societal contexts as a constant return of three mythical figures (the new life of Phoenix, the structures of Faust, the self-reflexivity of Narcissus); particular vocational meanings (education, critical or analytical thought, problem-solving); root metaphors (mechanism, machine, mosaic, arena).

2   In the late modern effort mentioned here, Kant, Husserl, and Whitehead made crucial contributions. One may also think of Gaston Bachelard's 'topoanalytic' efforts to reanimate the value-imbued and experiential nature of place against abstract space (Bachelard 1994; Buttimer 1980, 167). Casey also mentions Lefebvre's notion of 'differential space,' Foucault's idea of 'heterotopia,' and in progressive notions of dwelling (Casey 1993; 1998).

3   According to Jameson (who frequently invokes the theory of reification), Lukács strategically retranslated and expanded Max Weber's notion of *rationalization* as reification (*Verdinglichung*), betraying a stronger commonality with Weber than with Marx' theory of commodity fetishism (Jameson 1996, 62-63, 220; 2002, 82-85). For Lukács, reification operates, first, as a transformation of actual human interaction into material things with exchange value. Second, reification produces an existential fragmentation resulting from the triumph of isolated commodities and exchange value over the value-complexity and wholeness of social life. Despite his acknowledgement of temporality, and unlike the space-place dialectic emphasized by humanistic geographers and humanistic Marxists like Lefebvre, Lukács seems to accept a shallowly traditional view of space as *abstract* space.

4   For reasons of time, space, and philosophical partiality, my selection here is both limited and deliberate. Consequently, I am disclaiming any completeness in my discussion. As one reviewer noted, I have not considered the recent output of Actor-Network Theory, which

has gained increasing popularity in British geography and thinking about rhythms (Murdoch 1998). From this perspective, the world is seen as a dynamic amalgam of more or less durable networks, made up primarily by contacts, passages, and movements through their constitutive links. Modern technology, nature, culture, and other erstwhile separated entities, are implicated in the construction of all sorts of hybrid forms and flows. On a more philosophical level, this has affinities with 'non-representational theory,' which entails a radical questioning of traditional forms of representation (Thrift 1996). It trades on anti-epistemological theories of practice seeking to bear witness of the constant formation of events, through non-discursive modes of performance. It challenges what it sees as the anaemic political fantasies that go along with belief in the potency of representation. For another recent review of rhythms, which makes more of this (but takes little notice of place), see May and Thrift (2001).

5   For all the niceties in the extensive critique of time-geography, it remains important not to confuse Hägerstrand's original ideas with the far more radically dehumanized cartographic reason of the so-called quantitative revolution in Anglo-American geography. Hägerstrand's ideas of the everyday, quotidian, and profoundly material time-space existence of people and the human body simply cannot be understood without their deep concern with issues of social justice (the politics of welfare distribution in Sweden) and, as Buttimer conceded, humanity's biophysical environment. After all, how can one engage oneself with humans and social life without considering basic material conditions? While working on this introductory chapter, Torsten Hägerstrand assured me that this probably remains one of the most frequently misconstrued and ignored dimensions of his work, resulting from a more general ahistorical reading of 'applied' Swedish geography (see also Hägerstrand 1983).

6   At least in this respect (and I am certainly not denying fundamental epistemological and political discrepancies) there are quite interesting similarities between, on the one hand, Harvey's dialectic of Being and Becoming, and his ontology of place as permanence in process, and, on the other hand, Buttimer's rendering of the everyday existential quest for stability as well as change, home and range. Obviously, where Buttimer pulls attention to the way this dialectic expresses the body-subject's relationship to the rhythms of space and place, Harvey insists upon its functional relationship with the rhythms of capitalism. Despite his musings on the subjective complexity of the Modern, in Harvey's hands Being and Becoming seem to be readily exchangeable with the permanences and processes of capital. Buttimer is less prepared to draw such conclusions and seems to have little patience with functional explanations of this kind. Both, however, hold on to a relational theory that strongly resists the reification of time, space, and place, and their inherent dynamic.

7   In contrast to his work on space and the quotidian, Lefebvre's rhythmanalysis has attracted few commentaries and often very limited interpretations. Rob Shields (1999) has little to say about it in his otherwise informative intellectual biography. Mike Crang (2001, 192-193, 199-201) portrays Lefebvre's theory of rhythms as an attempt of unpacking the phenomenology of places as objects, which remains guilty of spatializing time (a claim based on Henri Bergson's more general critique of *différence*). I believe it is more edifying to comprehend rhythmanalysis as a dialectical method of representing a dynamic world of phenomenally apparent and hidden relations. The 'spatialization' of time is persistently criticized in Lefebvre's work on space and the quotidian. Crang also stakes the contentious theoretical claims that Lefebvre's work on rhythm fails to understand performance, the enduring importance of different temporalities, and the imposition of routinization. McCormack (2002) connects certain aspects of rhythmanalysis to Actor-Network Theory. A more sustained commitment with Lefebvre's ideas of urban rhythms can be found in John Allen's *Lost Geographies of Power* (2003, 159-188). Still, I am not aware of any more thorough Anglophone study of how rhythmanalysis strikes a core note in Lefebvre's

humanistic Marxist theory of the production of space and everyday life. Such a systematic reading cannot, I would argue, be fruitful without incorporation of his extensive engagement with issues such as time *and* space, symbolic *and* material practice, politics, culture, and the taken-for-granted, and his meticulous critique of capitalism (cf. Dimendberg 1998; Gregory 1994; Highmore 2001; Merrifield 1993; Soja 1989).

8   Bachelard, especially in his *Dialectic of Duration* (2000; the French original appeared in 1932), developed a notion of rhythmanalysis that was critical of Henri Bergson's intuition-centered construal of time and space (which accorded privilege to time over space), and instead brought out the importance of practical work and experience. We can add that in Bergson's extended philosophical campaign against 'spatialized' time (the conception of reality as a series of instantaneous snapshots or fixed moments rather than a continuous flow), a notion of the different rhythms of lived duration was married with an uncritical conceptualization of space as abstract space. Bergson's temporocentrism thereby also erased place (Ansell-Pearson 2001, 28 ff.; Casey 1997, 211; Bergson 1920; 1990).

9   Epistemological break is Althusser's illustrious translation of Bachelard's concept of discontinuity, borrowed for the notoriously anti-humanistic argument that from 1845 onwards (with *The German Ideology* and *Theses on Feuerbach*), Marx would have broken with his youthful 'ideological' Hegelianism to develop a new, and this time 'scientific' philosophy (Althusser 1969). This is not to say that Buttimer or Lefebvre adhere to any simplistic dichotomous categorization of, for instance, lived time-space and mechanical, de-contextualized time-space. I like to think of these 'breaks' as openings, a bringing into view of the lived and social beyond the smooth surface of abstraction.

References

Adam, B. (1990) *Time and social theory*. Polity Press, Cambridge.
Adorno, T.W. (1996) *Negative Dialektik*. Suhrkamp, Frankfurt am Main.
Adorno, T.W. (1997) *Aesthetic theory*. Athlone, London.
Adorno, T.W. and Horkheimer, M. (1997) *Dialectic of enlightenment*. Verso, London.
Allen, J. (2003) *Lost geographies of power*. Blackwell, Oxford.
Althusser, L. (1969) *For Marx*. Penguin, Harmondsworth.
Ansell-Pearson, K. (2001) *Philosophy and the adventure of the virtual*. Routledge, London.
Bachelard, G. (1994) *The poetics of space*. Beacon Press, Boston.
Bachelard, G. (2000) *The dialectic of duration*. Clinamen, Manchester.
Barrett, W. (1962) *Irrational man: a study in existential philosophy*. Anchor Books, New York.
Bergson, H. (1920) *Creative evolution*. Macmillan, London.
Bergson, H. (1990) *Time and free will: an essay on the immediate data of consciousness*. Zone Books, New York.
Berman, M. (1988) *All that is solid melts into air*. Penguin, Harmondsworth.
Bhaskar, R. (1979) *The possibility of naturalism: a philosophical critique of the contemporary human sciences*. Harvester Press, Brighton.
Bhaskar, R. (1993) *Dialectic: the pulse of freedom*. Verso, London.
Buttimer, A. (1968) 'Social geography' in *Macmillan's revised international encyclopedia of the social sciences*. The Free Press, New York, pp. 134-145.
Buttimer, A. (1969) Social space in interdisciplinary perspective, *The Geographical Review* 59, pp. 417-426.
Buttimer, A. (1971) *Society and milieu in the French geographic tradition*. Rand McNally, Chicago
Buttimer, A. (1972) Social space and the planning of residential areas, *Environment and Behaviour* 4, pp. 279-318.
Buttimer, A. (1974) Values in geography, *Commission of College Geography Resource Paper* 24, Association of American Geographers, Washington, DC.

Buttimer, A. (1976) Grasping the dynamism of lifeworld, *Annals of the Association of American Geographers* 66, pp. 277-292.

Buttimer, A. (1978) On people, paradigms, and 'progress' in geography, *Rapporter och notiser* 47, Department of Human Geography, Lund University, Lund.

Buttimer, A. (1980) 'Home, reach, and the sense of place' in Buttimer, A. and Seamon, D. (eds.) *The human experience of space and place.* St. Martin's Press, New York, pp. 166-187.

Buttimer, A. (1982) Musing on Helicon: root metaphors and geography, *Geografiska Annaler* 64B, pp. 89-95.

Buttimer, A. (1983) *The practice of geography.* Longman, London.

Buttimer, A. (1986) Classics in human geography revisited: author's response, *Progress in Human Geography* 29, pp. 517-519.

Buttimer, A. (1987) A social topography of home and horizon: the misfit, the dutiful, and longing for home, *Journal of Environmental Psychology* 7, pp. 307-319.

Buttimer, A. (1989) 'Mirrors, masks, and diverse milieux' in Boal, F.W. and Livingstone, D. (eds.) *The behavioural environment.* Routledge, London, pp. 253-276.

Buttimer, A. (1990) Geography, humanism, and global concern, *Annals of the Association of American Geographers* 80, pp. 1-33.

Buttimer, A. (1993) *Geography and the human spirit.* The Johns Hopkins University Press, Baltimore, MD.

Buttimer, A. (1998) Landscape and life: appropriate scales for sustainable development, *Irish Geography* 31, pp. 1-33.

Buttimer, A. (2001) 'Sustainable development: issues of scale and appropriateness' in Buttimer, A. (ed.) *Sustainable landscapes and lifeways.* Cork University Press, Cork, pp. 7-34.

Buttimer, A., Brunn, S.D. and Wardenga, U. (eds.) (1999) *Text and image: social construction of regional knowledges.* Institut für Länderkunde, Leipzig.

Carlstein, T., Parkes D. and Thrift, N. (eds.) (1978) *Timing space and spacing time* (three volumes). Edward Arnold, London.

Casey, E.S. (1993) *Getting back into place: toward a renewed understanding of the place-world.* Indiana University Press, Bloomington and Indianapolis.

Casey, E.S. (1997) *The fate of place: a philosophical history.* University of California Press, Berkeley.

Casey, E.S. (1998) The production of space or the heterogeniety of place, *Philosophy and Geography* 2, pp. 71-80.

Collier, A. (1994) *Critical realism: an introduction to Roy Bhaskar's philosophy.* Verso, London.

Cosgrove, D. (1984) *Social formation and symbolic landscape.* Croom Helm, London.

Crang, M. (2001) 'Rhythms of the city: temporalised space and motion' in May, J. and Thrift, N. (eds.) *TimeSpace: geographies of temporality.* Routledge, London, pp. 187-207.

Debord, G. (1995) *The society of the spectacle.* Zone Books, New York.

Dimendberg, E. (1998) Henri Lefebvre on abstract space, *Philosophy and Geography* 2, pp. 17-48.

Durkheim, E. (1965) *The elementary forms of religious life.* The Free Press, New York.

Entrikin, N. (1976) Contemporary humanism in geography, *Annals of the Association of American Geographers* 66, pp. 615-632.

Evernden, N. (1985) *The natural alien: humankind and environment.* University of Toronto Press, Toronto.

Foucault, M. (1977) *Discipline and punish.* Penguin, Harmondsworth.

Foucault, M. (1981) *The history of sexuality.* Penguin, Harmondsworth.

Foucault, M. (1986) Of other spaces, *Diacritics* 16, pp. 22-27.

Giddens, A. (1990) *The consequences of modernity.* Polity Press, Cambridge.

Glacken, C. J. (1967) *Traces on the Rhodian shore: nature and culture in western thought from ancient times to the end of the eighteenth century.* University of California Press, Berkeley.

Godlewska, A. (1995) Map, text and image. The mentality of enlightened conquerors: a new look at the *Description de l'Egypte, Transactions of the Institute of British Geographers* 20, pp. 5-28.

Gosden, C. (1994) *Social being and time.* Blackwell, Oxford.

Gregory, D. (1981) Human agency and human geography, *Transactions of the Institute of British Geographers N.S.* 6, pp. 1-18.

Gregory, D. and Urry, J. (1985) *Social relations and spatial structures.* Macmillan, London.

Gregory, D. (1994) *Geographical Imaginations.* Blackwell, Oxford.

Gren, M. (1994) *Earth writing: exploring representation and social geography in-between meaning/matter.* Department of Geography, University of Gothenburgh Series B 85.

Gurjewitsch, A.J. (1978) *Das Weltbild des mittelalterlichen Menschen.* Verlag der Kunst, Dresden.

Habermas, J. (1987) *The theory of communicative action* (volume two). Beacon Press, Boston.

Hägerstrand, T. (1983) 'In search for the sources of concepts' in Buttimer, A. *The practice of geography.* Longman, London, pp. 238-256.

Hägerstrand, T. (1991) 'Tidsgeografi' in Carlestam, G. and Sollbe, B. (eds.) *Om tidens vidd och tingens ordning.* Byggforskingsrådet, Stockholm, pp. 133-142.

Harvey, D. (1990) *The condition of postmodernity: an enquiry into the origins of cultural change.* Blackwell, Oxford.

Harvey, D. (1996) *Justice, nature and the geography of difference.* Blackwell, Oxford.

Heidegger, M. (1962) *Being and time.* Blackwell, Oxford.

Highmore, B. (2001) 'Henri Lefebvre's dialectic of everyday life' in Highmore, B. *Everyday life and cultural theory.* Routledge, London, pp. 113-144.

Hobsbawm, E.J. (1975) *The age of capital 1848-1875.* Weidenfeld and Nicolson, London.

Hobsbawm, E.J. (1999) *Industry and empire: from 1750 to the present day* (revised edition). Penguin, Harmondsworth.

Jackson, P. (1989) *Maps of meaning.* Routledge, London.

Jameson, F. (1991) *Postmodernism, or, the cultural logic of late capitalism.* Duke University Press, Durham, N.C.

Jameson, F. (1996) *The political unconscious.* Routledge, London.

Jameson, F. (2002) *A singular modernity: essay on the ontology of the present.* Verso, London.

Jameson, F. (2003) The end of temporality, *Critical Enquiry* 29, pp. 695-718.

Kern, S. (1983) *The culture of time and space 1880-1918.* Harvard University Press, Cambridge, Mass.

Lash, S., Quick, A. and Roberts, R. (eds.) (1998) *Time and value.* Blackwell, Oxford.

Lefebvre, H. (1976) Reflections on the politics of space, *Antipode* 8, pp. 30-36.

Lefebvre, H. (1978) *The survival of capitalism. Reproduction of the relations of production.* Allison and Busy, London.

Lefebvre, H. (1984) *Everyday life in the modern world.* Transaction Publishers, London.

Lefebvre, H. (1991a) *Critique of everyday life* (volume one). Verso, London.

Lefebvre, H. (1991b) *The production of space.* Blackwell, Oxford.

Lefebvre, H. (1992) *Elements de rhythmanalyse: introduction a la connaissance des rhythmes.* Edition Syllepie, Paris.

Lefebvre, H. (1995) *Introduction to modernity.* Verso, London.

Lefebvre, H. (1996) 'Seen from the window' and 'Rhythmanalysis of Mediterranean cities' in Kofman, E. and Lebas, E. (eds.) *Writings on cities.* Blackwell, Oxford, pp. 219-240.

Lefebvre, H. (2002) *Critique of everyday life volume 2: foundations for a sociology of the everyday.* Verso, London.

Ley, D. (1977) Social geography and the taken-for-granted world, *Transactions of the Institute of British Geographers N.S.* 2, pp. 498-512.

Ley, D. and Samuels, M.S. (1978) *Humanistic geography: prospects and problems.* Croom Helm, London.

Liedman, S-E. (1997) *I skuggan av framtiden: modernitetens idéhistoria.* Albert Bonniers Förlag, Stockholm.

Longhurst, R. (2001) *Bodies: exploring fluid boundaries.* Routledge, London.

Lukács, G. (1971) *History and class consciousness.* Merlin Press, London.

Malpas, J. (1998) Finding place: spatiality, locality, and subjectivity, *Philosophy and Geography* 3, pp. 21-43.

Malpas, J. (2003) Bio-medical *Topoi* – the dominance of space, the recalcitrance of place, and the making of persons, *Social Science and Medicine* 56, pp. 2343-2351.

Marx, K. (1954) *Capital* (volume one). Progress Publishers, Moscow.

Marx, K. (1973) *Grundrisse: foundations of the critique of political economy*. Penguin, Harmondsworth.

Massey, D. (1993) 'Power-geometry and a progressive sense of place' in Bird, J., Curtis, B., Putnam, T., Robertson, G. and Tickner, L. (eds.) *Mapping the futures: local cultures, global change*. Routledge, London, pp. 59-69.

Massey, D. (1994) *Space, place and gender*. Polity Press, London.

Massey, D. (1999) 'Spaces of politics' in Massey, D., Allen, J. and Sarre, P. (eds.) *Human geography today*. Polity Press, Oxford, pp. 279-294.

May, J. and Thrift, N. (2001) 'Introduction' in May, J. and Thrift, N. (eds.) *TimeSpace: geographies of temporality*. Routledge, London, pp. 1-46.

McCormack, D.P. (2002) A paper with and interest in rhythm, *Geoforum* 33, pp. 469-485.

Meinig, D. (ed.) (1979) *The interpretation of ordinary landscapes*. Oxford University Press, Oxford.

Mels, T. (2002) Nature, home, and scenery: the 'official' spatialities of Swedish national parks, *Environment and Planning D: Society and Space* 20, pp. 135-154.

Merleau-Ponty, M. (2002) *Phenomenology of perception*. Routledge, London.

Merriam-Webster (1993) *Webster's third new international dictionary of the English language*. Könemann, Cologne.

Merrifield, A. (1993) Place and space: a Lefebvrian reconciliation, *Transactions of the Institute of British Geographers* 18, pp. 516-531.

Mitchell, D. (1996) *The lie of the land: migrant workers and the California landscape*. University of Minnesota Press, Minneapolis MN.

Mitchell, D. (2000) *Cultural geography: a critical introduction*. Blackwell, Oxford.

Mitchell, W.J.T. (2000) Holy landscape: Israel, Palestine, and the American wilderness, *Critical Enquiry* 26, pp. 193-223.

Murdoch, J. (1998) The spaces of actor-network theory, *Geoforum* 29, pp. 357-374.

OED (1989) *The Oxford English dictionary* (second edition). Clarendon Press, Oxford.

Olsson, G. (1978) 'Of ambiguity or far cries from a memorialising mamafesta' in Ley, D. and Samuals, M.S. (eds.) *Humanistic geography*. Croom Helm, London, pp. 109-120.

Olwig, K. (1984) *Nature's ideological landscape: a literary and geographic perspective on its development and preservation on Denmark's Jutland heath*. George Allen and Unwin, London.

Olwig, K. (1996) 'Nature: mapping the ghostly traces of a concept' in Earle, C., *et al.* (eds.) *Concepts in human geography*. Rowman and Littlefield, Lanham Md, pp. 63-96.

Parkes, D. and Thrift, N. (1979) Time spacemakers and entrainment, *Transactions of the Intitute of British Geographers NS* 4, pp. 353-372.

Parkes, D. and Thrift, N. (1980) *Themes, spaces, places: a chronogeographic perspective*. John Wiley, Chichester.

Perec, G. (1997) *Species of spaces and other pieces*. Penguin, Harmondsworth.

Pickles, J. (1987) *Geography and humanism*. Geobooks, Norwich.

Pile, S. and Thrift, N. (eds.) (1995) *Mapping the subject: geographies of cultural transformation*. Routledge, London.

Pred, A. (1981) Social reproduction and the time-geography of everyday life, *Geografiska Annaler* 63B, pp. 5-22.

Pred, A. (1984) Place as historically contingent process: structuration and the time-geography of becoming places, *Annals of the Association of American Geographers* 7, pp. 279-297.

Priest, S. (1998) *Merleau-Ponty*. Routledge, London.

Relph, E. (1976) *Place and placelessness*. Pion, London.

Relph, E. (1981) *Rational landscapes and humanistic geography* Croom Helm, London.

Rifkin, J. (1987) *Time wars: the primary conflict in human history*. Henry Holt, New York.

Rodaway, P. (1994) *Sensuous geographies: body, sense and place.* Routledge, London.
Rose, G. (1993) *Feminism and geography: the limits of geographical knowledge.* Polity Press, Cambridge.
Said, E. (2000) Geopoetics: space, place, and landscape, *Critical Enquiry* 26, pp. 173-192.
Sartre, J-P. (1949) *Nausea.* New Directions, London.
Sartre, J-P. (1956) *Being and nothingness: an essay on phenomenological ontology.* Philosophical Library, New York.
Seamon, D. (1979) *A geography of the lifeworld: movement, rest and encounter.* Croom Helm, London.
Shields, R. (1999) *Lefebvre love and struggle: spatial dialectics.* Routledge, London and New York.
Soja, E.W. (1989) *Postmodern geographies: the reassertion of space and time in critical social theory.* Verso, London.
Thrift, N. (1983) On the determination of social action in space and time. *Environment and Planning D: Society and Space* 1, pp. 23-57.
Thrift, N. (1996) *Spatial formations.* Sage, London.
Tuan, Y-F. (1977) *Space and place: the perspective of experience.* University of Minnesota Press, Minneapolis, MN.
Tuan, Y-F. (1978) 'Space, time, place: a humanistic frame' in Carlstein, T., Parkes, D. and Thrift, N. (eds.) *Timing space and spacing time* (volume one). Edward Arnold, London.
Wallerstein, I. (1998) The time of space and the space of time: the future of social science, *Political Geography* 17, pp. 71-82.
Whitehead, A.N. (1929) *Process and reality: an essay in cosmology.* Cambridge University Press, Cambridge.
Williams, R. (1989) 'Metropolitan perceptions and the emergence of Modernism' in Williams, R. *The politics of Modernism.* Verso, London, pp. 37-48.
Wright, G.H. von (1986) *Vetenskapen och förnuftet.* Bonier, Stockholm.
Young, M. (1988) *The metronomic society: natural rhythms and human timetables.* Thames and Hudson, London.
Zahavi, D. (2003) *Husserl's phenomenology.* Stanford University Press, Stanford.
Zerubavel, E. (1981) *Hidden rhythms: schedules and calendars in social life.* The University of Chicago Press, Chicago and London.

PART II
REANIMATING PLACE AND DISPLACEMENT

Chapter 2

Sense of Place: Its Relationship to Self and Time

Yi-Fu Tuan

Sense of place would seem clearly a function of time: a period of time must lapse before one can have a sense of place. Yet this is not quite right for, as we shall see later, we can identify with a place immediately. More true is this: place must stop changing for a human being to be able to grasp it and so have a sense of it. Some places change so slowly that, from a human perspective, they are timeless. Large natural features – mountains, forests, and rivers – are outstanding examples. People come and go, generations pass, but the mountain or river stays much the same. Some old human habitations seem changeless. Of course, they have had a history, but that history – history as development – came to a stop, or seems to have come to a stop; and thereafter, human beings see it and remember it as changeless. A key characteristic of modern times, as we all know, is the rapidity and ubiquity of change. What was scrubland is now a shopping mall. Our hometown is barely recognizable after an absence of ten years. A whole mountaintop can be removed in a matter of weeks to gain access to coal. The last event is perhaps the most disturbing, for if even the hills – those 'eternal hills' of old eloquence – are not fixtures, what are? How can we develop a sense of place – of any place – if nothing stays put?

Even if places stay put and change little with time, human individuals do not. They age. The child sees the mountain or village one way, the adult another. At what period in our growth is our sense of place fixed? Not in childhood when every year brings about a new way of seeing and understanding. The answer would have to be maturity – a phase in life conceived as a standstill of some duration in the human life cycle, somewhat analogous to the solstice in the passage of the sun. In the course of this standstill, a firm sense of place develops that alters little thereafter. Our sense of place thus stabilized, we count on the material places themselves to be stable, especially those that are important to our emotional well-being. Commercial streets and towns can grow; their growth may even be welcomed as evidence of prosperity. But our home and hometown should stay the same. Now, why 'should'? How come a moral command has slipped into the language? I suggest that a 'should' or 'ought' slips in – and that it does so naturally – because we tend to confuse stability with integrity. What is stable may or may not be good, but what is integral and whole is good and should therefore stay that way. It is thus that we see our neighborhood, hometown, or countryside; and it is thus that we see our mature selves.

I have broached three principal ideas. The first is the necessity of pause – of time that has stood still – for a sense of place to take hold. The second is the emotional quality of 'sense' in the sense of place – a quality that is commonly assumed to be positive. I shall explore the positive emotion, even though I am well aware that the emotion can also be strongly negative: some places exude an air of evil for good reason, others appear inexplicably sinister. The third idea is the intimate tie between place and self: the language used of the one is much the same as the language used of the other; in both, aesthetic and moral terms abound (Sack 1997, 127-152). An example that I have already offered is integrity – how, in people's minds, the integrity of place and the integrity of self tend to merge. As I delve into these three sets of ideas, I find that what I had thought was a narrow and manageable topic is in fact quite broad, that in trying to understand sense of place, I am also trying to understand what it means to be human.

Home

Sense of place can be a passing emotion, like a fragrance; and it can be rich, deep, and enduring, with as many elements and layers as those that constitute a human personality. Home is the outstanding example of a strong and rich sense of place. It is so for reasons that are fairly obvious. Home caters to our basic biological needs. It is redolent of freshly baked bread and unwashed linen, dust and waxed furniture. It is where we find security and store food, where we recover from sickness under the solicitous eyes of kinsfolk – a touching arrangement that, by the way, is absent among other primates (Washburn and DeVore 1961, 101). Home is the scene of sexual congress, and, in the past, of birth and death. Home is a social world in which human relations are intimate and can be exceptionally intense. In modern times, home provides private space to which we can withdraw, freed from biological needs and social distractions, to recollect our thoughts and dream dreams. Also in modern times, we may have lived in several homes by late adulthood, but the place where we grew up is home in a special sense, with a foundation of meaning not available to other homes, for childhood is the period when our senses are most acute and our imagination most vivid.

The place – the home – to which one can become strongly attached, exists at different scales. To a young child, his bed or play room, rather than the whole house, may be his real home: he misses it the most when he is away at camp. To adults, the entire house and its grounds are home. Beyond that, in progressive figurative extension, home is neighborhood, city, countryside, nation, and, ultimately, the earth. Our sense of home in this extended sense weakens as we move beyond the house. But the gradient is not smooth, for arguably, our feelings about nation are stronger than our feelings about neighborhood or city. Uncomfortable as it may be to leave our familiar neighborhood for another town, it is nothing like the wrenching we feel when we abandon our citizenship and live in another nation. Loyal as we may be to our neighborhood, we are not obliged to defend it with our lives against the city council's bulldozer, as we may be obliged to defend our nation with our lives when it is attacked by enemy tanks.

Patriotism

Patriotism is not commonly considered an example of a sense of place. But why not? Patriotism may indeed be the most fervent sense of place now in existence. It is fervent, it can be passionate, because it can draw on all the inarticulate feelings that we have for our house with the picket fence and augment them with articulated images, ritual, rhetoric, and music. Note a peculiar bias among the well-educated, which include you and me, in our willingness to endorse an emotion or admit a feeling. We readily accept as our own the feeling about place that nature and pastoral poets evoke. We resist, however, the rhetoric of poets of patriotism. Is it because we consider it flag-waving jingoism? Maybe. Even so, are such lapses more reprehensible than the sentimentality that poets of nature and the rustic scene are all too prone? Neither failing is, in any case, inevitable. Wordsworth is not sentimental in his depiction of haunted villages and lakes, and Shakespeare is not jingoistic when he describes England as 'This happy breed of men, this little world, this precious stone set in the silver sea' (1984, act 2, scene 1).

Is size the cause of differing emotion? Are we naturally sentimental toward the small and patriotic toward the large? Not necessarily. Consider the planet earth. We can feel affectionate, but also proud, toward it. Shakespeare's eloquent words apply not only to England, but (with some modification) to the entire planet, which may well be described as a precious stone set in boundless space, home to happy breeds of creatures.

Stability

Inherent to the idea of home, at whatever scale, is stability. We expect home to be recognizably the same when we return to it after a period of absence. Our own sense of self depends on such stability. When the neighborhood we grew up in is demolished, we feel as though a part of our own personality is undone; and this would still be true even when the old neighborhood has been improved. We recognize the improvement, but what has it to do with us at the deepest level? A city can change through demolition and construction at the center, and through expansion at the edge. My Minneapolis has done both since I left it in 1983. I cannot return to my Minneapolis. Whatever sense of it I still retain is no longer supported by tangible reality. But, interestingly, I can return to Minnesota. Its political borders have remained the same, and although many internal alterations have occurred, these are swallowed up by the state's ample space.

Can the same be said of the earth? It still has the same borders, and whatever alterations have occurred on its surface are modest by comparison with the size of the planet. But this, though true through much of human history, is no longer true. Astronaut Frank Culbertson, from his window in the international space station, tells us in September, 2001 that our planet has changed notably for the worse – more dust and smoke, less forested areas – since his last mission in 1990 (*New York Times*, September 1, 2001). I translate this to mean: after ten years, Culbertson sees a different earth and feels that he cannot return to the same planet again. Seeing the planet from space as astronauts do, and as we do through photographs, should persuade us to make a greater effort to preserve areas that are still more or less natural. Any number of sound ecological reasons exist for preservation, but the reason I wish to emphasize here is the human need for stability – for a place where time seems to stop, a place to which we can return, a place whose wholeness and integrity confirm our own.

Love at First Sight

Home, as I have said, has the richest meaning. Less rich, but not necessarily less powerful, are certain places that we have seen in passing and may periodically revisit. The difference between them, emotionally, is like the difference between a love that has grown over the years, and a love that hits one like a bolt out of the blue. The first love is multi-layered and almost impossible to express without drifting into sentimental vagueness. The second love, perhaps because it has fewer layers, is more easily expressed, but this does not mean that it is any less intense.

The kinds of place to which one can form immediate attachment are many – almost as many as there are cultures and individuated human beings. Let me tell you about mine. It is the desert, which I encountered for the first time in 1952. Desert was never a part of my childhood experience, yet when I opened my eyes early one morning and saw the landscape of Death Valley in California bathed in subdued sunlight, I was not only astonished by the strangeness, but also consoled by its familiarity. The strangeness is understandable, but how can I account for the familiarity – for the sense that I have come home? Upon reflection, my answer is that I saw in the desert landscape, as I could not see in any other kind of landscape, the objective correlatives of my deepest values and beliefs: simplicity, clarity, purity, openness, generosity that is as encompassing as the sky, and space that is free of the lurid dramas of survival.

If I have fallen for the desert, other people have fallen for the mountain, the forest, or the seashore, to itemize just a few examples. Note that I am not now considering these landscapes as habitat. As habitat – and even the bleakest desert can be habitat for some people – they will have the richness of meaning common to home that I have referred to earlier. No. I am now considering landscapes beyond home that one may have encountered more or less by chance, and with which one forms, unexpectedly, a deep and immediate bond. Falling for a landscape or place at first sight is not, however, quite the mystery that I seem to be making it, because culture may have prepared one for the romance at a subconscious level. Consider what happened in Europe around 1700. For a variety of economic and technological reasons, cultural attitude reversed itself so that a mountainous landscape, formerly shunned, is seen in a positive light (Nicolson 1959). Yet just because European culture had learned to favor mountains did not mean that all Europeans came under their sway. Moreover, those who did might have been following a fashion rather than revealing their selves' deepest affinity. Following a landscape fashion can make for pleasant experiences in that landscape, but it is not the same as a sense of place that is rooted in the core of one's being, which can haunt one's dreams and drive one to return to a place, either in actuality or in imagination, again and again.

When I return to the Arizona desert, I see a landscape that is much as I remember it. Time has stood still, and I am grateful. The same can be said of other wildernesses. By remaining the same, they tell us that, for all the loss of hair and accretion of weight that make us almost unrecognizable to ourselves in the mirror, our basic values and selves remain more or less intact. Can a rural landscape dotted with small farms perform the same service? Apparently it can; and the reason is that, to many of us who are not farmers and do not actively transform the land, such a landscape can seem natural rather than humanly made. We yearn to protect it, freeze it in time as we do wildernesses, so that it can remind us of who we are; or rather, who we like to think we are – decent and hardworking people, immersed in the fundamentals of livelihood, close to nature and close to the earth.

Paintings

So far I have used the word 'place' as geographers understand it. I would like now to extend its meaning, for I believe that such an extension enriches geographical place by providing it with – as it were – a horde of 'cousins,' twice and thrice removed. Who or what are these cousins? They are the works of art. A landscape painting, for example. The real desert may be threatened by human encroachment, but the painting of it in my possession is not. My painting is immune to change; it is where time has come to a stop. I can always return to the painting for sustenance, provided, of course, that it is truly a work of art, able to capture some of the qualities of the real desert. The painting, at one level, is purely visual and so cannot be expected to stimulate all my senses: standing before it, I do not feel the sun warming my face while my back remains cool, a common enough experience in the desert. Yet this is not quite right, for the visual is synesthetically connected with the other senses; the better the artist, the more powerful is the synesthetic connection and effect such that when I look at a painting I do not see a flat surface covered in paint but rather a three-dimensional world that I can enter and, while there, be comforted or inspired.

Photographs

Photographs are another sort of surrogate place. Compared with paintings, photographs are deficient – lacking in certain aesthetic qualities – largely because the artist's control over them is incomplete. On the other hand, precisely because they are beyond the artist's total control, photographs exude a pathos peculiar to the real that paintings do not have. A photograph is not just a representation of reality, it *is* a bit of reality transmitted to and retained by a coated surface through physical and chemical means. When we see a bright patch on the photograph, we know that it is put there more by a bare cliff deflecting sunlight than by the artful manipulations of the photographer. When we look at the smile of a child in a faded photograph and smile back, we know that we are responding not just to a painter's illusionary skills, but to the smile of a real child – now probably an old woman or dead – who once stood five feet from the camera.

Unplanned elements in a photograph also contribute to a sense of reality. In real life, as in a photograph, the beer can on top of the fridge, the tumbleweed wedged between two boulders, are not put there by design: they are there either accidentally or through the operation of physical laws. Paintings, by contrast, typically exclude the accidental: everything in a painting is where it is for an artistic reason. Awareness of this fact makes even the most realistic painting a little unreal. Another difference between a painted picture and a photograph is this. Whereas both enable the viewer to contemplate process, the painting does not tell the viewer that an end or consummation is imminent. The photograph does. John Keats famously says that the lovers painted on the Grecian urn will never kiss, and we know he is right. However, if we see these lovers in a photograph, we know as we do when we see them in real life that, unless heaven falls, their lips will touch.

Early photographs were confined by their technical limitations to recording immobile objects. As technology improves, more and more pictures are taken of things on the move and even in rapid motion. Through exposure to both kinds of photographs, we develop the habit of dwelling imaginatively not only on stable objects – houses, neighborhoods, and landscapes, but also on objects in transit – the train pulling out of the station, the hand raised to wave good-bye. Objects so caught give the places in which they occur a vibrancy or poignancy that these places would not otherwise have. Photographs thus promote our sensitivity to place – our sense of place. They are also surrogate places in that we can visit and revisit them whenever our spirit needs rest or rejuvenation.

Movies

Later than the photograph in establishing a hold on people's consciousness is the movie; hence one might think of it as a more distant cousin of place. But why consider it a relation of place at all? On first thought, the movie cannot be place in any sense because it embraces time. In a movie, one event follows another. Even in its quiet moments, when time seems to slow down, a neighbor may still be shown mowing the lawn, a cat chasing a ball. The movie reminds us too much of the restlessness of real life, even exaggerating it, making us yearn for a pause, for a place where we can forget time and motion. Yet one reason we go to the movies is to take a breather from the demands of time – from having to do things now rather than later. Movies, it turns out, are as much bounded space – a world or place – as it is bounded time. Indeed, movies are more place than time because, whereas we can never go back and live in an earlier time, we can return to a movie. And that's how we treat our favorite movies. If we see them more than once, it is not so much to understand their plot better as to return to worlds that, for some reason, have the power to enthrall.

Movies, I say, are places. But maybe I speak in metaphors. It certainly is true that movies can evoke place. They do so with images and sounds: normal images of the sort that people experience at ground level and bird's eye views that are no part of daily human experience; normal sounds such as those of shrieking kettle indoor and the swoosh of passing cars outdoor, and sounds that are no part of daily life, by which I mean the music that the movie director adds to a scene, subtly in the background or loudly in the foreground, to create or enhance a mood.

Let me now make a confession. I have seen the movie *Gone With the Wind* at least a dozen times, the first time in 1942, when I was a child. My English was not good then. I could not understand much that was said. The Civil War, as history, meant nothing to me. The South, as geography, meant little more. So, why did the movie have this kind of grip? Could it be the war itself, always exciting to a child? The plot line? Yes, in part. But these excitements could not outlast the first viewing, and yet I returned to the movie again and again. I returned to it as I would to a fond place. For, after several viewings, *Gone With the Wind* became a place for me. The movie implanted in my awakening imagination a South of red earth, cotton fields, plantation houses, afternoon siestas and elegant parties. But it also awakened in me, through simplistic images and story lines that a child could understand, longings for romance, gallantry, noblesse oblige, courage, and indomitable will that remain with me unassuaged to this day, making me slightly ridiculous in the eyes of sophisticated friends who know the real, far darker history of the South.

Poems

We human beings may be primarily visual animals, but even more distinctive of us is language. Our language intertwines with place, its construction and maintenance, at every level, from the practical to the aesthetic and moral. Consider just the aesthetic, remembering all the while that the aesthetic entails the moral, that the one often merges seamlessly with the other. Sense of place is an aesthetic sensibility, and something we all have – more or less. If poets are different, it is because they have an exceptionally strong sense of place, and, more important, they can articulate it, so that their sense and sensibility can be shared with others. When I go to the Lake District or to the Wye River where the ruins of the Tintern Abbey stand, I would want to bring Wordsworth's poems with me. Reciting them quietly to myself can make what is diffuse take form and so be both more memorable and easier to remember. Wordsworth's poems may also embody the image and spirit of places that no longer exist, or have been radically altered. These poems act like old photographs: they become place substitutes that we can revisit in the imagination. I have revisited London of two hundred years ago, at the Westminster Bridge, many times; and can say with Wordsworth that 'Earth has not anything to show more fair,' and that 'Dull would he be of soul who could pass by/ A sight so touching in its majesty.'

Captured in this poem is not just the geography, but also the mood. It is the preternatural calm and silence of early morning in a great city filled with (as Wordsworth puts it) 'ships, towers, domes, theatres, and temples' that I wish to drink in again and again. This mood, transfixed by language, is for me a haven, a shelter, an oasis. I would go a step further and say that mood poems as such, even those with little specific geography in them, are places for me. They may not nurture my body, but they do revive my spirit. They do this by objectifying and clarifying my own inchoate feelings and longings.

Stories and Novels

Now, what about stories and novels? Are they also a type of place, related to place, cousins to place? As with movies, one might be inclined to say 'no,' because stories have a clear temporal dimension not only in the sense that they take time to tell or read, but that they have a narrative line or plot that depends on some sort of temporal sequencing. In other words, stories simulate life, and life is process and change; or, to put this a little more dramatically, life is borne forward by time whether we will or not. This relentless forward propulsion makes us want it to 'stop,' so that we can get off and stay put to recollect, savor, and enjoy. But life in its forward propulsion is also exciting, full of adventure, as are many stories. In life, as in stories, we want to know what happens next. Here, then, is another rub. Once we know, the excitement ends. Most of us, I believe, do not want to live our life all over again. Much as we fear death, repeating our life, if it were possible and offered, could seem more a punishment than a privilege, unless the first run is totally forgotten. Yet, as young children, we like to hear the same story over and over again; as older children we trust our favorite author to recreate the same world in book after book; and as adults we revisit our favorite novels as we would our alma mater or hometown. A similar response applies to stories and novels. Obviously, we do not go back to them for adventures whose outcome we already know. We go back to them, then, for the evocation of smells, sights, and sounds, and for a certain moral quality that diffuses through them, as it does through real places, like a subtle quality of light.

Music

Lastly, in this enumeration of arts, consider the supreme art of music. Since it occurs in time, it might be expected to remind listeners of time's pitiless passage. Yet this is not the case. Music seems able to annul time, to convert it into a sort of atemporal presence, or place. One age-old device used to achieve this illusion is to make the fundamental rhythms of life – the heartbeat and breathing in-and-out – not only audible but also hypnotically insistent. Noting the beats and mentally adding them up is certainly a type of time awareness; yet subjectively, if the beats have the period and invariance of a biological rhythm, they can have a calming effect and nullify the sense of time. Another conception of time, again taking nature as a model, is the circle or cycle. Everything in nature seems to move in circles – from the migratory path of birds to the migratory path of the sun. Life itself is a circular process of birth, life, and death, repeated over and over again. This conception of time is restful, for everything goes around and nothing is permanently lost. Time so conceived easily translates into place. The illusion is given that we can return to an earlier time, if we just wait long enough, as we can return to place.

Music, primordially, is the beat – reassuring even when it is the loud, repetitive 'boom' of rock. At a more sophisticated level, music aspires to the restfulness, the classicism, the sense of completion of a circle. Mozart's music still has this form. It starts somewhere in the world, rises to a culmination point – the sublime and serene adagio – and then returns to the world. Beethoven revolutionized Western music by creating a heroic style, at the core of which is a sense of striving and struggle toward a goal and final resolution. The last movement – the goal – acquired an importance that it never had before. The influence of Beethoven's heroic style is such that ordinary listeners have ever since learned to expect art music – classical music – to be a cumulative process, the spatial image of which is not the circle, but the arrow (Burnham 1995).

How does one appreciate such music? A sophisticated listener may strive to follow the process – usually imagined as sounds moving from left to right like words on the page – and say something like, 'we are now in the bridge to the second theme group.' Listeners who lack the technical knowledge to follow may feel inadequate; and yet, who enjoys the music more? More precisely, who enjoys more Beethoven's music – say, his Third Symphony – as it should be enjoyed, as the composer himself might wish it to be enjoyed? The technical listener stands outside the music and follows the development step by step, noting the marvels of transition as they arise. The non-technical listener – or the listener who deliberately sets aside his technical knowledge – enters the music as he would a place – an uncanny world. The musicologist Scott Burnham asks, 'Why do we keep listening to our favorite music? ... Do we really return to the music we value in the hope and expectation of hearing something new each time?' (1995, 164). His answer is no. We return to the same music again and again, not for additional intellectual stimulation, but to be exposed to a presence, to be in the midst of a magical place that provides us with nurture, self-knowledge, and inspiration.

Sense of Place and Self: Old and New

Place, I have been saying, slows down time, stops it, and can even – by offering the possibility of return – reverse it. We are comfortable in place, but not in time, as though we are atemporal creatures, inexplicably ensnared by the temporal stream. We catch ourselves saying, 'How he's grown!' and 'How time flies!' as though a universal form of our experience were again and again a novelty (Lewis 1958, 138). In other cultures and in earlier periods of history, people have not found time quite so problematic or burdensome. There was certainly less talk about it, less explicit awareness of it as a personal problem. A variety of reasons explain this apparent lack of concern. First is the absence of a sense of an individual self apart from the group. Only the individual human being is truly subject to time, for only the individual lives and dies. The group, by comparison, is immortal. Second, the material place of pre-modern peoples stays much the same from one generation to the next. Neither their natural nor their built environment contains obvious and constant reminders of time's relentless, one-directional passage. Third, when time consciousness does intrude, it is obliterated by sacred rites, during which the dead reappear to mix with the living, and the atemporal status of human beings is reaffirmed. I am saying, then, that people everywhere do not just roll along with the temporal flow in total acceptance and innocence, as animals presumably do. I would also say that place is the means most readily at hand to overcome this sense of ceaseless flow.

Take up, again, home – the traditional home. It has lost ground as place; it no longer is quite the repository of memory, affection, and meaning that it was in the past. Greater mobility is a major cause of this change. In a traditional society, men might spend their entire life in one place and women in two places, for they customarily leave their birth home and move to their husband's home upon marriage. In a modern society, men and women might move three, four, or more times before settling down in a Florida condo. No depth of memory can attach to any of these residences. What, one might wonder, is the psychological cost? If the bond between place and self is as close as I have intimated, then the cost ought to be considerable. Yet I do not think it at all proven that the mental health of mobile, searching people – cosmopolites ancient and modern – is any less robust than that of sedentary villagers. How come?

Let us admit, first, that excessive mobility can do psychological damage. People constantly on the move, treating places as mere rest stops and artworks as mere diversion, entertainment, or investment, can seem – for all their worldly success – not quite real, superficial. But this need not happen when the mobility is within reason and we are alert to its risks. So I ask again, How come? Well, I have already given my answer. This chapter is it. Let me, however, make one more stab at restatement and clarification, and then raise a question that each of us will have to answer for him- or herself.

Return to the commonly accepted notion that people's emotional engagement with home is profound. Home is where they grew up, where their sense of self developed along with their sense of place, the integrity of the one meshing with the integrity of the other in the subtlest of mutual influence. Once this is said, however, we need to go on in the interest of seeing the picture whole to recognize home's negative aspects and effects as well. What are they? For a start, home can be too confining, and this is so even when the meaning of home is extended beyond the house to include neighborhood, village, and town – a larger entity that might be called homeplace. The trouble with home or homeplace is that whatever integrity it helps a person to achieve may be achieved too soon, thus stalling the possibility of further growth. Second, it is just possible that the integrity is more illusion than reality. My reason for saying so is that, from any number of case studies, we know that hardly any home, village, or town, however outwardly unified and tranquil, is without internal divisions and contradictions. This means that the self – in the interwoven duality of place and self – cannot be quite the whole it seems or claims to be. Third, the sense of home and the sense of self that develops with it, because it is not only dense but contains unacknowledged contradictions, is frustratingly beyond the clarifying powers of articulation, whether it be language, gesture, or music.

Consider, by way of contrast, modern man and woman. Their home is never the cocoon, with all its advantages and some disadvantages. Modern man and woman are mobile and easily wander into exotic worlds far beyond home. In doing so, they do not necessarily lose a sense of self and a sense of place. To the contrary, in their travels, they may encounter landmarks and landscapes that reveal hitherto unknown aspects of themselves. As I have noted earlier, I see myself – my need for simplicity and clarity – mirrored in the Arizona desert. Someone else might see himself – his need for strength and elegance – mirrored in the twin towers of the World Trade Center. I do not live in Arizona and he does not live in Manhattan. Yet, for both of us, these places now have to exist in all their integrity if our own sense of self, enlarged by these recognitions, is to be sustained.

Self-discovery also occurs in encounters with works of art. These havens and power houses of renewal, like geographical places, enable human beings to forget, on the one hand, the boredom of repetitive cycles or routine, and, on the one hand, the relentless flow of time that carries them – that carries all of us – into oblivion. Home is or can be a temporary stay against time. To a literate person, home is as much *David Copperfield* as Peoria or Madison. Music's hold can be even stronger. Pablo Casals is deeply attached to Spain, his country of birth. Yet Bach is his emotional center and real home. For decades, Casals lived in self-exile from Spain. It is inconceivable that he can live in self-exile from Bach and still be Casals.

A Spiritual Quest

And now, as a parting offer, I turn – with some trepidation – away from ordinary people like you and me that social science studies to take a brief look at man and woman in spiritual quest who, although they have lived in different places and times, share certain characteristics. Typically, they lack a real home and do not particularly want one, for they are lured by the idea of an ultimate home. Like everyone else, they want time to slow down so that they can savor place, acquire a sense of it as well as a sense of self. But, in the end, they want to opt out of time altogether and be in eternity. They seek to be whole and to have a coherent sense of self, but not just yet and not at the expense of openness to the gifts of world, nature, and supernature that continue to transform the self. Man and woman in spiritual quest recognize themselves as restless desire, temporarily assuaged in various homes and landscapes – the literal ones of geography, and the figurative ones of art – grateful for the refreshment and self-recognition at each, but not content to stay.

'We are restless,' says Augustine, 'until we rest in Thee.' For Augustine, this Thee is God. For Plato, it is the Good. For the Hindu or Buddhist, it is the still center of the Rotating Wheel, or Nirvana. What is it for modern man and woman? Without an ultimate goal beyond time, hinted at in the security of home and in the ecstatic revelations of nature and art, will not the search for place end in a mere collection of pictures like those in a glossy tourist brochure? And will not the search for self-end in fragmentation? Each of us will want to retire to a quiet corner, reflect, and answer in his or her own way. But note this: we cannot have an answer, we cannot even have the questions, without the support – the nurture and stimulus – of place.

References

Burnham, S. (1995) *Beethoven hero*. Princeton University Press, Princeton.
Lewis, C.S. (1958) *Reflections on the psalms*. Harcourt, Brace New York.
Nicolson, M. (1959) *Mountain gloom and mountain glory*. Norton, New York.
Sack, R.D. (1997) *Homo geographicus*. Johns Hopkins University Press, Baltimore.
Shakespeare, W. (1984) *King Richard II*. Cambridge, Cambridge University Press.
Washburn, S.L. and DeVore, I. (1961) 'Social behavior of baboons and early man' in Washburn, S.L. (ed.) *Social life of early man*. Aldine, Chicago, pp. 91-103.

Chapter 3

The Stranger's Lifeworld: The Chinese Diaspora and Immigrant Entrepreneurs in Canada

David Ley

According to the *Web of Science*, 'Grasping the Dynamism of Lifeworld' (Buttimer 1976) is Anne Buttimer's most widely-cited publication, written early in a career whose international achievements and reputation were appropriately crowned by appointment as President of the International Geographical Union in 2000. 'Lifeworld' appeared in the notable June 1976 number of the *Annals of the Association of American Geographers,* an issue that included the work of leading scholars including Leslie King, Robert Sack, Yi-Fu Tuan, and David Ward. The editor, John Hudson, placed Tuan's and Buttimer's papers beside each other, and together they offered an impressive manifesto for a newly-developing humanistic geography, a position that was taking shape to counter the reduction of human agency both in the prevailing tradition of spatial analysis and also in the forceful interpretations of a structural Marxism, that for all its intellectual vigour and claims to political progress was even more dismissive of the role of human agency outside the prescribed task of class struggle.

In contrast to positivism and structuralism, Buttimer's essay looked to phenomenology and existentialism for its philosophical guidance. The breadth and depth of its proposals are remarkable, providing seminal, if not always acknowledged, precedents to the cultural turn in human geography in the 1990s. Among these seeds for a future harvest is the essay's discussion of self-consciousness, alerting the reader to the need for reflexivity and the inevitability of standpoint knowledge, while the concept of intersubjectivity points to the inherently social nature of all knowing. The pervasive presence of values and valuing, explored more fully in a short monograph two years earlier (Buttimer 1974), guarantees the centrality of meaning and human experience in geographical exposition. The concept of body-subject, drawn from Merleau-Ponty, anticipated the discussion of geographies of the body that animated feminist work in the 1990s. Part of the current reticence to acknowledge the intellectual foundations provided by these humanistic themes may be due to the individualism and subjectivism that Buttimer noted within some work in the philosophies of meaning. Such a predilection certainly seems disconnected from the overwhelming emphasis on power in the human geographies of the past 15 years. But power, at least as an interpersonal force, is only one expression of the broader family of constraints that surround the thinking and feeling subject, and these constraints (and opportunities) comprised an essential part of humanistic theorizing, at least in geography. The early monograph, *Society and Milieu in the French Geographic Tradition* (Buttimer 1971) made the agency-structure interdependence apparent in its title, and a few years later her essay 'Charism and context' (Buttimer 1978) pondered the redefinition of *milieu* that would be required for an interpretation following the French School to be made of the industrial city. Integral to such a task, she reflected, would be reference to industrial capitalism as the formative ideology and practice of such a city. My own contributions to this literature at the time also insisted upon both the structuring contexts of intersubjectivity, providing a social embrace of the individual, but also the broader enveloping 'spatial, temporal, societal and intellectual context ... introduc[ing] various degrees of influence and authority relations to the ongoing emergence of action' (Ley 1978). Indeed a textbook published shortly afterwards framed geographies of everyday life within conceptualizations of intersubjectivity and informal and institutional power (Ley 1983). In light of such unequivocal orientations towards constraint and power a tendency to see humanistic work only within the register of individualism and subjectivity reveals a curiously selective reading.

In 'Lifeworld,' Buttimer recognizes the limitation of treating the relation between people and their environment in terms of the fixed, familiar and stable world represented in studies of pre-industrial communities:

> Does 'home' always coincide with residence? Could a person be 'at home' in several places or in no place? Could the gestalt or coherent pattern of one's life-space not emerge from mobility as a kind of topological surface punctuated by specific anchoring points? (Buttimer 1976, 284).

The rhythms of commerce introduce a discontinuity to the more predictable rhythms of nature. The restlessness of mobility introduces new experiences of place, now strange not familiar, global not particular, stimulating not enervating, novel not traditional, punctiform not continuous, pragmatic not stewardly. As Buttimer notes, the taken-for-granted dimensions of the life-world are invariably problematized by life as a stranger, and she alludes to the essay on the stranger by the social phenomenologist, Alfred Schuetz (1944). As an interwar scholar in central Europe, the project Schuetz pursued was the integration of phenomenology with the sociology of action associated in particular with Max Weber. For Schuetz, like his counterpart George Herbert Mead in the United States, identity emerged socially and action too was predicated both upon the 'in order to' motives of the I as well as the 'because of' motives of the socialized me. In his short essay, 'The Stranger,' Schuetz meditated upon the precarious but essential task, following displacement, of assuming both a new identity and assembling a new stock of knowledge for navigating everyday life. The erosion of social, economic and cultural capital means that in social interaction the stranger's status and identity may well be devalued. Moreover, everyday life depends on the deployment of cultural recipes, well-tested responses to typical situations, and the 'crisis' for the stranger is that her recipes are geographically circumscribed and no longer relevant. With a sense of disorientation the stranger learns that his 'thinking as usual,' his habitual schemes of interpretation have no currency. Consequently, social life becomes, in Schuetz's telling words 'not a shelter but a field of adventure ... not an instrument for disentangling problematic situations but a problematic situation itself and one hard to master' (Schuetz 1944, 506). The meaning of place for the stranger becomes 'a labyrinth in which he has lost all of his bearings' (ibid. 507).

The Stranger Personified: The Chinese Diaspora in Canada

Alfred Schuetz's essay on the stranger was autobiographically informed, inspired by the severing of his own taken-for-granted world, arriving as he did in North America as a political refugee from Europe. His argument profits from its generality, and Schuetz referred to the similar witness of monographs written by William Thomas, Robert Park and other members of the Chicago School that had also illuminated, some in remarkable detail and poignancy, the experience of the European immigrant in America. In the remainder of this chapter I shall carry forward this genre by reflecting on the experience of place and society among contemporary economic immigrants from Hong Kong and Taiwan who have landed in Vancouver, Canada, in recent years and have found life as a stranger unexpectedly vexing.

Many of these migrants arrived in Canada by taking advantage of specific initiatives of the federal government to lure wealthy immigrants with the prize of citizenship following a three-year residency requirement. The Business Immigration Programme was developed to consolidate a pool of experienced, successful and resourceful entrepreneurs in Canada, offering entry and citizenship in return for economic enterprise. It comprises three immigrant categories, which have been subject to detailed fine-tuning since their inception (Nash 1996). The terms of the *entrepreneur* stream, founded in 1978, required during the 1990s[1] that

> an entrepreneur must establish or buy a business in Canada ... The entrepreneur is expected to participate actively in managing the business. The business must contribute to the Canadian economy and create one or more jobs in Canada in addition to the jobs created for the entrepreneur and his family. The entrepreneur is admitted on the condition that these requirements are met within two years of landing, and is expected to meet regularly with an immigration officer to monitor compliance with the terms and conditions (CIC 1999a).

A critical threshold in this procedure has been the two-year deadline for achieving programme objectives. In Schuetz's terms, immigrant entrepreneurs had to demonstrate economic success while, still strangers, they had barely mastered the recipes of the new society. Specifications for the *investor* stream have become more demanding in response to perceived and actual abuses since its introduction in 1986. As of April 1, 1999

> Investors must invest a minimum amount in approved projects in Canada. CIC [Citizenship and Immigration Canada] will act as an agent on behalf of the provinces and territories, which will decide where to invest the money. All investors must provide a minimum investment of $400,000 [Can.] and have a minimum net worth of $800,000. Provinces and territories secure the investment against loss (CIC 1999a).

The investor stream allows more passive entrepreneurialism, but at the same time anticipates a high level of risk as funds have normally been dedicated in the past to venture capital enterprises. The third and smallest category of business entry is the *self-employed* stream that permits entry of a range of skilled individuals, including artists, professional athletes, and also wealthy individuals. The specifications here are more general: 'Self-employed applicants must be able to establish or buy a business in Canada which will provide employment for themselves and will make an economic or cultural contribution to Canada' (CIC 1999a).

The business programme grew steadily to the mid-1990s when it accounted for over ten per cent of annual landings, reaching a peak of over 32,000 arrivals in 1993. Between 1981 and 2000 some 314,000 immigrants entered Canada through one of the three business streams. From 1984 to 1998, Hong Kong was the primary source of business immigrants to Canada, accounting for between a third and a half of the total each year. By the 1990s, Taiwan was adding another 15-20 per cent of arrivals. The most popular destination of business immigrants has been the Pacific coast province of British Columbia, which has received a third or more of all landings, with 85-90 per cent of these selecting a home in Greater Vancouver.[2]

With the requirement of past business success and significant available capital to invest, a popular stereotype has arisen in Canada concerning the wealth and economic power of this business immigrant cohort. Data published by the federal government department, Citizenship and Immigration Canada (CIC), in its annual report declare programme outcomes and have been much quoted by the press, by policy-makers and by immigration scholars. Tables list funds invested through the programme plus jobs created for each province. In the investor stream alone, for the period from its inception in 1986 until 2000, over $5.75 billion (Can.) had been invested in commercial ventures with over 40,000 jobs created or maintained (CIC 2002). The score-card for the entrepreneur stream was equally impressive in overall outputs. In the four years from 1992 to 1995, new immigrant-owned businesses had created or maintained 23,773 full-time jobs and close to 10,000 part-time jobs with a declared investment of over a billion dollars. British Columbia, the province of choice for business immigrants, had done particularly well. Entrepreneur immigrants had invested $540 million (Can.) in small businesses over the four-year period, creating or maintaining 12,400 full-time jobs and 5,450 part-time jobs (CIC 1996). More recent data for the entrepreneur stream in British Columbia show the benefits over the 1990-1998 period as over a billion dollars of business investment with the creation or maintenance of over 25,000 jobs (CIC 1999b).

We are speaking here of big money. Declarations of personal wealth to immigration officials indicate that household assets close to $20 billion (Can.) were available for expenditure by business immigrants in Greater Vancouver arriving between 1991 and 1996 alone (Ley 1999). The availability of substantial funds from new arrivals and off-shore investors has impacted the housing market significantly as immigration became by far the major source of metropolitan population growth between 1986 and 2001. By the late 1980s, inflationary Vancouver house prices had become the most expensive in Canada and have remained in that position to the present. Indeed over the 1971-1996 period, annual immigration and house price levels shadowed each other closely, with a correlation as high as 0.96 (Ley and Tutchener 2001).

The Business Immigration Programme in Practice

Data like the CIC annual score-card invoke the certainty of numerical measurement and the authority of official statistics. But their facticity obscures entirely from view their socially constituted origin for a socio-political purpose of programme justification (Ley 2003); they provide a perfect illustration of the intertwining of fact and value alluded to by the old phenomenological credo that values are not a layer set atop the facts but rather facts are condensed values. The score-card appears to demonstrate a remarkable process of adjustment, extolling successful learning of the recipes for doing business in a new nation. But they are far detached from the everyday life-world of the immigrant entrepreneur, even though claiming to illuminate it. They are a projection of a bureaucratic rather than an immigrant life-world and set of interests. Interviews permit a much closer examination of the experience of immigrant entrepreneurs on the ground.

Following an initial focus group that included a dozen business immigrants from Hong Kong and Taiwan, 24 in-depth interviews were conducted with economic immigrants living in Vancouver's affluent west-side neighbourhoods during 1998 and 1999. Interviews were conducted in English, Cantonese or Mandarin according to the preference of the respondent and conversations were tape-recorded.[3] Thirty people were interviewed: seven interviews were with men, eleven with women, and five with couples, while one man invited a male friend to accompany him at the interview. Seven of the women were members of 'astronaut' families (Waters 2002), with their husbands working overseas, while two of the couples had recently terminated an astronaut arrangement. Six years was the median length of residence in Canada among the group, with a range from three months to ten years, and a single outlier of 19 years. The median age at landing in Canada had been 40 years. Eight of the group had landed through the investor category, eight as entrepreneurs, two as self-employed, and six as skilled immigrants. The status of the group as economic immigrants conveys the same message of experience, skill, and entrepreneurial acumen as laid out by the business programme.

Respondents were recruited by a number of means: through the listings of an immigrant-serving NGO, through membership in an English as a second language programme, and through clients suggested by co-ethnic business professionals and a co-ethnic church minister. I approached these interviews expecting to meet a population of vigorous entrepreneurs, as I shared the preconception of popular images and government statistics. To my surprise, the first six interviews, located through three different entry points into the immigrant community, revealed that neither the household head nor any immediate family members were working in Canada. For sure, the investors had made their financial commitment, and others had passive stakes in real estate or the stock market, but none of the interviewees was gainfully employed. Some had taken early retirement, some were astronaut households, and others were waiting for a business opportunity that had not yet materialized. Subsequent interviews were weighted towards economic immigrants who were gainfully employed, but nonetheless of the 24 households, only one-half contained a member working in Canada. Five respondents owned a store or fast-food outlet, three others were in a trans-Pacific import-export business, three had part-time jobs, and one was an office manager.

This performance level fell so far beneath the expectations of economic immigrant scenarios that some time was taken in the interviews trying to understand the decision-making and circumstances that had contributed to these modest outcomes. There was unanimous sentiment among all respondents that economic success in Canada, even limited success, was extremely difficult to achieve. Mr. Chen, highly qualified with a British advanced degree, had owned a very successful business in Hong Kong and moved to Canada in his mid-forties. He has not been involved in economic activity in Vancouver, other than some passive real estate investment. He saw an uninviting business culture with significant barriers to entry:

> What really surprised me and is still surprising me is the tax situation. Not surprising, it's shocking compared to Hong Kong ... Be prepared not to be too optimistic regarding finding a job or making money or building a new business venture or enterprises. [Immigrants] should not be too optimistic and be prepared to eat up your capital. I think that's pretty serious ... In Hong Kong when we invest, we look at three, five, seven years at most. But here you have to look long term. If you fail in one business, we say 'Well, let's do it again.' Maybe one of these days I can strike the jackpot, but in Hong Kong if you don't hit the jackpot once or twice then forget it because time is short. It was before 1997. So the mentality is that we cannot look at things long term, keep on paying tax and keep on trying, trying ... And the rate of return, if we have less than ten per cent, then we say 'Forget it, it's a waste of time.' But here if you have ten per cent return a year, 'Wow, that's great!' So it is a different mentality.

High levels of taxation and low levels of return have discouraged or delayed entrepreneurial activity by potential business people. Mr. Chen told the story of a very wealthy acquaintance, a billionaire, who had planned to construct a large factory producing batteries outside Vancouver, but who relocated his investment to Mexico in the face of what he regarded as interminable environmental studies.

Two of the interviewees had themselves been manufacturers prior to moving to Vancouver – exactly the types of entrepreneurs sought by the Business Immigration Programme – but both had been discouraged by the regulatory environment they encountered. Mr. So had been in the dyeing and finishing sector of the textile industry in Hong Kong with a factory employing a hundred people. He had concluded that pollution laws, labour regulations, and start-up costs made the business not feasible in British Columbia. Upon his arrival, Mr. So took early retirement. In contrast Mr. Yee had been co-owner of a garment manufacturing business in Hong Kong and had tried to enter the same sector in Canada. But the business had failed due to a combination of problems with unreliable suppliers and buyers as well as tough labour relationships. Mr. Yee had made a career change and became a suburban greengrocer; after 14 years of persistent financial loss in Canada he was finally making money, at the time of the interview a net profit of some $3,000 (Can.) a month, for a work week of 70-80 hours:

> For me the best thing is, yeah, because I lost a lot of money in Canada, it's the first time I make money in Canada. This is very, very pleasant, because we really make a profit. This is the best thing in my life (laughs). Because, since the day I came, before I start this job I always lost money.

As a long-term resident (he was the oldest arrival, landing in 1980) who had tried a number of career paths before opening his shop, Mr. Yee's reflections on the business options open to newcomers like himself are illuminating:

> Yes, I change, a 360 degree change ... To make a living either it's being employed or being self-employed ... Self-employed there's only a certain kind of business you can make a living. There's a café, small restaurant, and produce. So I choose produce because, you know, it generates more people and more cash flow.

A significant career change in a society where the business culture and language are unfamiliar is, as Josephine Smart (1994) has suggested, a prescription for failure, and both the experience and the fear of failure weighed heavily on the respondents. John Kim had been in Canada only three months at the time of the interview. Landing through the entrepreneur stream, he felt the pressure to establish a business to meet the immigration requirement of quick achievement. But there were significant barriers. Like Mr. Yee he had realized that a career change was necessary; enquiries had shown there were no openings in Vancouver in merchant shipping, where he had risen to a senior management position in South Korea. His old recipes for doing business did not hold. A friend, also in the entrepreneur stream, had opened a shipping agency three years earlier, but was now losing substantial money from it. Prospects are daunting:

> [We] have much concern because I see other Korean immigrants are gloomy. They are idling, they are eating only ... I heard of, from some of the Koreans, who were successful in business. But only a few, that we can count very easily with one hand. Only a few. Most Koreans lose money. I visit one Korean grocery yesterday who immigrated here about nine years ago. He was my senior in the same department in our company ... He's now doing his own business as grocery in New Westminster. I visit there and hear about their family's last nine years from his wife. They also when they land, landed here, they made some contracts, buy some small business here from some other people. But later on they realize that ... it was not honest. He broke the contract. So he lost some money ... Not making money. Only getting money for his labour. They cannot hire somebody. If they hire, no money. My friend is working 365 days a year ... They go to the store eight o'clock, close 11 pm. I think they cannot make any profit, any money, only their living costs.

For Mr. Kim pressures are intense. Both the Korean and the British Columbian economy were in recession, his fellow countrymen in Vancouver were losing money in their businesses, despite more experience and better English than his own. His wife was becoming impatient, and family feuds had broken out. The clock was ticking toward the two-year deadline for meeting his entrepreneurial terms to establish a business employing a Canadian, but he had no idea of which way to turn. Here is the dilemma of the newcomer: his thinking as usual no longer bears any credit, he is 'a man without a history' (Schuetz 1944) in his new setting. Mr. Kim exemplifies all too well the typical status of the stranger: 'the stranger discerns, frequently with a grievous clear-sightedness, the rising of a crisis which may menace the whole foundation of the 'relatively natural conceptions of the world' (ibid.).

Investors are not under the same deadlines to establish a business. But waiting does not make the entrepreneurial decision any easier. Mr. and Mrs. Tsai have been landed for four years but during this time have not been employed or started a business. With two sons doing well in school, and living in a million-dollar house in a good neighbourhood, they have made a commitment to Canada. But they are living off savings, and the prospects of earning money in Vancouver are not good. Mr. Tsai, owner of a small installation company in Hong Kong, makes the comparison:

> Well we have a lot of experience ... Not only my experience. We always speak with the other immigrants who are from Hong Kong or from Taiwan. We have a lot of experience also from them ... 95 per cent of female immigrants like here very much ... 95 per cent of male immigrants feel upset, feel embarrassed, feel very bad ... middle age. Around 35 to 50 ... men in this age usually feel upset because they have lost their self-confidence ... their job, lost their respect because when they are in Taiwan and Hong Kong, maybe in the highest position in the company ... very high level position. After moving here they live with the family, wife and children go to school, the husbands stay home. Nothing to do. No work to do. It is very hard to look for a job same as they are in Hong Kong, They can only have their job in very low class job, very low class.

Mr. Tsai's observations rang true in the experience of virtually every respondent, who had invariably faced downward mobility and deskilling in Canada. With Schuetz (1944) they have learned that a stranger 'may lose his status, his rules of guidance and even his history.' The asymmetrical gender impact, penalizing men, is noteworthy, though it can be savagely reversed in the case of some stranded astronaut wives, or 'widows' as they may call themselves (Ong 1999; Waters 2002). Male identity challenges are frequently associated with additional family quarrels, and in the case of one respondent led to divorce, as her husband returned home with the failure of his company, while his wife chose to remain in Vancouver, living off payments from investments in a separate family business in Hong Kong. Another respondent told of the stress that overcame a friend's husband, formerly a successful accountant, who had been unable to secure employment in Canada. After four years he entered a deep depression, threatened his son with a knife, and suffered a break down. Adding to the trauma, the parents decided to return to East Asia, but their two teenage children refused to accompany them.

Inactivity, then, becomes a response to perceived weak business opportunities. Another novel response to an unfamiliar cultural and economic setting is early retirement, for respondents all agreed that if British Columbia's economy could not compete with Taiwan or Hong Kong, its quality of life was unequalled. Mr. and Mrs. Leung have four other family members living in Canada, and two in Hong Kong, all of them bearers of a Canadian passport. At the age of 52, Mr. Leung landed in Vancouver in the investor stream, and began his retirement. He had earlier worked for a multinational corporation in Toronto before returning to Hong Kong, and had a keen sense of business opportunities in the two nations:

> Frankly, you can't earn any money here. [If] you have your own money you can come. Just stay here and relax. You cannot earn good money here because the tax here is so high. So you must have earned enough money to come. And then if they come they must expect their lives to be totally different especially when they are working. Even though they have

money in Hong Kong, they earn a very good salary in Hong Kong, when they come here they may not be able to find a job, or the salary would be about one-third or even less that of Hong Kong.

In Mr. Leung's circle, those over the age of 50 were retired. 'Those who come recently, they, we, are all retired people.' A barrier to economic activity in Canada was the high tax rate and other costs like employment benefits that removed any incentive from doing business. He estimated that even returns from property investment were no better than three per cent. Aside from his mandatory business investment, Mr. Leung's funds are all in the stock market.

The apprehensions of immigrants, hesitant while outside business participation, are repeated and thereby confirmed by respondents who are doing business in Vancouver. Almost every one of them found the going very difficult. Mrs. Chi, who entered Canada from Taipei with her husband as an investor, has an import-export business. She is disillusioned as a result of the conditions she has encountered, the disqualification of her expertise acquired elsewhere:

> When we first came here, we were told this country is open for us in terms of business. When we got here we found that we were not needed ... Our 20-30 years' working experience are not needed here, they are not recognized.

Mrs. Chi has advised relatives not to migrate to Canada, and 'they threw away the application form.'

> We had to adjust to the situation of being unemployed. We had to start our business from scratch. We could hardly face it...We just tried to make enough to cover our expenses ... It's more stressful here ... We really don't know what to do, we are kind of, at a loss, we are confused.

Because of language difficulties and business experience, the couple is working within an ethnic enclave economy of fellow Chinese immigrants. The enclave economy is an attempted solution to the irrelevance of familiar recipes in Canada. Restore a Chinese social space in Canada and the cultural patterns that led to success at home may be repeated. But despite a market of 300,000 ethnic Chinese in Greater Vancouver, this strategy has foundered through over-subscription. The ethnic enclave economy is overcrowded, and competition is fierce (Wong and Ng 1998; Ley 2002). With a business that is not profitable, the Chis continue to live on long-term savings six years after landing.

Mrs. Seng had also been in Canada for six years at the time of the interview. Entering through the entrepreneur stream, she and her husband opened a health food store. A teacher in accounting at a business school in Taipei, like Mrs. Chi she had high hopes initially for economic success. Inability to speak English has been a greater problem than they expected, and adjustment has been difficult and has led to family arguments. Like the Chis they are also finding that heavy competition in the ethnic enclave economy is crippling their business: 'Here people have to "steal" customers from each other. We are not able to make a reasonable profit.' They are clearing only a little over a thousand dollars a month, business is getting worse, and despite a depressed housing market her husband is trying out real estate as an additional revenue stream. Similar in some respects, Mr. Chow is a tailor operating from a small store-front. He had been working in Beijing before he entered Canada in 1991 as an independent immigrant with only an elementary school education. His two children, however, are both university graduates. Mr. Chow was able to find work quickly and later opened his own business; despite low returns on his business and long hours (over 60 a week) he is content with his move to Canada.

Mrs. Wang is much less satisfied. Unlike Mr. Chow, who considered only Canada as a destination and was uncertain of the success of his application, Mrs. Wang's family studied immigration options carefully, checking the United States, Singapore, and Western Europe, and visiting Vancouver three times before submitting their application. Perhaps it is this care in the application process that has contributed to her current unease. In Taiwan they were successful entrepreneurs, owners of a company manufacturing telephone components. Qualifying in the entrepreneur stream, they brought $2 million (Can.) with them. The family owns two homes in British Columbia, and their son is in boarding school in the province. They set up a business as required by the terms and conditions of their visa, but it did not bring in sufficient income. Importing products from their Taiwanese company for the North American market, they could not make their prices competitive and the business has become inactive. Mrs. Wang's husband was unable to find an alternative, and returned to Taiwan. Mrs. Wang's disappointment has been fuelled by the failure of a second initiative, an investment in a local printing company, which subsequently declared bankruptcy. These reverses are hard to take for a family that has proven its entrepreneurial skills in Taiwan, but Vancouver and Taipei are separate spheres with credit lines it seems that are scarcely transferable. Like virtually every respondent, Mrs. Wang appreciates the quality of life in Vancouver. But the economic liabilities are substantial. To survive, a new immigrant needs 'enough huge money, because the expenses are high, the tax is very high. And everything you have to spend money on, you cannot make money here.' Within two or three years even that 'huge money' would be seriously depreciated. The prospects for finding work are limited. In Taiwan Mrs. Wang, a university graduate, was a successful business woman:

> I think I'm still young, and have enough energy and have very good work experience to find a job here, but they don't agree with your previous work experience. So we're just wasting our experience. We have more than twenty years' experience. I think I'm a really good businesswoman but I cannot do anything here.

Once again, the stranger faces the abolition of personal history and geography as past achievements that brought deserved rewards elsewhere, have no purchase here.

It is interesting to note that once the probationary terms and conditions for entrepreneur status were lifted, the Vancouver business, already struggling, was allowed to languish, and out of necessity Mr. Wang became an astronaut, working in Taiwan and returning every three months to see his family. A similar transition has occurred in the Nam family who entered Canada in 1991 as investors. Initially unemployed, Mr. Nam subsequently rekindled business interests in Taiwan. But the astronaut option is a difficult one. With the disruption of family roles, routines and competences in an unfamiliar culture, women can get lonely and depressed at the burden of managing a family (Waters 2002). With the dissolution of conventional rhythms, men can become disconnected from their families. Affairs occur; marriages break up; children become unruly.[4] Some of these circumstances showed up in the Tse family, who landed as entrepreneurs in 1992. Neither of them had post-secondary education, and neither spoke English. Mr. Tse had a business as an electrical contractor in Hong Kong and looked for a similar business opportunity in Vancouver. His linguistic limitations confined him to the Chinese ethnic economy. He worked long hours, and came home late, leading to suspicions by his wife of his whereabouts. Returns on the business were 'very, very low' and as the Hong Kong enterprise had not been wound up, Mr. Tse resumed these activities, becoming an absentee, part-time owner of his Vancouver operation. Mrs. Tse became depressed and fearful in her home in his absence: 'We do miss him a lot. But it's been like this for so many years, we are used to this situation now … it is tough … not an easy kind of life *(in tears)*. If I don't have the religion I don't think I could handle my life.' From this testimony we are inclined to answer Buttimer's question – 'Could a person be "at home" in several places or in no place?' – negatively.

Consequently just as readily as families fall into the astronaut option, they fall out of it. This was true of two of the married couples interviewed together. Both had young children, the wife spoke little English and felt isolated and insecure. 'When he leave Vancouver … we cry in Vancouver and he cry in Taiwan.' But movement out of the astronaut lifestyle solved one problem only to produce another, for the couples had no answer to the dilemma that had prompted them to economic transnationalism in the first place, the lack of an employment income source in Vancouver. Both families were demoralized with no clear strategy for economic improvement. One was surviving through rentals on two Vancouver houses they owned and some stock market investment; the husband was fearful of starting a business that he knew to be a high-risk option. The second family was spending its way through its dwindling savings fund and Taiwanese investments. They estimated their savings would last a total of 3-4 years. In these distressing conditions, immigrants use among themselves the telling metaphor of imprisonment or immigrant jail (cf. Findlay and Li 1997). One respondent described the length of his term in Canada as 'three years, eight months,' adding this was also the duration of the Japanese occupation of Hong Kong!

These families may well be candidates for return migration following the achievement of citizenship (Mak 1997). Inevitably business immigrants who have left Canada have fallen outside the sampling net of CIC statistics and are absent from consultants' and academic surveys. But their numbers are substantial though imprecisely known. Estimates in the early 1990s located some 25,000 Canadian passport holders in Hong Kong, but in the wake of 1997 these figures have been revised upwards to in excess of 200,000. Undoubtedly, immigrants in the business programme who have experienced demoralizing economic returns in Canada would be strongly represented in this group. All of my respondents had friends who had made the return move:

> I have quite a number of friends that have already returned to Hong Kong in 1997 and this year. It's about six or seven families that are already gone for different reasons. The most common reason is that they cannot find a job.

Once her children have graduated from university, this respondent will also return permanently to Hong Kong to reunite with her husband whose professional status has not permitted him to find an equivalent position in Canada.

Of the 24 families that were interviewed only three could be said to have made a satisfactory economic transition to life in Vancouver. The Lin family decided to leave Hong Kong because of its intense business culture that crowded out family life. After visiting a number of nations in the late 1980s, Mr. and Mrs. Lin selected Vancouver as their destination. Mr. Lin left behind a senior position as a civil engineer in project management, while his wife was a health care administrator. The family had significant savings and treated their first year in Canada as an extended vacation. Employment, however, proved a problem and Mr. Lin ran through several positions before he accepted the challenge of owning a fast food franchise. The business seems to be doing quite well, and the family now employs 5-6 staff. While the position represents a considerable social demotion from their former careers, they are satisfied with the additional time it offers for family and religious activities.

Mr. Chun is the success story in the sample. An investor, he has established a successful ginseng store at a high rent location within two years of landing in 1997. At the time of interview the store was generating an income of $6,000 (Can.) a month. Undoubtedly a previous business selling herbal remedies in Hong Kong had prepared the Chuns for this market niche, aimed at the mainstream as well as the ethnic market. The So family has also made a relatively easy transition. Mr. So is Vice-President of a Taiwanese development company. He entered Canada as an entrepreneur and established an import-export business sending hot tubs to construction companies in Taiwan. This carefully planned venture, maintaining the umbilical cord home, will allow the terms and conditions of his visa to be met. But he is also preparing the ground for his Taiwanese company to enter the development sector in Canada. Within six years of landing he is commanding a Canadian income of some $50,000 (Can.) a year.

Conclusion

This chapter has addressed one of the questions in Anne Buttimer's seminal 1976 paper: 'Could a person be "at home" in several places or in no place?' In this essay I have reviewed the uprooted experience of Chinese immigrant entrepreneurs and their various strategies to address the failure of their thinking as usual.

Their strategies have typically involved attempts to extend the longevity of recipes learned in their homeland. In a variety of ways they seek to make over the new land into the old land, in order to sustain the viability of established recipes of practice and expectation. The ethnic enclave is an attempt to insulate themselves from a harsh economic environment where too much is strange and difficult to navigate; the astronaut household abandons this attempt, and the breadwinner returns to an economic home where the path to success is familiar; return migrants complete the break altogether with a Canada where they cannot feel economically at home. Retirees by-pass the need to participate in Canada's menacing economic market-place, carrying their wealth transnationally from past achievements elsewhere. Among those who are truly in the market place in Greater Vancouver, disappointment and very modest success are the norm; some await the passage of their three-year 'prison sentence' and the award of citizenship and, perhaps, return migration. Others struggle on and day by day create a new measure of familiarity and sense of place (Waters 2002).

While Schuetz was preoccupied with the rupture of personal history in the status of the stranger, more pronounced I think is the dispossession of a familiar geography. Place matters, as is made abundantly clear by the respondents' constant juxtaposition of 'here' and 'there' in their commentaries. Mrs. Lam, a business immigrant recounted:

> So we started the business, we thought it was easy like Hong Kong. There we had a retail shop. If you start a business, people will come, and we will have business. But no, here it's completely different. So we had the same experience, we lost quite a big sum. From that time on, we are very cautious, very careful with what we have left. We dare not venture into another business. We try to keep our savings as much as possible.

It was not so much history as geography that separated the spheres of success and failure, status gain and status loss, relevant and irrelevant recipes. Indeed history could be reversed with the astronaut option or return migration. But life on opposite shores of the Pacific remained a conundrum of incompatibilities, of negotiating the abrupt transition from insider to outsider. Routines, rhythms, cultural and economic competences that allowed personal progress in one setting, became in the other 'not an instrument for disentangling problematic situations but a problematic situation itself and one hard to master' (Schuetz 1944, 506).

Consequently the tenor of these interviews was often melancholy. Avoidance became the rational strategy in addressing uncertainty; avoidance of business through retirement, through passive investment, or through hunkering down until residential eligibility for passport status had been secured; avoidance of Canada through the astronaut option or return migration. Many of those who had started a business were facing self-exploitation in terms of hours worked for meagre returns. A common experience was financial loss. When an English as a second language class of 30 Chinese economic immigrants was asked how many had lost money in Canada, every hand was raised. At a focus group, a Taiwanese immigrant in the investor stream lamented the common shock of losing his invested funds: '... the first experience to us of business is loss of money. We are afraid of ... We have $150,000 (Can.). I mean most of us lost this kind of money, 80 per cent. That's our first experience.' The business programme is predicated on a model of continuity, of an unproblematic transferal of proven business skills to the Canadian market. But this myth of the bureaucratic life-world misses the more problematic spatial reach of recipes and competences. Here is not there, and the sureties of home are not readily attainable when one is away.

Notes

1   Legislative change in 2002 introduced a more significant reconfiguration than the continuing but minor revisions of the 1990s.
2   Including secondary migration, principally out of Quebec, 37 per cent of business immigrants who landed in Canada between 1980 and 1995, and completed a tax return in 1995, were living in British Columbia that year (CIC, 2000).

3   Ten interviews were conducted in Cantonese or Mandarin by Hugh Tan and Andrea Tang.
     Mr. Tan was also present at the focus group meeting.
4   Ong (1999) places the divorce rate as high as 20 per cent among wealthy East Asian
     immigrants in the San Francisco region. Two of the women in my sample had divorced
     since moving to Canada.

References

Buttimer, A. (1971) *Society and milieu in the French geographic tradition*. Rand McNally, Chicago.
Buttimer, A. (1974) *Values in geography*. Association of American Geographers Resource Paper
     24, Washington DC.
Buttimer, A. (1976) Grasping the dynamism of lifeworld, *Annals of the Association of American
     Geographers* 66, pp. 277-292 .
Buttimer, A. (1978) 'Charism and context: the challenge of *la geographie humaine*' in Ley, D. and
     Samuels, M. (eds.) *Humanistic geography*. Croom Helm, London, pp. 58-76.
Citizenship and Immigration Canada (CIC) (1996) *Business Immigration Division, Program Statistics
     1995*, Ottawa.
CIC (1999a) *Immigrant services*. Ottawa (http://www.cic.gc.ca/english/).
CIC (1999b) *Vancouver admissions*. Business Unit CIC, Vancouver.
CIC (2000) *The interprovincial migration of immigrants to Canada*. IMDB Profile Series CIC, Ottawa.
CIC (2002) *Immigrant Investor Program: fully paid subscriptions by province, 1986-2000*. Ottawa
     (http://www.cic.gc.ca/English/bus-stats2000.html).
Findlay, A. and Li, F. (1997) An auto-biographical approach to understanding migration: the
     case of Hong Kong emigrants, *Area* 29, pp. 34-44.
Ley, D. (1978) 'Social geography and social action' in Ley, D. and Samuels, M. (eds.) *Humanistic
     geography: prospects and problems*. Croom Helm, London, pp. 41-57.
Ley, D. (1983) *A social geography of the city*. Harper and Row, New York.
Ley, D. (1999) Myths and meanings of immigration and the metropolis, *The Canadian Geographer*,
     43, pp. 2-19.
Ley, D. (2002) *Immigrant entrepreneurs: indicators of success* (unpublished report). Department of
     Geography, University of British Columbia, Vancouver BC.
Ley, D. (2003) Seeking *homo economicus*: the strange story of Canada's Business Immigration
     Program, *Annals of the Association of American Geographers* 93 (in press).
Ley, D. and Tutchener, J. (2001) Immigration, globalisation and house prices in Canada's
     gateway cities, *Housing Studies* 16, pp. 199-223.
Mak, A. (1997) Skilled Hong Kong immigrants: intention to repatriate, *Asian and Pacific Migration
     Journal* 6, pp. 169-184.
Nash, A. (1996) 'The economic impact of Canada's Business Immigration Program: a critical
     reappraisal of theory and practice,' paper presented to the Symposium on Immigration and
     Integration, University of Manitoba, October.
Ong, A. (1999) *Flexible citizenship: the cultural logics of transnationality*. Duke University Press,
     Durham NC.
Schuetz, A. (1944) The stranger: an essay in social psychology, *American Journal of Sociology* 49,
     pp. 499-507.
Smart, J. (1994) 'Business immigration to Canada: deception and exploitation' in Skeldon, R.
     (ed.) *Reluctant exiles*. M.E. Sharpe, Armonk NY, pp. 98-119.
Waters, J. (2002) Flexible families? 'Astronaut' households and the experiences of lone mothers
     in Vancouver, British Columbia, *Social and Cultural Geography* 3, pp. 117-134.
Wong, L. and Ng, M. (1998) Chinese immigrant entrepreneurs in Vancouver: a case study of
     ethnic business development, *Canadian Ethnic Studies* 30, pp. 64-85.

Chapter 4

Softly Heaves the Glassy Sea: Nature's Rhythms in an Era of Displacement

Edmunds Bunkse

Ak, don't be so close to the earth! At least not on Sundays.
On Sundays vapors from the bog turn into a white cloud.
Imants Ziedonis

Giant Footsteps

'Those are not the footsteps of a giant,' the boy kept telling himself. 'There are no giants. It is the sound of trees bending in the gale, brushing against one another. It only sounds like giant footsteps.' Yet the moment he had thus convinced himself of the logic of the situation, there they were again – slow, rhythmic, gigantic footsteps, coming from the direction of the sea, coming ever closer, ever closer, ever closer ...

Twelve years old, the boy was thin, gangly. He had just been made a boy scout and he was standing guard, alone, over the encampment in a mixed deciduous and pine forest on a high shore above the Baltic Sea. He was standing beneath a tall pine, shivering in the wind-driven, sparse raindrops. His watch was between three and four in the morning, the darkest part of the night.

The hour was endless, but finally it did come to pass and he went to wake the next boy. He said nothing to him about the giant.

In the morning, when they all gathered for reveille, before the usual mad dash into the sea, barefooted, down the hard-packed clay path that in its coldness burned their bare feet, he learned that the boy he had awakened was found asleep. He had curled up in the protective warmth of a former German army field kitchen soup kettle on wheels.

Nature's Rhythms: First Thoughts

When invited to participate in this book, I immediately responded that I would investigate the works of Imants Ziedonis, a modern, Soviet-era Latvian poet, who managed through his nature prose and poetry to engender pride in forgotten virtues of work and responsibility. This was a safe and manageable path for a historian of geographic ideas for whom literature is a major source of study. What could I add to sensibilities about nature's rhythms that reached back to ancient Egypt, Mesopotamia, and Classical Antiquity? What could I illuminate about nature's rhythms that had not already been illuminated by Virgil in his *Eclogues* (1950), for that matter, by Henry David Thoreau in *Walden*, Henry Beston in *Outermost House* (1962), Annie Dillard in *Pilgrim at Tinker Creek* (1974), Barry Lopez in *Arctic Dreams* (1986). Or by the Latvian country writers that I had read and plays I had seen as a youth: Janis Jaunsudrabins, Anna Brigadere, Rudolfs Blaumanis, Edwards Virza; or just recently, Imants Ziedonis.

Some six months later, when it came time to write the essay, I felt strangely restless and mildly dissatisfied with the chosen theme. Having just finished a book manuscript in which I dared to delve into myself for observations of geographic experiences and perceptions, I was not ready to resume being a historian. But realizing that I am neither a farmer, fisherman, or hunter, and have never lived close to 'nature' more than a few months at a time, what insights could I possibly bring to the theme of nature's rhythms?

Then inspiration came from Gabrielle Colette. Had not she been a thoroughly urbanized person, given to illuminating her racy life among the urbanites of Paris, yet retaining her connections with her roots in the countryside? Roots that ran through her mother Sidonie into the French provincial landscape. She never let a day go by without noting the weather, where the wind was blowing from, what that direction meant for living things. Even on her deathbed she was still asking her servant to look out the window and tell her about the state of the atmosphere, and to not take it from a radio report.

I am not quite as urbanized as Colette, having been active in nature whenever life decreed or free time allowed it. I am not a systematic student of nature, such as many birding or stargazing enthusiasts are. But like Colette, I have preferred to 'sniff the air,' as my father would say. I watch the skies all the time, am always aware of the wind – where it is coming from, what it is like – and observe the night sky (especially keeping track of Venus, or *Auseklis*, the morning star in Latvian). I am also a walker and a sailor. Could not I explore how a modern individual who in everyday life is not directly connected with nature's rhythms – an individual who lost his homeland and had to develop attachments to foreign lands and different regions – manages to observe and be aware of some of them, however impressionistically and tenuously?

Thus the theme of this chapter came about, viz., how a mobile, displaced person perceives nature's rhythms in different places and landscapes. It represents musings about nature's rhythms by a humanist, who in humanistic fashion hopes that others might resonate when reading this. In the essay I ask questions of myself. What are nature's rhythms? How do we know them, how do we become aware of them? Does nature 'speak' to us directly or through a perpetual cultural filter? Do we need to be in place to know the natural rhythms of that place, or can we find them in spite of mobility?

Culture and Nature's Rhythms

In music a rhythm is a pulse. In nature is a recognizable, dynamic, recurring pattern that we perceive through any of the senses. The steps of the giant evoked at the beginning of this essay are a rhythm that I drew from my childhood. I do not know if this is unique or if others have heard similar forest sound rhythms (there are the

'Forest Murmurs' that Wagner composed, which is quite different). I certainly have never again heard anything quite like it, even though I have listened to forest and individual tree sounds ever since (more on that later). No doubt it had something to do with the fearful mystery of that stormy night for a twelve-year-old's imagination. It might also have been a unique configuration of the trees, among which there were many pliant birches. And there was also the psychological fact that this was a foreign landscape, albeit still on the shores of the Baltic Sea, on which I had experienced my first years of life. But this was Northern Germany, not Latvia. We were in our third year of exile, living in a Displaced Persons camp in the old Hanseatic city of Lübeck. Standing guard (against whom?) that night in a foreign land made it seem all the more dangerous. Whatever the case, I have always regarded that one hour in the blackness of the Baltic night as a unique, pure experience of nature.

But was it a unique experience, directly from nature? Yes and no. There was, after all, the idea of a giant – certainly not a figment of my imagination, but of fairytales that had been heard and read. Indeed it is hard to think of awareness of any rhythms in nature that we do not receive second-hand from our elders and from our cultural heritage. Thus, had it not been for my father, who awakened me from deep sleep one cold and clear winter night in Latvia, to go outside and watch the crimson-green-white glow of the aurora borealis, I would not know their hair-tingling magic until much later in life. The slow, pulsating rhythm of those lights was a pure fairy tale. It reached far beyond ordinary, daily reality. It made the earth and sky a mythical unity within which we humans lived. But I also saw it in the context of the time – wartime. It reminded me of a burning city that I had seen at night beyond the horizon (also shown by father), I could not escape from the foreboding thought that more cities would burn and more people suffer horribly. Only much, much later, in the peace of Northern Maine and Manitoba, has the aurora borealis lost its associations with human catastrophes.

The point is that it is nearly impossible for us to experience anything in nature without doing so through the prism of culture. Something in nature may indeed catch our attention without any prior prodding from the cultural milieu, but there is the tendency to humanize that experience and to surround it with associations. This is true of most rhythms of nature that a human being is aware of.

Seasonal, and diurnal rhythms are of course the most obvious. Lunar rhythms, expressed in menstrual cycles and tides, although not at the foreground of daily consciousness for everyone, are nevertheless important to women and those who journey in tidal waters. Seasons, though naturally self-evident, are laden with cultural associations that vary from place to place, society to society and have become clichés for organizing diaries, memoirs, newspaper and other mass media expressions, and, of course, for organizing work and leisure on massive scales. Even in cities – perhaps most of all in cities – there is awareness of changing seasons, especially here, in the Megalopolis on the Eastern Seaboard of the US, where summer can be stiflingly unbearable out-of-doors and winter snows are generally unwelcome, chaos-causing occurrences. We have culturally given rhythms built into a calendar year, some of which are connected to ancient pagan seasonal feasts and festivals, such as Easter or Christmas. For some, such as Latvians, Swedes, and other Scandinavians, celebration of the summer solstice is a direct inheritance form pagan times. For Latvians the connection is still strong enough to make *Janu diena*, as the solstice celebration is called, the most important feast day of the year.

But there are other pulses in our calendar year that have nothing to do with nature In the US Independence Day on the fourth of July marks for many the proper start of summer, while Labor Day, at the beginning of September, marks the beginning of the school year and the effective, if not real end of summer (J.B. Jackson thinks that in American small towns of the Middle West and West, these and other shared events may be more important in generating a sense of place than places in these towns. Jackson 1994).

Culture also informs us about rhythms in human nature, the time within. Early on we learn about life and death and about the human life cycle. We know that birth, youth, middle age, old age, and death are inviolate laws of nature. Science is also giving us insights into subtle biorhythms of the brain and body, such as circadian rhythms (the 24-hour clock). These rhythms correspond to rhythms in the universe and have 'many evidences: body temperature, excretion, brain activity, heartbeat, breathing, eye movement, menstruation, dreaming, growth, muscletone [sic], hormone production' (Lynch 1972, 117-118). Medical research has found diurnal variations in blood clotting and other micro-changes in the body and therefore best times for administering different medicines.

There are also great natural rhythms that are being uncovered by science: fluctuations in animal populations, hundred year flood cycles, twenty two thousand year sun spot cycles (Lynch 1972, 118) and the cyclical nature of climate. These latter rhythms are interesting, knowledge of them enriches one's life, but it does little to connect one to a particular place or landscape, to its season, its flora and fauna (at least not as John McFee, 1989, has demonstrated in the case of Los Angeles). We may be aware of these and other rhythms, but from a geographer's perspective the question is *what part do nature's rhythms play in our sense of places and landscapes?*

Looking into my own experience, I find that awareness of specific rhythms of nature in the landscapes and places where one grew up make an imprint that is difficult to erase later in life. Indeed, such awareness can be so strong emotionally and so place-specific that it may hamper making connections with other places in the world. In this context I will examine the seasons in general and in my life in particular. This will be followed by the rhythms of the sea, birdsong, and the music of trees.

The Seasons and Sense of Place

> Many an aspen, many an elm bowed and rustled overhead, and hard by, the hallowed waters welled purling forth of a cave of the Nymphs ... Lark and goldfinch sang and turtle moaned, and about the spring the bees hummed and hovered to and fro. All nature smelt of the opulent summer time, smelt of the season of fruit (Theocritus, cited in Edmonds 1960, 105).

The seasons are the most obvious grand rhythm of nature, a universal human experience representing both 'an environmental imperative' and a 'plastic mental construct' (Buell 1995, 250). No matter how humans try to predict a coming season, or how to influence it through rituals, a season comes in its own inevitable, unpredictable way. A myriad of associations, beliefs, and customs are linked to seasons. What one thinks and knows about seasons is based on ancient cultural heritage and the heritage of the region or culture group. For individuals the seasons are bound up with place and memory. With time spent in a place, they come to represent powerful emotional and intellectual associations with that place.

On the face of it, knowledge of the seasons in the modern world is not as important as it once was to people who lived directly from nature, such as hunters, gatherers, and fishing folk. They were bound most directly to the earth and its flora and fauna. No doubt the cycles of the sun, moon, and the stars were important, but more important were the rhythms of plants and animals. Hunters and gatherers had to pay careful attention to when and where particular game species would appear, and so too plants (Martin 1992). Their life rhythms were the rhythms of the living world around them. Thus the Cheyenne tribe of North America would live in small, dispersed bands for much of the year because the buffalo, the prime object of their hunt, were dispersed. When in autumn the buffalo congregated in large herds, the Cheyenne bands also came together and formed a nation that was almost city-like (Grinnell 1923; Hoebel 1978).

According to C.L. Martin (1992), the links between the seasons and place became weakened when plants and animals were domesticated and agriculture and cities invented. Now humans controlled the reproductive cycle and the lives of plants and animals. Instead of using the 'calendar' of living things on the planet, people began to look to the heavens for temporal divisions – to the sun and the moon and their cycles; or to the 'sky gods,' as Martin calls them. A priesthood came into being and made sure through various ritual functions that the seasons would arrive on time and were fully realized. Without pursuing this theme any further, I should add that seasons and place were further distanced, once the measurement of time became portable (Boorstin 1983).

Although in contemporary, technologically advanced societies only a small minority of people is directly dependent on the seasons for their lively hood, everyone is aware of seasons and each season is associated with a cluster of activities. People are perfectly capable of discerning the obvious changes in seasons, but in a consumer-oriented society mass marketing and mass media exploit seasonal changes to introduce products and to spur on their consumption. In the U.S., boats, swimwear, and barbecue gear are an obvious link with spring and summer; school supplies and clothing, football games and foliage-watching with the fall; skiing, snowboarding, and snow-mobiling with winter. It is astonishing how finely attuned to seasonal ailments is the pharmaceuticals industry. Here in the Northeastern US the different hay fever cycles are preceded by hay fever medicine advertisements, the cold and flu season with the same. In the fall one may be perfectly healthy, however the sudden appearance of advertisements for cold medicines makes you brace for the inevitable with a certain sense of dread of what is to come. The seasons may be a natural cycle, but they are defined largely through artificial cultural activities and artifacts.

There is a paradox in the seasonality of mass consumption. Like all consumption it is geared to world marketplace forces of uniformity and standardization, and as such it is inimical to the idiosyncrasies of places and landscapes (Zukin 1991) and their natural rhythms. Yet by emphasizing the seasons and playing them for all that they are worth in the game of marketing, they are, after all marking the seasons. Are they then the equivalents of the pagan feasts and festivals? In a vague sense, yes. But only that, for there is a profound difference with the paganistic celebrations of moon cycles and the sun god's yearly passage through solstices and equinoxes. The sun god in particular was keenly observed and honored in particular places with sacrifices, songs, dances, and other activities. At the very least it was a transcendental epiphany for the participants, especially young ones, with an awareness of the self in relation to vast presences in the universe. That certainly was the case with the one night of summer solstice celebration that I experienced in Latvia as a child, before we had to flee from our homeland. (There had been others, but I was too young to have much awareness of them.) I do not doubt that market place induced seasonality leaves fond memories of those seasons associated with particular places where recreational activities took place, but how profound they are is an open question.

Here I wish to examine how a season becomes important in one's life in a particular place and how that becomes mental baggage that one takes along into the world as a base of reference for life in new places. That season is winter. Others will have different fondness for a different season engendered by different life-paths in different places and cultures.

A Love of Winter

> Oh, you need to be here then [in winter]. It's the real Maine then. Everything's *some* different. The tourists, the leaf people, they've gone back ... It's all shut then. It's when I like it best (Richard Ford, 2001).

The Ancient Greeks had a proverb: 'Wherever you will go you will be a *polis*.' It meant that no matter where you may roam in the world, the gods of your city and its way of life will be within you. You can never lose them. I think this is true also of the sense of seasons: what you experience of them directly and learn from secondary sources in childhood stays with you for the rest of your life. The seasons may be among the most difficult obstacles in adapting to a new place and developing a strong sense of place. We carry within us the seasons of the places and landscapes in which we have sojourned for a long time, especially during childhood, when lasting impressions are created.

The seasons that I came to know in Latvia have stayed with me wherever I have lived, but none more so than winter.

Thoreau, whose *Walden* is a record of observed natural cycles, devotes almost two thirds of the book to summer (Buell 1995, 242-243). While he made sharp observations of winter, it was not his favorite season. Living in the semi-rural countryside between Philadelphia and Baltimore, I have heard similar opinions from the weather forecasters of these and other nearby cities. Yet I have deep affection for winter and try to ignore their predictions of the dire consequences that come with a two-inch snowfall (if snow comes at all). I love the summer too, with its great freedom for the body and its fecund life, just as long as I do not have to spend it in the oppressive heat of Pennsylvania and Delaware. Fall and spring are marvelous too, but winter holds my greatest affection and admiration.

It is not a simple matter of disliking oppressive heat. Indeed, I can readily imagine the good feelings that come from playing sandlot baseball on a hot July night. But having never experienced that as a child, I can only surmise those feelings intellectually. Not so winter. Its cold, ice, wind, and snow are integral parts of my being and I know them in their myriad manifestations. Much of it has to do with the fact that winter was the sociable season for children (of whom many were only children, the case with myself), when school prompted us to be together, as opposed to summer, when we might have played on the beaches in the supervision of a parent. In winter, during school-time, no matter the weather, we ran wild outside during intermissions, often without hats or gloves, without supervision. We built snow forts during free time, stocked them with snowballs and had fierce battles, back and forth between forts, all good-natured. There was cross-country skiing and wild ice-skating on ponds and lakes. Most of all there were endless games outdoors. I have never lost the positive associations that I have with frozen fingers and ears and damp clothing.

However, deeper, more thoughtful feelings about winter came from quiet contemplation and from Latvian folklore and all the arts.

First snow is one such experience. Long before I read about the magic of first snow in Latvian literature, I experienced it without any associations whatsoever. I was not yet of school age when I became aware of first snow. When I awoke one morning, something had changed. Outside sounds were muffled, distant and inside everything was illuminated by a surreal, soft light. When I looked out of the double-framed windows, what had been a gray, dismal autumn landscape was covered in white. However, what riveted my attention were the large flakes that were swirling down from an unseen source in the sky. I was fascinated by their ever different, fluffy shapes and by the fact that they kept coming down endlessly from a whitish-gray nothingness; that they seemingly materialized out of nothing. My soul was in a happy state. I do not know how long I watched like this. I do not remember becoming tired of it. Never again have I been able to watch any natural phenomenon this long. To this day, whenever it snows with large, swirling flakes, I am happy, no matter what might be going on in life.

There was also memorable literature that fostered a deeper understanding of winter. The first book I ever read was about an icebreaker. Obviously from then on I wanted to captain an icebreaker. (I have some regret for not having joined the Coast Guard, where such an opportunity was possible.) A grim short story that was discussed in fourth grade about a group of ice fishermen on the Baltic Sea, who were blown out to sea by a powerful storm, was a morality tale of life, death, meanness, selfishness, and nobility. It left a powerful imprint, for all of us knew what was involved. More powerful still was a story of a poor, outcast boy, who, after being humiliated during a snow-ball battle, goes into a spruce forest heavy with snow, curls up and eventually dies a dreamy, peaceful death in white beauty. It then seemed like something akin to an earthly heaven. From that moment on I knew that it was possible to cope with death. (I had already known the closeness of death when, at age seven, I was bleeding to death from a bullet wound inflicted by a soldier. A feeling of a marvelous – and dangerous – floating away into sleep was very much like in the story of death in the snow.) The seeming truth of this was reinforced, when one Christmas Eve father took us to a country Lutheran church in a hired sled. Sitting between my parents, covered by warm furs, sparse snow gently brushing against the face and tall, snow-laden spruces slowly drifting past on the narrow forest lane, I soon fell into blissful sleep. How often have I wandered in snowstorms since then, alone, in the Sierra Nevada, Vermont, and Maine, not in search of death (I always make sure to have a way to find direction), but to enter into that earthly paradise!

Winter was also magical in the way snow created a completely new kind of landscape. It literally covered up churned up November mud and even the human nastiness of war. But I also have no illusions about being a refugee in winter. We were such during the winter of 1944-45, one of the bitterest in Europe. However, the miseries of that winter were not the fault of nature and did not diminish my positive feelings.

Yes, winter is a very special season. Certainly I can also sing the praises of summer and the other seasons. Especially summer, with bright, clear mornings (one forgets the not so pleasant mornings), its scents of pitch in spruces and pines, its long days and twilight that stretched long into the night, the pleasures of swimming in the sea, in lakes, and streams. There is also the pleasurable feel of the wind and the sight of wind blowing waves across hillsides of rye and barley. But summer was not only warmth. At fifty six degrees of latitude, there was always the freezing from staying too long in the cold waters of the Baltic Sea, from being too long out in the rain. Summer does not seem proper if one has not been cold and then slowly getting warm again in the gentle sun, in the lee of a sand dune. With autumn I have few associations, at least from childhood. I remember mainly successions of bleak, overcast autumn days, when my parents would not let me outside, when the beach had nothing to offer, and the only outdoor activity was walks in the city with parents. There was, however, one magical aspect to fall – the fogs that occasionally floated in thin streams just above the ground. Early on it was explained to me that our ancestors (and some rural people even then) believed that these were *veli* (literally the rolling ones), the souls of departed folk, who come out in November and roam (roll) over the earth and visit with living relations. Spring was a season with more associations than fall. The most powerful impression of spring came from pussy willow branches with which the house was decorated at Easter and from the young birch saplings that father placed in each corner of the living room to mark the pagan rite of the coming of summer (19 May). The birches sent out a strong yet gentle aroma peculiar only to birches. (When the overwhelming smell of honeysuckle blots out all other scents here in Pennsylvania, I long for that gentle scent of birch saplings.)

These associations with seasons did not transfer very well to Northern California and Southeastern Pennsylvania – two places where I have lived for fairly long periods of time (of course it is not so simple and has also to do with the complex psychology of exile). Especially in California, where I never fully opened up myself to what I knew intellectually were fascinating seasonal rhythms and seasonal characteristics in different places. The sometimes-heavy rains of winter were fine, but the six or so month long dry season became so oppressive, that I would head north to British Columbia in search of rain. The hot, dry Santa Anna Winds, which blow from the east, were particularly depressing, picking up and swirling about accumulated dust, sand, and depending where you happened to be, bits of city trash of all sorts. The wind made your skin 'creep,' that is dry out. However, the intercalation of cold streams of coastal fog on hot July afternoons with the heat just a few miles inland was fascinating and felt delicious on one's hot skin. And how can I forget the fresh, pink beauty of tiny Japanese cherry blossoms at the end of February, graced with droplets of rain that shone diamond-like when the sun came out? Nor the silken waves of dry grasses that the wind drove up the steep hillsides in the Russian and Eel river valleys in late August? I perceived all this and more, but never truly connected emotionally with the rhythms of California landscapes. At that time (it was at the height of the Cold War) I was in the grips of painful remembrance of my lost homeland and missed horribly both the rhythms of its nature and those of my mother tongue.

And then there is Southeastern Pennsylvania, where I have lived for the last fourteen years, a beautiful, rolling, semi-rustic landscape. The seasons are pronounced. It has an additional quality in that the seasonality of other places is imported by air masses from the West and can interrupt the normal course of the season. A thaw can come in January from the Gulf of Mexico and an 'Alberta Clipper' – a fast moving cold front – can bring quite wintry conditions in late fall or early spring. For much of the summer Canadian air stays away and humid subtropical air can rule for days, even weeks on end. Known as heat waves, the humidity is oppressive, there frequently is so much ozone in the air that it has the grayish-yellowish color of phlegm and one dares not inhale deeply its thick consistency. Such times can be eerie; the landscape feels like an isolation ward for some strange plague. And so, if I can, I escape with my family to the cool freshness of a Maine North Woods lake, essentially an escape to Latvia; or to Latvia itself.

There are, of course lovely, contemplative late autumn times in Pennsylvania, after the leaves are gone and a Canadian High prevails, when after sunset there is a low afterglow from the sun on the horizon. Thus autumn has become a favorite season, while I dread summers and even the coming of spring with its exuberant outburst of fecundity.

No, the seasons do not travel very well, at least not in my case.

The Sea

The sea of my past is a gift to you,
Its depths scented in flaming amber.
Amanda Aizpuriete, *To tu pevaica vetrai* (Ask the Storm, 2002)

Just as one carries the seasons within oneself wherever one goes, there are the rhythms of the sea that also come along, if you happen to spend your childhood by the sea.

In childhood the first awareness of nature came from the Baltic Sea. I grew up in the small provincial port city of Liepaja, just a block away from the sea. The sea imposes its rhythms on you, some of which are so obvious that you do not recognize them until you go away and then return. It is a well-known fact that the rhythms of the sea enter one's being, whether you are on or in the sea, or merely by it. Even a brief walk along the sea – much more so an ocean – will bring the rhythms of the wind and waves into you.

At Liepaja, as along most of the coasts of Latvia, the sea is shallow. After an initial deepening to perhaps a half a meter, there will be a sand bar and a succession of sand bars further on, ever deeper. As little children we were not allowed to venture beyond that first bar. So here was a physiographic rhythm that we learned. Another such rhythm was the way that the fine white sand was shaped into tiny, hard wavelets that ran parallel with the shore. They were endlessly fascinating in the then crystalline water. In my child's mind I wondered about the force that created these and concluded that they must be linked to the endless rows of small, steep waves that were driving ashore from the sea. These too were endlessly fascinating, especially during storms, which were frequent.

A great pleasure was to decide to go in the water suddenly, run in with gleeful shouts, and feel the shape of the waves against the shins and the hard wavy sand against the bottom of the feet; and plant one's feet in such a way that sent sheets of spray up ahead and to the sides. On bright, sunny days water droplets would sparkle and small rainbows appear.

On windy days long lines of waves would be breaking far out in white foam and the air was dense with the sound of a continuous, loud and powerful hiss, not unlike that of a large waterfall. I was so used to this sound, which, of course could be heard at our house, that I never heard it. That is not until I returned to the Baltic Sea after many years absence and the hiss could be heard long before arriving at the shore.

Sometimes the sea was glassy smooth. Then it had a slow, rhythmic pulse that made a profound impression. All would be silent, then slowly, languidly a little, smooth wave would wash up on the beach and lap at it several times – swish … swish … swish – and all would be silent again. It seemed as if the sea was slowly heaving as a whole, big body. Although I learned early on (on my own?) that it was but some reverberation from the passing of a distant fishing boat or ship, there was a mystery to it that fascinated. I did not personify the sea, but it seemed to be a big, alive creature. A friendly creature.

These experiences of the Baltic Sea shaped my reactions to other seas and oceans. Like many who grew up by the sea, I am always drawn to it and become restless when I have to live far inland. But experience of the Baltic Sea also limited my feeling for the oceans. I have swum with reckless delight in huge combers of the Atlantic Ocean in September, but could never get quite used to how sharply the bottom dips down into the deep. The sublime power and presence of the Pacific always makes me stop. Many are the times that I have admired the widely spaced, smooth waves rolling in from the West like powerful beings in their own right. Pacific surf breaking on steep beaches, with their crests blown off by the strong northwest wind I can also watch for hours, just as I watched waves on the Baltic. Even sailing in thirty-foot storm waves one night just north of the Golden Gate had its fearful attraction. So too a powerful impression came from a 'conversation' with a lone seal in the high surf, which kept looking back at me, another lone being, on the beach. But all the many experiences of these two oceans have not won my heart. In fact, at times of great stress in life I have had dizzying nightmares about being in their ever more monstrous, green waves.

Not surprisingly, whenever I have been at one of the Great Lakes, I felt immediate warmth and affection for their waters and beaches, as well as vegetation, especially Lake Superior, but also Lake Ontario. Similarly, the larger lakes of North America, such as Lake Shuswap in British Columbia and Moosehead Lake in Maine seem familiar like home. I know how much I limit myself in this, but affection for waters and their rhythms cannot be turned on and off at will. The oceans offer great adventure and beauty, but I can only admire them, not take them into my heart.

Birdsong

Two summers and other, shorter visits to my maternal grandmother's farm in the uplands of Northeastern Latvia opened me to sensibilities that come from the rural landscape. Although allowed a feral existence from morning to night, I do not think it made me a country boy. What I know of the country is too impressionistic. There seldom were other children present, so I learned haphazardly, except for fishing in the stream that ran through the farm, which my uncle took time to teach. Other than a few hours of herding cows and helping to stomp down hay in the second floor of the barn during haying time, the work seasons of the farm were for me distant. The days went by in the timeless blur of childhood happiness, from bright, cool early mornings to the long, gray twilights of nightfall of that latitude. Nevertheless, there are impressions of natural rhythms that form a part of my sense of that place and also have helped me to connect with foreign places. Most of the impressions come from birds, probably because they were active presences in the landscape. Of these are not many, for we fled Latvia when I was only nine, but they left indelible impressions and gave me a strong sense of that place and landscape. They also helped me to connect with foreign environments.

One is the cuckoo birdcall in spring. It was a quiet, evening enveloped in a still grayness that I first heard it. The slow, leisurely call came from the far forest across rolling, open fields and meadows: 'Coo-coo ... coo-coo,' a long silence, then, again the same sound. In the great silence that reigned on that evening the voice of that bird was lonesome and plaintive. After a long interval, it came again. I learned to wait for the call. Although the rhythm of the song was always the same, the intervals in-between were unpredictable.

Uncle Francis identified the sound for me and asked if I had a coin in my pocket. No, I did not, whereupon he gave me one. 'You must always have a coin in your pocket when you hear the first cuckoo bird in the spring,' he said. 'Otherwise you will be poor all year.'

Now I realize why mourning doves have become dear to me here in Pennsylvania – they have the same plaintive cry as the cuckoo in Europe, the same, slow rhythm. They do not signal great distances, as does the cuckoo, they are invariably close by, but they connect one with a kindred spirit in nature. I never cease to admire the vocal gymnastics of the mocking bird, which sings the night away perched atop the sharp peak of our house or in the great honey-locust tree next to it, but I have no particular fondness for it. It is foreign, exotic. But the mourning dove seems like a friendly voice of home.

Another rhythm was radically different. I associate it with spring and early summer. It was a seemingly chaotic, rapid rhythm – the mad trilling of seldom visible skylarks high in blue skies, above fields and meadows. Many times the days were windy. The skylark embodied the carefree happiness of summer days.

For many years the memory of skylarks was buried in my subconscious. When I returned to Latvia after many years absence and one warm late spring day heard their trilling high in the sky, it was instant recognition. It was as if the world was an orderly place once again. As if I had come home.

The golden oriole (*Coriolus oriolus*) is another bird of which I am fond. I came to notice its particular voice before the onset of rain, which in turn I came to associate with a certain stillness, high flying, darting swallows and a happiness within. Why the happiness, I do not know. It is buried in some fortuitous condition of childhood. Its song is not one that I can readily summon forth in my mind; nevertheless it is there, part of that rural landscape.

These few birdsongs were a deep part of my being. Their songs gave color and depth to places and landscapes in which I grew up and they went with me into foreign lands. I missed them in northern California, especially in the redwood forest, which is a silent forest, seemingly devoid of bird life (and insect life). Here in Pennsylvania, as mentioned before, I have found companionship with the mourning dove. In the spring there is too the mournful, high-pitched whistle ('Old Sam Peabody, Peabody, Peabody') of the northward migrating white-throated sparrow (*Zonotrichia allbicolis*), which I meet also in the woods of Maine (though by midsummer its song has lost a syllable).

Although there are a few nesting loons in Latvia, I never heard one. On our lake in the North Woods of Maine there are always loons. Over the years I have learned to know their varied songs, from the wild 'laughter' in the spring, to mournful calls at night in August. The loon has made me feel at home in my new homeland. But other than the recently arrived coyote, few aspects of Maine have ever seemed foreign.

Tree Voices

This essay began with an account of a boy's fearful night in a forest on the German shores of the Baltic Sea. It is fitting to end it with a consideration of the different voices of trees.

Even before that stormy night I had been aware that the wind can be seen and heard in trees, and wherever I have been, I have looked and listened for it in trees. In the most alien landscapes I have found kinship with trees. Not only the voice if wind is in them, but in the feel of needles and leaves to the touch, in how branches break or bend, how a birch or an alder bends when you climb it. How trees change from spring through the summer I observed intimately, but not the fall, because by that time I was back in the city by the sea. As an only child on my grandmother's farm I had to create daily life and trees, and wood pieces were part of it. Uncle Francis showed me how to make little, square-rigged Viking sailboats from thick pieces of pine bark, a stick, and thin cardboard. I sent fleets of these boats down the creek on raids to distant lands. He also showed how to fashion flutes from bush-willow branches in spring, when a section of bark (skin) can be pulled off whole and fashioned into a flute chamber. When a big spruce full of pungent pitch was felled in the nearby woods, he showed me how to use a hatchet to pare its branches.

In school we also learned about trees, from folklore, fairytales, literature, and history in which forests and trees had been significant. The girl's, or woman's, folk rhyme, 'Through a silvery birch grove I walk'd,/Not a branch did I break' was the very first that I learned. The Medieval tale of the 'Hornblower of Talava' was fascinating. There he was, this observer atop a tall spruce in a spruce forest, far from the Talava castle, with his horn, on the lookout for enemies, ready to blow his horn in warning. On a dark and stormy night came Teutonic Knights. He blew his warning over and over, but from Talava there was no response. He kept blowing, even as the German axes were felling the tree. How our youthful minds were moved by that terrible scene in the dark forest!

We learned about the virtues of different trees: the birch as the ancient Latvian cult tree of white (*balts*); the oak and manliness, sturdiness; the linden tree and womanly virtues. And thus, for complex reasons, I became aware of trees and their 'voices.' The birch, tall and pliant, with leaves that shimmer in bright sunlight, rustles in the slightest of breezes. In a gale of wind it sways violently and gives a hissing sound like no other tree, not even its cousin, the aspen. It sounds like water of a mountain stream running over small pebbles. I do not know the sound made by oaks, but conifers I identified early on. Words cannot describe these sounds. A spruce forest or an individual spruce gives off a deep roar in a gale. In light wind it hisses deeply. Pines, especially individual ones, hiss with a higher pitch. There is something mysterious and ineffable in the sounds that wind makes in pines, something that makes one feel the closeness of eternity.

As with birds, these early experiences and associations were carried abroad. In California I did not connect with the chaparral until many years after I had left it and could delight in its delicate fragrances, particularly the myrtle. However, it had no voice to it, no perceptible rhythmic texture in the winds (which blow hard there.) But I connected with the eucalyptus, which had wind voices. The Eucalyptus grove on the Berkeley campus was a favorite place to be, not only because of the aroma, but also especially for the paper-like rustle of its leaves in light airs. It was not familiar, but nevertheless felt as if there was a certain kinship with birch groves. It also epitomized lazy, mild Sunday afternoons in California.

Coastal redwoods (*Sequoia sempervirens*) are provocative. As a lover of trees, I wanted to know them. At first they were incomprehensible, strange, primeval, out of any human scale whatsoever, and utterly voiceless, silent, for not a bird would sing. The strangeness was especially pronounced in the Eel River Valley of Humboldt County, where their trunks rise straight out from river silt. Together with their great height and straight, branchless trunks, they seem like architectural monuments. Bit by bit, over time, I got to know them, but they never engendered emotional warmth. However, they were curious trees and provoked trains of thought. Many a time, on moonlit nights, I wandered among them alone. At such times they were like a magnet. The way the moonlight would slant down among the black trunks was almost like an artificially lit stage setting. Walking through patches of moonlight and dark shadow made it like walking in a fairytale, except I had no imaginary figures to populate this landscape, no associations. Until one night I saw naked youths from our ranch leaping into the river from a huge redwood trunk, up high, that had fallen part way across the river. With the dark, towering mass of redwood trees on one side, the shimmering waters of the river in the center, and the brushy shore and prairie hills on the other, I was witnessing a scene thousands of years old. These were Native Americans. Thenceforth prehistoric man walked with me among these prehistoric beings.

After some time I found the voice of the redwoods. It was familiar and strange. In summer, when in late afternoons, almost like clockwork, it starts to blow a half a gale from the Northwest, it sounds like surf in the top branches of the redwoods. They jerk around violently, while below there is utter stillness. The massive trunks are immobile, as are the leaves of the poison oak that climbs partway up the trunks, and the ferns that grow in every patch of the forest floor not covered by the acidic needles of the redwoods.

As much as I admired the noble redwood forest, it was the one place in America where conscious thought occurred of not wishing to die there. There are many worse places, especially in subtopian landscapes, but this, after all was a coniferous forest. A primeval forest without snow.

Afterthoughts

> I doubt that character and conduct are much shaped by landscape, climate, or geography. We manage to breed saints, brutes, barbarians, and mudheads in all sorts of topographies and climates. But what country does to our way of *seeing* is another matter, at least for me. By and large I do not know what I like, I like what I know. I respond to the forms and colors and light I was trained to respond to, I acknowledge what revives my memory. But only when I have submitted to a place totally (Porter, Stegner, and Stegner 1981, 110).

This has been an exploration of nature's rhythms by an individual who, unlike Thoreau or Annie Dillard, did not spend a year or so with a self-conscious plan to observe nature. The essay is necessarily a *bricollage* created from different places and regions on two continents, in particular circumstances. And yet, there is a commonality to the aspects of nature that were recognized as having resonance with landscapes of one's childhood. Being deprived of one's homeland at an early age, I was in effect seeking to make a home in the world and found connections where I could. I liked what I knew, to paraphrase Stegner.

In 1830 Chief Arapooish ('Rotten Belly') knew that, 'The Crow country is in exactly the right place. Everything good is to be found there. There is no country like the Crow country' (quoted in Porter et al. 1981, 111). I too knew this about my country. And thus, without a country for much of my life, I have created another country, both through the vicissitudes of contemporary life and by choice. It is an extended country that reaches from Narvik beyond the Arctic circle, to Oaxaca near the tropics, from British Columbia to Maine, a collage country of bits and pieces of home, made up largely of northern rhythms that first revealed themselves in childhood.

The story may be particular but it represents a universal human propensity to select elements from past cultural contexts in order to shape the present (Lowenthal 1985; Allan 1986). It is more complex than I was able to present it here, because, as is well known, leaving a homeland for new environments creates ambivalence towards the present, past, and future (White 1995, 3). I suspect that it is also a universal story for many people who are highly mobile, who do not work or live close to nature for any sustained length of time, and therefore do not have much time to fathom it in the fullness of its natural rhythms.

The story makes the case for a postmodern condition of fragmentation of time and space, not as an arbitrary academic, literary, or architectural abstraction, but as a reality of contemporary existence. With one vital distinction: motifs of the past are not chosen arbitrarily. In the mystery of writing it (Lowenthal and Prince 1976, 130) I found 'new relationships between the past and the present' (Beitnere 2002, 154). The elements selected – the sea, seasons, birds, trees – are not cut off from the past, they form an organic continuum with it in the present. At least within the frame of a single life, which is, after all, a world.

References

Aizpuriete, A. (2002) To tu pavaica vetria, *Literaturas menesraksts Kargos* November, pp. 90-96.

Allan, G. (1986) *The importances of the past: the authority of tradition.* State University of New York Press, Albany, NY.

Beitnere, D. (2002) Vai visi latviesi ir tikai zemnieki? Sabiedribas pasapraksts un kulturas identitate, *Literaturas menesraksts Karogs* November, pp. 150-163.

Boorstin, D. (1983) *The discoverers: a history of man's search to know his world and himself.* Random House, New York.

Beston, H. (1962) *The outermost house.* The Viking Press, New York.

Buell, L. (1995) *The environmental imagination: Thoreau, nature writing, and the formation of American culture.* The Belknap Press of Harvard University Press, Cambridge, MA, London.

Dillard, A. (1974) *Pilgrim at Tinker Creek.* Harper and Row, New York.

Edmonds, J.M. (1960) *The Greek bucolic poets.* William Heinemann, London.

Ford, R. (2001) 'Richard Ford' in Lester, T.S. *Maine: the seasons.* Alfred A. Knopf, New York, pp. 107-110.

Grinnell, G.B. (1923) *The Cheyenne Indians: their history and ways of life.* Yale University Press, New Haven.

Hoebel, E.A. (1978) *The Cheyennes: Indians of the Great Plains.* Harcourt Brace College Publishers, Forth Worth, Tex.

Jackson, J.B. (1994) 'A Sense of Place, a Sense of Time' in Jackson, J.B. *A sense of place, a sense of time.* Yale University Press, New Haven and London, pp. 149-163.

Lopez, B.H. (1986) *Arctic dreams: imagination and desire in a northern landscape.* Charles Sribbner's Sons, New York.

Lowenthal, D. (1985) *The past is a foreign country.* Cambridge University Press, Cambridge.

Lowenthal, D. and Prince, H.C. (1976) 'Transcendental Experience' in Wapner, S., Cohen, S. and Kaplan, S. (eds.) *Experiencing the environment.* Plenum, New York.

Lynch, K. (1972) *What time is this place?* The MIT Press Cambridge, MA, London.

Martin, C.L. (1992) *In the spirit of the Earth: rethinking history and time.* The Johns Hopkins University Press, Baltimore and London.

McFee, J. (1989) *The control of nature.* Farrar, Strauss and Giroux, New York.

Porter, E., Stegner, W. and Stegner, P. (1981) *American places.* E.P. Dutton, New York.

Virgil (1950) *Virgil's works.* Random House, New York.

White, P. (1995) 'Geography, Literature and Imagination' in King, R., Connell, J. and White, P. (eds.) *Writing across worlds: literature and migration.* Routledge, London and New York, pp. 1-19.

Ziedonis, I. (1997) 'Viddivvarpa. Poema par maizi' (Poem about bread), in Ziedonis, I., *Raksti 7.* Nordik, Riga, pp. 197-237.

Zukin, S. (1991) *Landscapes of power: from Detroit to Disney World.* University of California Press, Berkeley, Los Angeles.

Chapter 5

Rhythms and Identity in
Boyne Valley Landscapes

Gerry O'Reilly

My central themes are journey and identity, storytelling and interpretation, individual
experience and societal context, contingent events and broader historical movements.
Anne Buttimer

Context and Definition

Each Autumn I take first year third level geography students on fieldtrips to the
Boyne, County Meath, with the aim of reintroducing them to this holistic discipline
emphasizing human-environmental relationships and sustainable development. The
majority has studied history to 15 years old, and a minority continues with it to
Leaving Certificate level. A vast majority of the students are bilingual in English and
Irish, and come from areas outside Dublin (Figure 5.1). In orienting the voyagers,
maps and time-slice landscape representations are juxtaposed with archaeological sites,
bocage farm-scape, Valley contours and seasonal rhythms of the flood plains. Different
knowledges are used: empirical data from physical geography, archaeology, history,
political economy and other systematic data, but also anthropology and tourist
interpretations. Without these voices including poetry, music or painting, much of this
landscape could become static museum-scape (Buttimer, 1980). Combinations of
Boyne sites are used, but narratives of human-environmental rhythms and power
relations can be traced in the landscape. Besides empirical data, students have
inherited images of the region through tale and legend. Boyne River rhythms and
seasonal migrations of eel and salmon call to mind the *Salmon of Knowledge* legend
retold to each generation. Romances of Gaelic *fleadhs* and *feiseanna* (celebrations and
competitions) from Ireland's Celtic Golden Age come alive at Royal Tara (Beresford
Ellis, 1994). Like Tara, the Bend of the Boyne (*Brú na Bóinne*) and Slane are associated
with St Patrick. Then come school-hood memories of 'invasions – us and them' –
Vikings, Normans and English, and defeat at the Battle of the Boyne (1690) resulting
in 'foreign' landlords, famine, revolution and eventually national independence with
renaissance in wealthy 'Meath of the pastures.'

Voyagers find echoes of Swift's *Gulliver* and the Duke of Wellington in Trim,
James Bond (007) alias Pierce Brosnan from Navan and Mel Gibson, Academy
Braveheart (1995) hero who put Trim Castle on Hollywood maps playing the thirteenth

century Scottish nationalist, William Wallace, disemboweled by English forces and shot on film in a Norman castle on the Boyne.

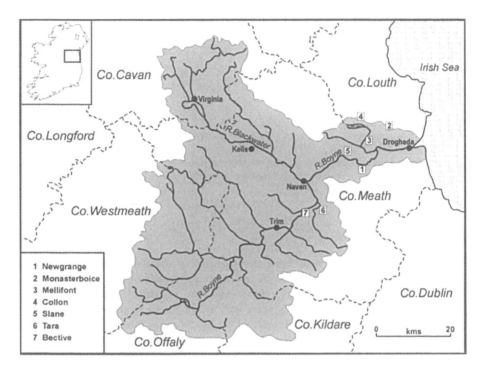

Figure 5.1 The Boyne Valley and catchment area
Source: Author.

Students commented:

> Statements in the landscape are awesome and create a sense of power and strength. One wishes to protect this landscape in order to carry on Irish culture and heritage. There is a great sense of wonder at the scale of these statements in the landscape (Aileen). I was overawed by what I experienced, especially Newgrange. They must have felt that there was an afterlife, much more important than this one on earth as illustrated by the light box at Newgrange with victory of light over darkness, life over death (Sorcha). Each part could not exist without the river, local natural resources and rhythms. There was a feeling of harmony with nature (Eileen).

Trying to Make Sense of Rhythms

In contrast to the flux and muddle of life, art is clarity and enduring presence. In the stream of life, few things are perceived clearly because few things stay put. Every mood or emotion is mixed or diluted by contrary and extraneous elements. The clarity of art – the precise evocation of mood in the novel, or of summer twilight in a painting – is like waking to a bright landscape after a long fitful slumber, or the fragrance of chicken soup after a week of head cold (Tuan, 1993).

Words try to catch essence, enhanced by transformational grammars of space and place echoing over millennia while the eye catches snapshots of Boyne tableaux, affected by flooding and weathering; agricultural schemes; house, road, heritage center construction; ever-increasing tourist numbers and archaeologists scraping. Interpretations are sought (Bourdieu, 1995; Price and Lewis, 1993; Duncan and Duncan, 1988; Cosgrove, 1985, 1989; Cosgrove and Daniels, 1988; Duncan and Ley, 1993; Barnes and Duncan, 1992). Students commented:

> The landscape had a type of power over me because of its beauty and the many stories it holds. Mellifont and Newgrange have many tales with evidence of our ancestors (Helen). When I arrived at Newgrange I realized that I had never been there before – I saw culture in reality and not just a text. I realized my sense of place in society (Paula).

While scrutinizing landscape, *pentimento* appears – this art term refers to reappearance in an oil painting of elements of drawing or painting that the artist tried to obliterate by over-painting. If pigment becomes transparent over time, ghostly remains show through. *Pentimento* indicates evidence in a composition such as landscape that original work has been changed. The pigment with which the artist covered unwanted beginnings or perceived mistakes, with time or cleaning can become transparent, revealing original intentions through the finished composition. Use of X-ray photography can uncover evidence of the creator's original intentions, as with Caravaggio's *Lute Player*, while in Michelangelo's Sistine Chapel *Last Judgement* panels, clothes put on his nude figures posthumously are still well in place. While the artist may be working for him- or herself, he or she may also be forced or sponsored to re-create. Original canvas may be planned, and fully or partly painted by one artist; different artists can change it at varying periods due to shortage of money, recycling material, or the painter's wish to give a different view. At Mellifont Abbey, in the nineteenth and twentieth centuries, *pentimento* of medieval biblical scenes with human forms came through the whitewashed walls of the chapter house, which had been over-painted by seventeenth century Puritan powers.

Author Lillian Hellman (1973) uses the term *pentimento* as metaphor in describing places and associated people she met on her life-path, now part of her mindscapes. Her metaphor snapshots are organic, making specific places and associated people alive in the mind of the author and reader, going beyond mosaic or patchwork metaphors. As a student stated:

> I saw something greater – landscape and society. I experienced the extension of innovations of ancestral creation, and rhythms that have shaped contemporary society and our spirituality. Something awesome happened that helped shape who I am today. I like to think I awoke and arrived at each picture with an open mind and gained a wealth of insight (Joan).

The Valley offers a gallery of *pentimento* views. Going from the whole view to component sections, *pentimento* becomes obvious. Definition is important avoiding over-simplified deconstructions.

The landscape was like a talking painting (Shauna). For me it told a thousand tales if you looked deep enough. It was shaped by earlier generations and was given to us, full of wisdom, life and cultural rhythms. The landscape gives us a sense of where we come from, and how we treat it shows us what kind of people we are (Eimear). There is a uniqueness and spirit about the Irish landscape as I have seen in the Valley (Joe). There is much sentiment and great cultural influence from many periods. It has many depths and should be given the respect it deserves (Marie). Landscape is who we are and where we come from (Maura). As we traveled through the *Brú na Bóinne*, Newgrange, Navan, Trim, Tara I was overwhelmed by the sense of historical rhythms – generation to generation all these places hold ... how people coped in the past ... how people had to adapt to nature and environment. The tombs were built so that they are still standing. These places show the strong sense of community and spirituality, and were passed down (Niamh).

Geographical Discourses and Generational Rhythms

During the fieldtrip, geography is described as a holistic science examining human-environmental relationships in areal or spatial patterns. Differentiations and continuities in contemporary landscapes are constantly (re)created in time-space contexts and paradigms. The Valley offers a perfect location for regional and thematic studies (Sauer, 1925; Hartshorne, 1939; Gregory, 1989; Hägerstrand, 1982).

Lifeways and values are expressed through all forms of activities in Boyne landscapes (Vidal, 1926; Jackson, 1984; Buttimer, 2001a). Historical geography presents evidence in place-names and documentary sources; rhythms, actors, processes and events that created Boyne landscapes come to life: Neolithic necropolis, Celtic raths, monastic settlements at Monasterboice, Donaghmore and Kells; Norman vestiges at Dowth, Dunmore and Trim; eighteenth century estates; 'planted' villages; geostrategic battle sites and so forth. Juxtaposed are the daily urban routines of Drogheda, Navan, Trim and Kells, and rural or tourist-related economies in complex combinations. Linear national narratives emphasize dates – Viking raids (eighth century) – but cultural processes that have shaped the landscape are too often subsumed. Transitions related to sedentarization, or Neolithic *pentimento*, may overshadow other innovative processes. Boyne landscapes are being (re)created with European Union policies and tourist industries in the name of sustainable development.

Maps and histories in Irish, Latin or English help bring to life the power of commissioners and artists of Boyne tableaux especially at Mellifont and Trim. Yet there is always the desire to fill in unexcavated gaps: Neolithic mounds or absent memorials. Constructed rational paradigms must leave space for phenomenological and humanistic interpretations (Buttimer, 1982; Tuan, 1974; Ley, 1980; Johnston, 1978).

No matter what changes ostensibly come in the future, we will always have our past in this landscape (John). It bears the marks of many civilizations, which came to Ireland and have left their stamp. The landscape shows evidence of innovations that each brought. Irish landscape bears evidence of the value of different civilizations and multi-cultural society (Helen).

Rhythms of Sustainable Landscapes and Lifestyles

Sustainability is premised on the healthy functioning of ecology and environment in nature's rhythms and appreciation of this passed from generation to generation through culture and identity.

> Sustainable development in the Valley refers to preserving its natural, physical features while simultaneously fulfilling social and economic needs, as future generations will depend on these as much as we do (Nano). It is where there is a balance of resources: cultural and economic in society which leads to long-term development instead of a short term quick fix. We must have something strong to hand on to the next generations. In order to achieve sustainability, bottom-up organizations like the Irish Country Women's Association must inform people and collaborate with GOs like the Heritage Service to avoid extremes (Clare). It is the change of environment through different periods – Neolithic to modern, as a result of people's influence and interference (Paula). With sustainability the Valley is alive, it is frightening to think that this landscape is in our generation's hands (Paul). Sustainability must be about more than physical survival and species diversity; cultural diversity and identities are all important (Jane). Our heritage is of great importance to our culture and to preserve it we must sustain our landscape (Dermot). Throughout the entire trip, each part of the landscape combined together to express feelings of life, pride, spirituality, awe, interest and patriotism in me – sustainability is about place, roots, rhythms, staying alive (Nora). A sense of spirituality remains in these places where the artifacts are preserved encouraging us to go on (Therese). If you don't sustain the Valley, then it dies out (Aibheann).

As the past helps define the present cultural landscape, 7,000 years of human impact on the Valley necessitates planning processes, which will mold legislation to achieve future lasting development.

As positive development aims at getting a balance between ecological, economic and socio-cultural needs where none of these domains suffer irreparable damage – from pollution, resource destruction, or over-rapid cultural change creating social vulnerability – modulating processes via technology, and consensual institutions facilitating top-down and bottom-up synergies is all important (O'Reilly, 2001; Buttimer, 2001a). Students commented, 'Our heritage is of great importance to our culture and to preserve it we must preserve nature and our landscape in a sustainable planned manner catering for interconnected natural and social rhythms' (Eoin). 'The landscape expresses Ireland's deep roots; it enables us to look back and forward' (Bernadette).

Within Boyne landscapes exist millennia of interconnected arenas (re)created via territorialities struggling for resources, defense and identities. Besides advantageous carrying capacity and geostrategic location – the Valley is pregnant with symbolism; a medium between humans and nature trying to sustain themselves and give live a meaning with echoes of Saint-Exupéry's *'donner un sens a la vie'* constructing the metaphorical *Citadelle* (1948). Cultural interpretations of nature, seasonality, mortality, eternity, identity and legitimacy have sustained continuous habitation of the Valley as evidenced at Newgrange necropolis.

> Newgrange is an expression of human belief and endeavor. The genius of the creations and purpose amazes me, even by today's standards (Anne). The Boyne has its own beauty

... we shouldn't destroy the evidence we have of previous ways of life just to contribute to future economic development. Too much tourist development would make the Valley loose its spirit (Fiona). While some sites have been preserved by government, when it established heritage centers, it is important that other sites don't need to be, or can't be rebuilt e.g., Mellifont, and which period could be used, twelfth or fifteenth century? ... Hill of Tara, should it be preserved and turned into a major tourist site or should it be left as it is? I don't know how this question should be answered but I believe that each generation of Irish people have the right to see it as it is (Una).

Even though thousands of years old, it's amazing how these features still impose ... innovation captured people's minds. They tried to make a better life, no matter the difficulties faced. An aura of nostalgia can be felt and I realize the extent of my inheritance (Paula). I felt awe and wonder at the magnificent skills, work and craft portrayed in the landscape. The buildings show how dedicated people were to their beliefs (Helen). It is clear that our ancestors were concerned with spiritual values (Cathal). To really see the landscape gave me a feeling of poetry and magic. Settlement upon settlement ... shows that the land was of great value to generations and generates a feeling of power and strength to me. To think powerful rituals – ceremonies took place here, kings of great power and wisdom ruled these lands ... this leaves a feeling of insignificance in relation to the Ireland of today. The culture and wisdom held in the palm of the Valley is a sacred asset ... giving us a sense of strength through our history (Jennifer).

The Physical Canvas

Legendary goddess *Boinn* gave her name to the river – only her husband Nechtain and his cupbearers could look upon Sidh's well; one day *Boinn* gazed in, water came out blinding one eye. She fled but the water followed her to the Irish Sea at Drogheda (Green, 1995, 1996) (Figure 5.1).

The Valley's terracing and gorges attest to glacial history 12,000 years ago (Meehan and Warren, 1999). The Boyne (113 km) has a catchment area of 2,670 sq km and gentle gradient (1.24 m/km) with fifteen tributaries, twelve being rich in salmon, eel and trout. With high banks, an uneven riverbed and canalized stretches, there is fast runoff during annual floods. Limestone with shales and sandstones are salient features. The Boyne system drains 67 per cent of Meath (567,127 acres) with rich pasturelands and level plain, nowhere rising above 260 m. The catchment includes parts of five other counties. Activities supported by the Boyne include provision of drinking water, fisheries and recreation. The river now comes bottom of the national list in relation to unpolluted waters with 45 per cent of the channel surveyed falling into this classification; only 31.8 per cent is considered unpolluted. Abstractions from surface waters for drinking impact on quality by reducing flow and diminish the river's capacity to assimilate discharges.

Concerning municipal water discharge, 31 agglomerations (population 111,930) discharge effluent, with eight generating 80 per cent of this load and Drogheda (population 56,000) accounts for half of total discharge. Industrial waster water comes from food processing and intensive agriculture based industries (e.g. piggeries, poultry), while tourism, mining, cement manufacturing and textiles also impact. Industries are concentrated towards the northeastern catchment area, the majority of which discharge effluent to watercourses rather than into municipal sewerage systems.

Four effluent discharges are regulated by the Environmental Protection Agency, and 13 have Local Authority licenses (Three rivers project, 2002).

Boyne Valley Cultural Landscapes and Rhythms

On the landscape, power *pentimento* strikes the observer, especially at Newgrange (Nolan, 1988; Aalen et al. 1997; Andrews 1967; Johnston, 1994; Brady and Byrnes, 2002; Devine-Wright and Lyons, 1997; Mitchell and Ryan, 1998; Moody and Vaughan, 1986-96; Smyth and Whelan 1988; Foster, 1989). This prehistoric necropolis projects views in the Bend of the Boyne UNESCO World Heritage Site with imposing megalithic monuments (3900-3500 BC). Natural heritage areas also exist there, including Crewbane marsh and woods, and associated migratory swans giving genesis to gods, legends and tourists to Newgrange: 60,000 visitors in 1980 rising to 230,000 in 2000 (Stout, 2002). Symbolically, at the winter solstice, the *Taoiseach* (Prime Minister), dignitaries and satellite television crews welcomed in the new Millennium as the rising sun made its annual penetration of the main burial chamber. On the same date in 2001 and 2002, people from the Valley protesting at plans for the construction of a waste incinerator at nearby Duleek congregated here (Figure 5.2).

Massive stone-kerbed mounds on the Bend's long ridge, and 40 satellite burial chambers, dominate Newgrange, Dowth and Knowth landscapes. At Newgrange, the corbelled roof stands on the main passage grave; a roof box allows sunlight to penetrate during the solstice giving *chiaroscuro* – obscure bright impressions. Here place, time, seasonality, rhythms of communal farming activities, life and death are intertwined. Outside, standing stones and shadow play with time; intense ritual activities were once practised here sustaining and giving life a meaning (Figure 5.3).

Near Newgrange, Neolithic or Early Bronze Age ceremonial enclosures exist along with ringforts and souterrains from the Early Christian period. In the twelfth century *Senchus na Relec* survey of royal cemeteries, individual monuments are listed; the 1837 Ordnance Survey identified these as a royal cemetery (Stout, 2002). During the Neolithic period, *Brú na Bóinne* was a ritual pilgrimage center for native, Romano-British and continental visitors located on the Royal Tara – Ulster route. Newgrange became part of Mellifont Abbey farmlands (1384) and the Bend was an epicenter of the Battle of the Boyne; afterwards Newgrange lands were transformed into an estate with enclosed field-scape. Newgrange House has been demolished; farmlands are owner-occupied, while the main prehistoric sites are state managed (Stout, 2002).

Knowth mound (3500 BC) with decorated stones aligned on a lunar standstill has graves in rhythm with the equinox. One curbstone can be read as a 365-day solar calendar; others indicate recurring measurements of time. Dowth graves are aligned with midwinter sunset and quarter days. With the demise of Dowth's usage as a necropolis, communal rituals continued in earthen henge monuments, which were replicated at Newgrange, Knowth and Monknewtown. In the Christian era, a monastic site developed, due to Viking attacks (ninth century) protective souterrains interconnecting burial chambers were constructed.

Figure 5.2 Newgrange, main ceremonial burial chamber
Source: Author.

Like Knowth, Dowth became a center for Celtic kings, and in the twelfth century both were incorporated into the ditch-protected Pale region of Anglo-Norman Ireland. A Norman church (1202) lies near the main megalithic chamber, on the site of an earlier monastery and a manorial village developed around the fortified mound. In the eighteenth century Dowth House and parklands, incorporating a tower house manor (fourteenth century), prehistoric tombs and henge was built. In the nineteenth century, further tenant land subdivision continued, as did quarrying of stones from the mounds. The state has acquired the lands in the immediate area (Stout, 2002).

> I always wanted to see Newgrange – it's spectacular. It's large, breathtaking (Tracy). Knowth expressed the eternity of the world to me, and people's insignificance in the face of this. We are merely passing through, the landscape is transient, yet defies the test of time. I feel that power is associated with the landscape (Tara). There is a type of spirit in the tombs as history tells us that there are many people buried there. You can see the pride people had in their area in the way they worked hard to build these structures. Respect: people had a love for their deceased relatives, building magnificent tombs to hold them. Power and determination: of a people to build these tombs and castles. Devotion: to gods so that they could spend their lives undisturbed (Catherine). Knowth showed the people were fearful of the gods and celebrated the seasons, which we still do (Aisling).

It is much easier to imagine our past and future when we come face to face with tombs and ruins. These features gave me a sense of value, that people are a sum of the whole past ... What a wealth of knowledge our ancestors passed down through the generations. They paved the way for us to follow. I feel that I am part of the whole bigger picture that is my heritage. America and other countries don't have these features from the past ... I have a sense of belonging (Elizabeth). To comprehend Newgrange is surreal; and how the builders knew how to live and build in correspondence with nature (Solstice). I almost felt the spirits in the tombs (Jacinta). I feel gratitude to have this landscape. The inhabitants of the Valley were highly organized, an intelligent society with a great respect of nature, life and the sun. I got a great feeling of freedom and wide-open space: a sense of historical and cultural rhythms (Marian). The landscape expresses feelings of anxiety when I see how easily it can be destroyed, moving nature to construct buildings (Deirdre). I felt fear of the dead (Linda).

Figure 5.3 Newgrange, entrance stone and light box of main chamber
Source: Author.

At the Bend and on the Blackwater tributary at Loughcrew, prehistoric *pentimento* illustrates agricultural innovation and human continuity. Political and religious elites have tried to legitimate their being by leaving their imprint on the landscape, in life and death. 'Graves evoke emotions within you and make a realization between yourself and your ancestors. The necropolis site itself is expressive and spiritual. The wealth of the tomb is installed in you or perhaps you are reminded of it, as it is part of all of us. All the monuments were at one with nature' (Pauric).

Celtic Pentimento

Tara, upstream from Newgrange was the most venerated religious-political site of legendary Celtic times (between the third and eleventh century) hosting High Kings. Only earthworks remain, yet megalithic monuments on the Hill, and events give this place a special symbolism. The Mound of the Hostages (passage grave, 2000 BC) is prominent near the earthwork Banqueting Hall (230x27 m); Neolithic in date, it may have been the ceremonial entrance to the Hill. Nearby is a three-banked ringfort, the Rath of the Synods; in 1899 British Israelites searching for the Biblical Ark of the Covenant excavated it (Bhreathnach, 1995; Bhreathnach and Newman 1996). The Royal Enclosure (245x290 m) with two ringforts hosts the Royal Seat (*Forradh*) and Cormac's House on which stands a stone marking a mass grave (1798 Rebellion). In its center lies the *Lia Fáil* (Stone of Destiny), a phallic symbol, associated with the coronation of Celtic kings. Legend places the British coronation Stone of Scone with it; housed in Westminster Abbey until recently before being reclaimed by Scotland. Nearby are King Laoghaire Fort, and Gráinne's Fort named for Cormac's daughter, heroine of the tragic love tale of *Diarmuid and Gráinne* studied by most Irish students (Green, 1995, 1996).

> Tara worked more feeling for me than any other place. Maybe it was the rain as a contributing factor, or the feeling of quietness. There was a spiritual mystical atmosphere. The way it is at the moment is the way I would like it to remain because if they build a heritage center, it will surely take away from its spirit. This is where we have to be careful in how to sustain such sites (Aoife). I felt a sense of gratitude for the landscape that we have. Tourists shouldn't erode Tara. I feel extremely fortunate to be able to experience what we have and sometimes this landscape is exploited by us (Orla).

Relating to Tara, St Patrick is said to have confronted King Laoghaire (432 AD) with the Christian Gospel, kindled the forbidden paschal fire on Slane Hill, confounded Druids, explained the Trinity mystery using a shamrock leaf, worked miracles and converted Ireland to Christianity.

> The lands of Meath expresses a feeling of a time when this was an area of great importance as with the royalty at Tara. There is a spiritual vibe from the area, both pagan and Christian with churches and Neolithic burial chambers (Colm). On the Hill, massive raths exist; they are almost overwhelming in their nature. The sheer impact the landscape had on us is difficult to explain (Georgina). I felt spiritual as I walked through Tara as I felt the area where St Patrick once walked (Kathy).

When entering Tara, with its pasture and sheep, the typical nineteenth century Anglican, Church of Ireland church and high spire located within the graveyard is what strikes most. Evidence of Celtic, Norman, Reformation and colonial ascendancy imprint on this church and grave space – in ruins, tombs and commemorative plaques to landlord families; their association with religious elites and Empire wars are encrusted on walls inside the church.

Due to declining Protestant congregation numbers, the church is now used only on St Patrick's Day (17 March), for religious services. Otherwise it serves as an interpretative center imaging Celtic Tara on a mobile screen over the altar. State agencies are responsible for upkeep of this symbolic space. Recently a white statue of St Patrick on a pedestal has been erected outside the churchyard gate.

> I feel the landscape is sacred and gives us an idea of how our ancestors live ... developments through time. Tara is imbued with a magical mystical atmosphere, feelings of history, gods, immortals, St Patrick and Christian ceremonies, fires, passage of souls transcending space to new places (Noreen). I felt Ireland's roots at Tara (Pauric).

During bicentennial celebrations of the 1798 Rebellion, controversy surrounded plans for taking soil from Tara where rebels fell and bringing it in cortege to the 1798 Memorial – Heritage Center, Wexford. This project was abandoned, especially in the context of the Peace Process (1998).

> I realized how symbolic the land was to our ancestors and in the struggle for independence. Standing on Tara, I got to see what it was like being Daniel O'Connell addressing his crowd. I saw the ditches surrounding the Hill and felt impressed by the ability of our ancestors ... they held a high value for religion whether pagan or otherwise (Sorcha).

Dan O'Connell (1775-1848) the 'Liberator' chose Tara for one of his 40 'monster' meetings attracting 100,000 people during his Catholic Emancipation Campaign with discourses on *le peuple,* nation and democracy. He used similar methods attempting to repeal the Act of Union of the United Kingdom of Great Britain and Ireland (1801); protests were overtaken by Famine (1845-1848) and Rebellion (1848). Tara's landscape became part of nationalist revival traditions from the 1880s on. During De Valera's intensive nation-state building campaigns (1928-1966), his son carried out archaeological work, which yielded little treasure. Attempting to revive the Irish language in the Anglicized Pale, De Valera created *Gaeltacht* – Gaelic-speaking areas – in the Valley, distributing estate lands to Irish speakers from the West.

The demise of landlordism followed the Famine (1845-1848). Tara Estate landscapes including Celtic sites, was broken up into owner-occupied farms, while the demesnes now serve as a golf club. Neighboring Dowdstown Estate's demesnes became the domain of the Colomban Fathers, Missionaries to China, wishing to rekindle the spirit of the early Celtic missionary church, but now becoming part of Britain's imperial venture. By association, the word Tara became a popular first name for girls and houses, especially in the Irish Diaspora or 'scattering' lands, as fictionalized in Margaret Mitchell's *Gone with the Wind* (1939) and Scarlett O'Hara's struggle to save the family plantation of Tara during the American Civil War (1861-1865).

Christian Images

Monasterboice hosts St. Boetus' monastic settlement (sixth century) plundered by Vikings (968) and destroyed in 1097. The churchyard encloses ruins of Celtic and

Norman churches, round tower (33.5 m high) and three Celtic crosses, two adorned with sculptured panels of biblical stories – Adam and Eve, Crucifixion and Last Judgment – and Celtic motifs. Local families continue to bury relatives here and the multiplicity of gravestones makes this cemetery a living site. 'Monasterboice evokes feelings of loneliness, solitude, spirituality, independent living, and sensitivity' (Susan).

At Mellifont Abbey (Figure 5.4), the eye is struck by the lavabo and cloister with many (re)constructions (twelfth-fifteenth centuries). Mellifont, on the Mattock tributary, was founded by Saints Bernard and Malachy, and reforming Cistercians from Clairvaux (1142) predating the Norman invasion (1169). They introduced innovations in monastic lifestyles, architecture, agricultural management and fishing technology diffusing their know-how upstream to Bective Abbey (1145). Within 50 years 23 houses had been founded from Mellifont despite internal power struggles between pro-Norman factions supported by the English king and opponents. Power-games surrounding Mellifont, and Cistercian bases in France and England, hold echoes of Umberto Eco's *Name of the Rose* (1980). Royal patronage tied Mellifont to the English monarch who dissolved the abbey in 1539. Cistercians returned from France in 1831, and purchased the Oriel Temple demesnes that contained 1,000 acres of old Mellifont lands, and established New Mellifont Abbey.

> The spiritual magic of Mellifont is evident as soon as you begin to explore the ruins. The area is busy, yet quiet and pervasive as if it is not just rubble or stone, but a permanent reminder of the daily rhythm of work and prayer. You can imagine the original structure standing tall in all its glory and spirituality. Every piece of rock, or raised ground in the Valley inspires you to believe that there is a continuing story behind it all (Deirdre). Mellifont is like a scrapbook showing different stages in development over a 300-year period; but associated with the nearby ruins of an early Celtic settlement (Michelle).

Bective on the Boyne was rebuilt as a manor house (1600) blending Romanesque and Tudor styles. Like Mellifont, this space passed to elites who had supported Reformation monarchs drawing the Valley into the estate system, capitalism and Empire markets. Besides Bective's *pentimento* and agricultural activities, the Big House (eighteenth century) lies in the background. Writer Mary Lavin, in *Tales from Bective Bridge*, captured the area's scenic tranquility, while dramatic events are portrayed in the abbey's cloister in scenes from the film *Braveheart*.

Norman Influences

Upstream from Bective, Trim (*Ath Truim*, Ford of the Elder-trees) is a market town whose skyscape offers rich images: King John's Castle (twelfth-thirteenth century) Keep (21m), towers and barbican gate reinforced by symbolic religious ruins. These, with their actors' stories in Latin, Gaelic, Norman French and English link Trim to the Norman empire and Crusades (Duffy, 2000; Foster, 1989).

Despite traffic-jams and weekly marketing, in Trim town center's medieval streets can be seen solid eighteenth and nineteenth century buildings interspersed with multicolored houses and shops. Talbot's Castle (1415) built with stones from St. Mary's Abbey (1368) became the property of Esther Johnson ('Stella') (1717), friend of local man Jonathan Swift (1667-1745). Observing Ireland's dual society his political tracts brought governmental threat that he circumvented with metaphorical *Gulliver's Travels* (1699-1726).

Figure 5.4 View of Mellifont Abbey
Source: Author.

During Talbot's Castle period as a school, Arthur Wellesley (1769-1852) received his education there, before becoming Duke of Wellington, victor over Napoleon at Waterloo (1815) and UK prime minister (1828-1830) at the period of Catholic Emancipation (1829). Wellington's imperial statue towers on Trim's skyline, but lower than the Catholic Church steeple overshadowing the nearby dour limestone Workhouse built for Famine victims (1845-1848). Though recycled as a maternity hospital and senior citizens' nursing home, unmarked graves remain vivid. Trim Heritage Center's multi-media exhibition, *The Power and the Glory*, projects images of Celts and Normans, with no mention of Swift or Wellington, but emphasizes human continuity with rhythms of the Boyne and market town life. The filming of *Braveheart* and tourist boom of the 1990s drew local and state funded restoration of the castle (Figure 5.5).

> Trim invokes conflict, exchange and battle (Susan). Trim Castle shows a sense of order and superior planning brought by the Normans in contrast to Viking towns like Wexford (Anne). Trim Castle ... once the capital, brought feelings of conquer of ancestors and that

is what they would have been, to then become part of my ancestry (Joan). These cultural feelings express values and sentiments of nostalgia and a sense of the meaning of the words being Irish (Paul). Our heritage sites are being used in films. Our landscape today shows that we citizens of Ireland are a team. Communities work to preserve what we have and build centers to show our culture to others (Orla). These cultural features emphasize the power and wealth of many families throughout history (Sinead). It is still a living landscape; it expresses the need for defense and fortification with the castle at Trim showing how important defense was for our ancestors, and the use of high ground for defense at Tara and Newgrange also. As we moved from one area to another, the importance of land was obviously necessary for power, wealth and growth (Tracy). I particularly enjoyed viewing the castles; the large defense batteries instilled a sense of power that excited me, the large forbidding walls a testament to past defensive design. I found it interesting that these defenses are still standing; they gave me the impression that they are still waiting for an attack (Paul).

Figure 5.5 Trim town skyscape
Source: Author.

Battles for Territories, Power and Souls

Protestant and Counter Reformation theological discourses often became subsumed into elite power-territorial agenda. When juxtaposed with nation-state building strategies they had major politico-cultural consequences throughout Europe, including Britain and Ireland. Due to coincidence of Reformation and colonization in Ireland, Irish cultural perceptions see both processes as synonymous.

In the seventeenth and eighteenth century conquest and Plantation – due to the Boyne's geostrategic location as a defensive barrier north of Dublin – decisive battles took place here but are not marked by monuments. However Boyne battle *pentimento* still bursts through on Belfast and Derry street murals struggling to legitimate contested identities and spaces.

Historian and documentary maker (*History of Britain*) Simon Schama, recently sparked controversy by referring to Cromwell's massacre of 3,000 unarmed soldiers at Drogheda (1649) as a 'war crime … atrocity so hideous that it has contaminated Anglo-Irish history ever since. Cromwell's army were intent on crushing Royalist forces' as well as ridding Ireland of 'papal barbarism.' He is held to have massacred 4,000 people at Drogheda, although revisionists question such events. When Cromwell's death mask was displayed in Drogheda's museum in 2000 it caused tensions; protesters arrived with tomato ketchup. The director of the *History of Britain* series stated, 'It is incredibly important to challenge myths … History should not be the happy chronicle of the unfurling story of a great nation. It should have a very strong streak of scepticism and rigour … Cromwell was an imperialist bigot but not the genocidal lunatic that Irish folk memory has turned him into. We want to separate the true from the false' (Burke, 2001).

Near Drogheda, the Battle of the Boyne area coincides with the *Brú na Bóinne*.[1] Areas used by the armies, especially at Oldbridge, on the southern bank of the river occupied by James' forces, are highly symbolic. After his victory, William III transferred the lands of Ireland to his supporters, engendering the landlord estate system, planned demesnes, parklands and ornamental gardens, planted villages including Collon and Slane along with intense capitalist production. These villages now with advertising on shop fronts, still display eighteenth century geometrical logic in their morphology and buildings; imposing iconographic location of churches, village green in Collon and *pentimento* of Georgian architecture. Slane Castle built in Georgian style came into the ownership of the Coyninghams after 1690; their stewardship is marked by annual rock concerts hosting 80,000-120,000 fans that view celebrities including Queen and Bruce Springsteen.

While there are no monuments on the battle site, Unionist parades and songs – *On the Green Grassy Slopes of the Boyne* – celebrate William's victory in Northern Ireland every July 12th; Orange parading at the Boyne site was discontinued in the 1970s. Perceived as triumphalism by nationalists who resent partition of Ireland (1921), such anniversaries lead to violence, debilitating the Northern Ireland Peace Process (1998). In the Republic, no independence day is formally celebrated as such.

In 2000, the Oldbridge Demesnes (500 acres) on the Battle site was purchased by the state that plans to create a Visitor Center. A mansion, parks and riverine canal was constructed on the site in the eighteenth century and the Protestants of Great Britain and Ireland erected a memorial obelisk, which was blown up in 1923. Landscape architecture and management on the estate reflected eighteenth century classical formalism, before being impacted on by wilder Romantic nineteenth century fashions. Eventually, tenants became owner-occupiers. The demesnes remained largely intact, before being bought by the state.

I felt awe when I saw places where events in Ireland and beyond took place. I saw exactly where the Battle of the Boyne was. It showed the path that Ireland followed (Therese). A person without heritage is like someone with amnesia, who doesn't know his or her own name. We can't move forward if we don't know where we are coming from (Deirdre). The cultural features gave me a great sense of identity and belonging to the Irish nation. This sense of Irishness is drawn out of the landscape ... a type of historical mosaic. This filled me with a sense of identity, belonging, knowing that I am Irish and therefore directly part of this country's evolving landscape (Emma).

On the southern banks, a local NGO has organized a tourist heritage trail, and the state is undertaking restoration of Oldbridge House for which little information is available. Different identities and traditions must receive parity of esteem in Northern Ireland according to the British-Irish Belfast Agreement (1998). Like paradigms surrounding organization of bicentennial celebrations for the 1798 Rebellion, discourses manufacturing consent have yet to be heard on Boyne battle sites (Chomsky, 1989; Herman and Chomsky, 1994; Hooson, 1994).

I feel a certain degree of discontent towards the landlords who lived in their vast walled estates living off the sweat of peasant people. This social gap existed and still exists today. Driving through the countryside, viewing old Manors gave me a sense of sadness as I saw how tightly the land was controlled and in what grandness the owners lived. However I did enjoy the architecture of the 'Big Houses' – they are a cultural monument to Ireland's past and a reminder as to where we have been. The domineering buildings in the planned Slane village show the sense of segregation that existed with ten per cent of the population reigning over the rest. All these things shape us. Seeing the shabby houses and huge mansions of the past I feel lucky to be living in the wealthy era of the Celtic Tiger (Eamon).

150 Student Voices Speak on Boyne Valley Landscapes

In challenging parameters, postmodernist discourses try to promote 'other voices' (Soja 1989). In what follows, neophyte geographers have their say. Students were requested to circle a keyword or add one that best described the Valley's landscape. The results were: symbolic (43 per cent), followed by spiritual (18 per cent), national (16 per cent), iconographic (nine per cent), power (nine per cent) and religion (five per cent). The remaining keywords included heritage, historical, beauty, spectacular, ancestors and identity. The most common collocations relating to nature or environment experienced were beauty, unique, tranquil, water, unspoiled, unchanged, green, spectacular, fertile, sacred, symbiosis. Student descriptions of landscape features related to the following sites in rank order: Newgrange, Dowth, or Knowth (100 per cent); Trim (33 per cent); Tara (31 per cent); Mellifont (22 per cent); Slane or Collon (2 per cent) Monasterboice (1.5 per cent); field systems and farms (0.5 per cent). Concerning sustainable lifestyles, viewpoints include (note that quotation marks separate anonymous voices):

'Development is a way of keeping our Irish heritage alive with tourism and economics.' 'It relates to the preservation of important features such as Tara, Slane Castle ... without too much restoration or change ... to accommodate tourism.' 'It refers to the worthwhile development of the Valley which when done correctly can benefit it: tourism equals money, money equals more sustainable development.'

Students wrote about values or feelings the landscape may express to them; identity, pride and positive feelings are the sentiments most expressed (95 per cent).

'There is a history of recurring activity and rhythms in the landscape going back to the Neolithic ... my heritage. Newgrange and Knowth have a mystical spiritual quality ... accentuated by nature and locations in relation to the sun, moon, and solstice. Tara generates a magical mythical atmosphere with the host of historical stories attached to everything: religion, warriors, kings, and annual pagan festivals – Bealtine and Lunasa. I felt that water was very important in the whole life of the Valley as the river means more than just water, but seasonality; rhythms of continuity, and ... linkages to the sea and foreign lands.' 'I felt a sense of mystery at all the Celtic signs and symbols.' 'It was good to ... see the real side of Ireland. Respect is impressed upon me for what we have available to us in the landscape ... a crucial reminder we owe it to the landscape and future generations to look after it. I feel inspired by those who left it to us. I hope that in years to come the uneven spread of economic growth does not destroy it.'

Landscape and Feelings

'This landscape expresses sacredness, a feeling of awe and wonder.' 'The area has a spirit and personality.' 'It generates spiritual and cultural meaning as it gives me an insight into past generations.' 'I experienced pride and a sense of place.' 'It expressed a strong religious and community spirit. There were solid structures. The landscape mirrors how traditions are passed down.' 'The local community obviously had and has great respect for the area.' 'Family values and ancestors shone through. The value of religion is the main thing expressed by this landscape.' 'I think that it is important to preserve these sites in their original form; it is close to sacrilege to destroy them, as they are symbols of our growth as a nation and ... our way of life throughout the ages. This is clearly seen at Knowth where the mound was first used as a burial site, later a residential site and today as a tourist attraction.' 'A patriotic feeling came to mind, and pride in ancestors. I had a great feeling of ... the gods.'

Identity and Place

'The landscape is our background. The Valley is something awesome that helped shape the who, of who I am today.' 'The landscape is part of who I am ... where I live, what I do, what I believe.' 'It gave me a sense of nationality and heritage, because it reminds me that before the time of Christ ... our ancestors, had a very advanced know-how of nature's rhythms and architecture.' 'The landscape makes me feel proud of where I am from and appreciate the history of our small island. It's unique to us as Irish people and we should maintain it, so generations to come can experience it as we have.' 'I felt our ancestors.'

Heritage

'It is wonderful to see how the Irish worked together as a community to build something unique which distinguishes us from other nations. I think these feelings still exist today; everyone is proud of their heritage and therefore wishes to sustain it.' 'Many feelings were evoked through the power of historical settlements. Each portrayed its own sense of character and left a feeling of mystery as to what it was like thousands of years ago. It shows the power that our small country held.' 'It allowed me to focus on the people who lived and worked the land before me. Connection between the present and past evokes a sense of power and community in me.' 'I could feel a sense of history seeping from the ground.'

Landscape and Nation

'Heritage is not just about artifacts but also a feeling of belonging to a particular people and culture.' 'I got a sense of awe, amazement that our ancestors were capable of such wondrous achievements … I feel respect that our nation has not abused these sites e.g., burial chambers, burials in the fetus position.' 'The landscape gives us a feeling of being Irish, it is what makes us Irish and different from any other country.' 'People still have respect for the somewhat strange cultures of our ancestors. I felt pride as this area belongs solely to the culture of Ireland.' 'I felt a sense of ownership, not personal ownership, but more something that is unique to Ireland. How do we sustain it, who decides?' 'It helped me connect with the geography in the past and present.' 'I discovered Ireland, the Emerald isle, from the top of a megalith. It expresses feelings of allegiance to my country … my ancestors worked to build a community together.' 'I felt proud that the people who constructed these places are Irish.' 'It gave me a sense of belonging. I am a part of Irish culture just like the people of the Valley were.' 'The landscape made me appreciate where Ireland and I come from.'

Some Conclusions

The holistic approach to geography presented here, attracts students with a wide range of interests. My fieldtrips through the Boyne Valley aim at appreciating the discipline by reading landscapes and rhythms that call on a variety of approaches. However, what students experienced – *pentimento* and metaphor – is principally what they wrote about and may impact on their future career paths, in this case mostly in education.

Students scored high on tests based on physical geography and definitions of various branches of geography including development. Questions and discussions about the Valley based on economic geography including agriculture and tourism interested students in a general way, while sustainable development discussions stimulated more interest, especially with the aid of landscape snapshots. Concerning sustainable development, interest was shown in ecological-human facts, but students were most interested in culture and identity perspectives, such as human relationships to the Valley, aesthetics and emotional significance to Irish identity, and contemporary reality in landscape and mindscape (cf. Buttimer, 2001b). This was especially evident in reactions to the Newgrange complex and cosmology, and the Hill of Tara. These sites evoked much use of the words 'our,' 'nation,' 'gods,' 'awe,' 'mystical,' with students wanting to read their autobiographies in the mirror landscapes. Little enthusiasm was evident at the Battle of the Boyne site, nor its associations with Unionism and contested identities. Questions asked about future (re)writing of this heritage space were generally met with procrastination – too real perhaps.

Students wished to enjoy and experience Boyne landscapes and in their own words wrote mostly narratives of ancestors, architectural and survival achievements. They wrote about spirituality, humanity and environment, but not formal religion. Awe and pride in landscapes and heritages were used interchangeably. Love of life, place, landscape, and country was rendered devoid of traditional nationalist discourses. Self and group-identity thoughts evoked by Boyne landscapes linking the prehistoric directly to the modern were most salient in student presentations as they constantly spoke of sustainable heritage and Irishness with its daily and centennial rhythms through the generations.

Note

1   England's 'Glorious Revolution' (1688) deposed James II, the last Catholic English King, and enthroned Mary his daughter and her Dutch husband, William III, Protestant Prince of Orange, ruler of the Dutch Republic. To distract William from his French wars, Louis XIV backed James II and his Irish allies. William was most popular in Ulster, while James in his attempt to strengthen his hold on Dublin and southern provinces made the Boyne his defense line – 60,000 men in two multinational armies went to battle at the Bend. Some 150,000 Jacobites and 50,000 Williamites were killed.

References

Aalen, F., Whelan, K. and Stout, M. (eds.) (1997) *Atlas of the rural landscape*. Cork University Press, Cork.

Andrews, J. (1967) 'A geographer's view of Irish history' in Moody, T.W., and Martin, F.X. (eds.) *Course of Irish history*. Cork University Press, Cork, pp. 17-29.

Barnes, T. and Duncan, J. (eds.) (1992) *Writing worlds: discourse, text and metaphor in the representation of landscape*. Routledge, London.

Bartlett, T. and Jeffery, K. (eds.) (1996) *A Military History of Ireland*. Cambridge University Press, London.

Beresford Ellis, P. (1994) *Dictionary of Celtic Mythology*. Oxford University Press, London.

Bhreathnach, E. (1995) *Tara: A Select Bibliography*. Discovery Programme and the Royal Irish Academy, Dublin.

Bhreathnach, E. and Newman, C. (1996) *Tara*. Stationery Office, Dublin.

Bourdieu, P. (1995) *Language and symbolic power*. Polity, Cambridge.

Boylan, H. (ed.) (1998) *A Dictionary of Irish Biography*. Gill and Macmillan, Dublin.

Brady, C. and Byrnes, E. (2002) 'A pilot archaeological survey for the site of the Battle of the Boyne, Oldbridge Estate, County Meath,' paper abstract, Second Irish Post-Medieval Archaeology Group Conference, Dublin.

Burke, J. (May 6, 2001) *The Observer*. Schama's TV trial of 'war criminal,' Cromwell.

Buttimer, A. (1980) 'Home, reach and sense of place' in Seamon, D. *Human Experience of Place and Space*. Croom Helm, London.

Buttimer, A. (1982) Musing on Helicon: root metaphors and geography, *Geografiska Annaler* 64B, pp. 89-96.

Buttimer, A. (1993) *Geography and the human spirit*. Johns Hopkins University Press, Baltimore.

Buttimer, A. (ed.) (2001a) *Sustainable landscapes and lifeways: scale and appropriateness*. Cork University Press, Cork.

Buttimer, A. (2001b) 'Humanistic geography' in Smelser, N. and Baltes, P. (eds.) *International Encyclopedia of the Social and Behavioural Sciences*. Elsevier, Amsterdam.

Chomsky, N. (1989) *Necessary illusions: thought control in democratic societies*. Pluto Press, London.

Cosgrove, D. (1985) Prospect, perspective and the evolution of the landscape idea, *Transactions of the Institute of British Geographers* 10, pp. 45-62.

Cosgrove, D. (1989) 'Geography is everywhere: culture and symbolism in human landscapes' in Gregory, D. and Walford, R. (eds.) *Horizons in human geography*. Macmillan, London.

Cosgrove, D. and Daniels, S. (eds.) (1988) *The iconography of landscape*. Cambridge University Press, Cambridge.

Devine-Wright, P. and Lyons, E. (1997) Remembering pasts and representing places: the construction of national identities in Ireland, *Journal of Environmental Psychology* 17, pp. 33-45.

Duffy, S. (2000) *Atlas of Irish history*. Gill and Macmillan, Dublin.

Duncan, J. and Duncan, N. (1988) (Re)reading the landscape, *Environment and Planning: Society and Space* 6, pp. 117-26.

Duncan, J. and Ley. D, (eds.) (1993) *Place/Culture/Representation*. Routledge, London.

Foster, R. (1989) *The Oxford History of Ireland*. Oxford University Press, Oxford.

Green, M. (1995) *Celtic goddess: warriors, virgins and mothers*. British Museum Press, London.

Green, M. (1996) *The Celts*. Routledge, London.

Gregory, D. (1989) 'Areal differentiation and post-modern human geography' in Gregory, D. and Walford, R. (eds.) *Horizons in human geography*. Macmillan, London, pp. 67-96.

Hägerstrand, T. (1982) Diorama, path and project, *Tijdschrift voor Economische en Sociale Geografie* 73, pp. 323-39.

Hartshorne, R. (1939) 'The character of regional geography' in Hartshorne, R. *The nature of geography: a critical survey of current thought in the light of the past*. Association of American Geographers, Washington, DC, pp. 436-44.

Hellman, L. (1973) *Pentimento*. Little, Brown and Co., Boston

Herman, E. and Chomsky, N. (1994) *Manufacturing consent: the political economy of the mass media*. Vintage, London.

Hooson, D. (1994) *Geography and national identity*. Blackwell, Oxford.

Jackson, J.B. (1984) *Discovering the vernacular landscape*. Yale University Press, New Haven.

Johnston, N. (1994) *The Norman impact on the medieval world*. Colourpoint Books, Newtownards, Northern Ireland.

Johnston, R.J. (1978) Paradigms and revolution or evolution? Observations on human geography since the Second World War, *Progress in Geography* 2, pp. 189-206.

Ley, D. (1980) Geography without man: a humanistic critique, *School of Geography Research Paper* 24, Oxford University, Oxford.

Meehan, R. and Warren, W. (1999) *The Boyne Valley in the Ice Age*. Geological Survey of Ireland, Dublin.

Mitchell, G.F. and Ryan, M. (1998) *Reading the Irish landscape*. Town House, Dublin.

Moody, T. and Vaughan, W. (eds.) (1986-96) *A new history of Ireland*. Oxford University Press, London.

Nolan, W. (1988) *The shaping of Ireland*. Geography Publications, Dublin.

Ó Fiach, T. and Forristal, D. (1975) *Oliver Plunkett: his life and letters*. Fourcourts, Dublin.

O'Reilly, G. (2001) 'Scaling democracy and sustaining development in an Irish context' in Buttimer, A. (eds.) *Sustainable landscapes and lifeways: scale and appropriateness*. Cork University Press, Cork, pp. 287-317.

Price, M. and Lewis, M. (1993) The reinvention of cultural geography, *Annals of the Association of American Geography* 83, pp. 1-18.

Saint-Exupery, A. de (1948) *Citadelle*. Gallimard, Paris.

Sauer, C.O. (1925) The morphology of landscape, *University of California Publications in Geography* 2, pp. 19-54.

Smyth, W. and Whelan, K. (eds.) (1988) *Common ground: essays on the historical geography of Ireland*. Cork University Press, Cork.

Soja, E.W. (1989) *Postmodern geographies: the reassertion of space in critical social theory*. Verso, London.

Stout, G. (2002) *Newgrange and the Bend of the Boyne*. Cork University Press, Cork.

Three Rivers Project (2002) (www.threeriversproject.ie/mainpages/boyne/).

Tuan, Y-F. (1974) Space and place: humanistic perspective, *Progress in Geography* 6, pp. 233-46.

Tuan, Y-F. (1993) *Passing strange and wonderful: aesthetics, nature, and culture.* Island Press, Washington, D.C.

Vidal de la Blache, P. (1926) *Principles of human geography.* Constable, London.

PART III
REANIMATING URBAN
LIFEWORLDS

Chapter 6

Temporality and the Rhythms of Sustainable Landscapes

Edward Relph

Introduction

The argument in this essay has its origins in two comments by Anne Buttimer in *Sustainable Landscapes and Lifeways* (2001) with which, in a friendly way, I will take issue. The first comment is that 'issues of sustainable development ... inevitably involve nature and its multiple rhythmicities' (Buttimer 2001, 7). While this is certainly so, I argue that a human response to these rhythms requires an underlying sense of time that can inform sustainable development in a consistent and coherent manner.

The second comment is the conclusion where Buttimer remarks that 'the knowledges which the twentieth century has produced in academic and applied settings are apparently not of the kind needed for guiding imagination beyond the current contradictions' (2001, 378). The contradictions to which she refers are those implicit in objective scientific observation, for this is necessarily filtered through cultural practices and is really not objective at all. I agree that a different kind of understanding is needed if we are to imagine ourselves out of the present situation, but I think the foundation for this can be found in phenomenological insights about existence and the experience of time, many of which were articulated in the twentieth century as a sort of counter-knowledge to prevailing scientific methods. Phenomenology understands time as it is experienced, with complex links between recollection, the present moment and anticipation. This understanding is known as 'temporality,' and it is temporality that underlies and must inform sustainability.

Sustainability (or sustainable development – in this essay I regard them as the same) is an enigmatic idea because it has been exploited as a description of short-term economic growth, even though this is based on drawing down the natural capital of the earth and hence is manifestly unsustainable. As I understand it, sustainability is above all a concept about time. It indicates a form of change, whether in ecosystems, economies or cities, that can endure indefinitely. Attitudes toward time, and the forms of change that follow from them, are expressed in cultural landscapes. Thus, reverence for the past is demonstrated in stylistic revivals and the preservation of old things; celebration of the future is evident in modernist designs that self-consciously embrace innovation. Sustainability requires a composed attitude toward time that balances past, present and future in the interests of continuity. Where time seems to be regarded as a commodity and landscapes are filled with unrelated fragments of heritage and futuristic gestures, sustainability remains distant. There is substantial evidence that this is the current circumstance. Conversely, temporality is revealed in landscapes that display an easy coexistence of old and new, and here sustainable practices are either already present or are possible.

The Necessity of Sustainability

The historian Eric Hobsbawm, in his sweeping survey of the twentieth century (Hobsbawm, 1994, 558) concluded that it was ending in 'problems for which nobody has or even claimed to have solutions.' When he turned his attention to the future he saw environmental problems, the huge gap between the rich and the poor, and the evaporation of clear norms and common values that could serve as the basis for solutions (Hobsbawm, 2000, 162-167). A no less chastening view of the current state of things is expressed by the Nobel laureate economist Amartya Sen. He writes:

> We live in a world of unprecedented opulence, of a kind that would have been hard to imagine a century or two ago ... And yet we also live in a world with remarkable deprivation, destitution and oppression. There are many new problems as well as old ones, including persistence of poverty and unfulfilled elementary needs, occurrence of famines and widespread hunger, violation of elementary political freedoms as well as basic human liberties, extensive neglect of the interests and agency of women, and worsening threats to our environment and to the sustainability of our social and economic lives (Sen 1999, xi).

Sen's hope is that rethinking basic ideas, and treating development as a process of achieving freedom from poverty and injustice, might reduce or resolve many of these problems. This will, of course, require a huge change of perspective about the purpose of development; his argument is intended to begin the process of bringing about this change.

Although huge shifts of social perspective in short periods have occurred before, for instance, with the abolition of slavery, and more recently with women's liberation, there is a serious difficulty posed by the sorts of problems that Hobsbawm and Sen describe. It has been suggested that they are so complex, difficult to understand and to resolve, that solutions may exceed human ingenuity. One of my colleagues at the University of Toronto, Thomas Homer-Dixon (2000) calls this 'the ingenuity gap' and sees evidence of it, for example, in global climate change, which demands sophisticated and expensive research, multilateral international agreements, the co-operation of countless agencies, changes in individual and group behavior and a shift of political will and economic investment. His hopeful prognosis is that the ingenuity gap can be closed by working simultaneously from both its sides. From one side, improved research and education could lead to more ingenious solutions (but these are precisely what Anne Buttimer considers inappropriate and not up to the task). On the other side of the gap, we can try to slow down change, and look for simpler, more adaptable ways of doing things that will reduce the need for ingenuity. This is the way of sustainable development.

Temporality and the Townscape of Markham

Intentionally or unintentionally, the appearance of built places and landscapes reveals a great deal about what a culture considers important because it involves substantial investments of effort and money. A culture's attitude to time, no less than anything else, is expressed in the landscapes that it makes for itself.

In 1976 Hugh Sloan, a graduate student I was supervising, wrote a thesis about the manifestations of time in the townscape of Markham, a small town near Toronto, that has subsequently developed some unusual policies and practices with regard to heritage. Sloan adopted a distinctive perspective to time and townscape. All attempts to explain time are, he wrote, inadequate and inconclusive when compared with the intricacies of everyday 'temporality.' Temporality is the lived-experience of time that precedes any notion of quantitative clock time; it is the dense association of memory, present awareness and expectation that, among other things, integrates us into landscapes. Kevin Lynch describes it in *What Time is this Place?* (1972, 124-125) as 'an elastic flow within an intermittent present, moving now fast, now slowly, according to biological rhythms of which we are only half aware; and changeable future and past ... in which there are peaks and valleys, rhythms, eras and boundary zones.' Landscapes and townscapes are simultaneously the contexts of temporal experiences and subject to temporality. They are the settings for diurnal, weekly and seasonal patterns of human activity, the backdrops and reference points for recollections and expectations. They are an essential component of the geography of memory. And in a manner broadly similar to that of human life, albeit at many different and overlapping tempos, landscapes have rhythms of creation, change and decay. A street is planned and surveyed, buildings along it are constructed, all of them new at once; the street endures, perhaps for centuries, but its buildings are modified differently, some are maintained with few alterations, others are deeply renovated or deteriorate and are completely replaced. Eventually the street and all its buildings, like the cities of the ancient Mayas, may be abandoned and disappear beneath the slow and relentless forces of natural decay.

Sloan drew an analogy between music and townscape. He noted that rhythm in music gives order and structure to a composition, with the individual moments of rhythm contributing to the total character of a piece of music even as they pervade harmony, tone, melody and intensity. Similarly, he suggested, the temporal quality of townscape pervades all aspects of townscape experience. The manifestations of time in activities, objects, and buildings contribute to an overall temporal pattern of decay and renovation, of newness and oldness. In a landscape that is well composed, the past, present and future are interwoven by this rhythm.

The town of Markham, which was then about 15 kilometers beyond the edge of the built up area of Toronto, provided Sloan with a typical North American case study for the investigation of temporality and townscape. Its streets were laid out in the early 1800s, but many of the buildings had been constructed in the late-nineteenth century and its general Victorian character remained. Some of the original buildings had been replaced by supermarkets and gas stations, and there were some new subdivisions on adjacent to the old areas.

Sloan carefully examined Markham's townscape for different expressions of attitudes to time. He considered ways in which the past was manifest through restorations and preservation, he looked for landscapes that seemed to be lodged in the present, and he explored intimations of the future in such things as signs that announced plans for development. Markham's main street of shops he described as showing a time of 'coexistence.' In this he saw a 'coincidence of transience and permanence' in which objects and buildings from several generations were rhythmically integrated as though in a visual equivalent of a musical phrase that linked different melodies. These, he suggested, are the qualities of a temporal townscape. In contrast, where new subdivisions adjoined Victorian areas there was a 'time edge' – an abrupt juxtaposition – where it was possible, in effect, to step from the late-nineteenth century into the late-twentieth century. And Markham's pioneer museum, with a collection of old artifacts, barns and isolated buildings of doubtful significance, was, he thought, a 'significant time deception.' Both it and its elements were out of context and discordant.

Disrupted Temporality and Heritage

Since 1976, when Sloan completed his study, Markham has experienced intense urban development; and it has been enveloped by the conurbation of Toronto. Its municipal boundaries, which previously bore some relationship to the geographical entity that most people would have recognized as a town, have been expanded to include several hamlets and villages and a substantial surrounding agricultural zone. Between 1996 and 2001 the population of this new Town of Markham grew from 160,000 to 207,000, with most of the increase in pleasant subdivisions of single family, detached houses with vaguely Victorian or neo-Classical facades. Markham has also become a major center for high-technology in Canada, with one of the major North American hubs for IBM, and many other electronic and pharmaceutical facilities housed in buildings designed according to the best principles of utilitarian modernism. In reaction to this rapid development, Markham has developed some remarkable, innovative approaches to heritage and in 2000 it was the recipient of the first Prince of Wales prize for Stewardship of Built Heritage.

When Sloan wrote in 1976, the concept of 'heritage' was in its infancy and he scarcely mentioned it. Since then it has become an essential part of the discourse for things old and considered valuable. It is not an unproblematic idea. Its primary intent is to maintain connection with the past in the face of remorseless change, but in landscapes its chief manifestations seem to be time edges, discordances and disruptions of the rhythms of temporality. This is no less true for Markham than for anywhere else. Consider, for example, the following three local landscapes.

Heritage Estates

Current Canadian approaches to suburban development require the land surface to be scraped flat so that impediments to construction are minimized. Old buildings get in the way of this simple process. The most innovative approach to heritage in Markham, and the one that probably was instrumental in its winning the Prince of Wales prize, has involved the relocation of old houses to Heritage Estates, a subdivision of 'last resort' for buildings that would otherwise have been demolished.

In Markham it is possible to buy a 150-year-old house for a dollar, on condition that you uproot it, put it on the back of a truck, move it to Heritage Estates, renovate it and live in it for at least two years. The total cost of relocation is about the same as buying a new house.

Heritage Estates has 38 lots and in 2002, 30 of these were occupied by relocated houses with the rest waiting for customers. It is subject to strict site planning controls. The exteriors of the houses have to be restored to their original appearance, though they are mounted on new concrete foundations and the interiors have been deeply renovated. The only garages allowed are in old barns and sheds, also relocated. The houses are arranged mostly by the order in which they arrived and without regard to age or style. They have standardized setbacks and rather oddly face onto bulbous culs-de-sac that have to be about triple the normal road width to allow the house-carrying trucks room to maneuvre.

The intention behind Heritage Estates is commendable. It aims to maintain as much of the built past as possible in the face of rapid urban change. But it has the sanitary feel of a retirement home for old houses that once had separate existences on farms, and have now been uprooted and pushed into uneasy proximity. Heritage Estates is adjacent to the pioneer museum of old wooden barns and farm buildings that Sloan regarded as a time deception, and like that is the outcome of a sense of time in which fragments of history can be transported, rearranged, segregated, and otherwise turned into commodities.

Market Village/ Pacific Mall/ Heritage Town

Market Village and Pacific Mall are parts of a single shopping mall located a few miles from Heritage Estates. They have a complex, brief and intensely post-modern history. In the late 1980s a giant crafts and garden center was built on this site; it was housed in a massive barn constructed, in a sort of precursor to heritage preservation of last resort, out of lumber recycled from local barns that had been demolished. With old farm machines as ornaments, fresh baking and a petting zoo, the garden center exuded rural nostalgia. It was so financially successful that a contiguous development called Market Village was begun. This was a disneyfied blend of simulacra of the architecture of small town Ontario with a medieval townscape of winding, outdoor pedestrian streets. Market Village was a financial disaster, probably because it was completed in the early 1990s just as an economic depression began, but also because open-air shopping malls are not very sensible options for Canadian winters.

By 1990 Markham had become a popular destination for Chinese immigrants, especially from Hong Kong. Market Village was acquired by a Chinese developer who promptly enclosed the pedestrian streets by the expedient method of installing a flat metal roof without regard to the fake facades of buildings, parts of which face onto the now enclosed streets and parts of which stick up above.

In its new, enclosed, Chinese form, the stores were quickly rented and Market Village prospered so much that about 1997 another Chinese developer bought the adjacent crafts and garden center, demolished the barn and constructed Pacific Mall. This consists a two-storey industrial shed with the ground floor divided into a grid of glass-walled, retail cubicles about five meters square, with the ductwork and structural components all exposed. Upstairs, however, there is a very different style of retail

area, identified as 'the largest Heritage Town in America.' It is entered through an
elegant gateway of ceramic sculptures of dragons and lions, with ridge end details that
are replicas of those in the Forbidden City in Beijing. Inside there are narrow
walkways lined with booths decorated with lanterns and intricate signs and selling the
same food and goods that are available in the cubicles downstairs.

The Future is the Past at Cornell

About two miles east of Heritage Estates is Cornell, one of the largest new-urbanist
projects under development in North America. When completed it will have a
population of about 30,000. Planned by Andres Duany and Elizabeth Plater-Zyberk,
the pre-eminent new-urbanist planners of North America, it has reduced-width
streets, garages situated along rear laneways, porches on the fronts of the houses and
architectural details that are loosely derived from the Victorian parts of Toronto.
Behind these old-style facades, every house is fully networked with a dedicated
Internet connection on each floor. The aim in Cornell is to create a new type of
development that reproduces the pedestrian and community-oriented qualities of old
towns, while accommodating the latest technologies. A photograph in the
promotional sales materials shows the developer Larry Law shaking hands with
Andres Duany (Figure 6.1). The caption in large letters across the top of the
photograph reads 'I have seen the future and it is the past.'

Figure 6.1 The future becomes the past. New urbanist Andres Duany shaking
            hands with developer Larry Law in publicity material for the new
            urbanist project at Cornell in the suburbs of Toronto
Source: Publicity material for the Cornell project.

These three examples from Markham reveal deep confusions in contemporary attitudes to time. Heritage Estates treats old buildings as movable commodities, Market Village and Heritage Town reduce history to an instantly replaceable, exportable set of decorations, and Cornell blends past and future into an eerie timelessness. There is nothing exceptional about these temporal confusions. When I visited the Acropolis there was a crane inside the Parthenon and its columns were being sprayed with chemicals to protect them from weathering. In Glasgow, Vancouver, San Antonio, and Sydney I have seen old facades propped up with steel beams, waiting for new back-buildings. On a wall of the exhaust-stained, concrete perimeter of the Barbican project in London is a plaque that declares 'The probable site where on May 24 1734 John Wesley "Felt his heart strangely warmed." This experience of grace was the beginning of Methodism.' The sign is surrounded by towering apartments, dead spaces, roads, and traffic. Lever House on Park Avenue in New York, completed in 1952, is an icon of modernist and futurist design; it married the angular, undecorated, aesthetic of glass and steel with forward-looking corporate, capitalist notions of progress. In 2001 it was extensively restored. I have seen the future and it is the past.

Figure 6.2 The closure of temporality: a farm on the edge of Toronto
Source: Author.

In the last quarter of the twentieth century there has been a proliferation of time edges and time deceptions. Heritage projects preserve fragments of the past with little regard for historical or geographical context, and the future seems to have become increasingly difficult to imagine without a veneer of something old. I suppose that in some limited sense there is a time of coexistence in the landscape of Heritage Estates, with its old-new houses and the empty lots waiting for more; there is also a sort of time of coexistence in the complex, superimposed structures of Pacific Mall and in the future-past visions of Cornell. But it is difficult to see in any of them temporal rhythms that possess harmony, tone and melody. On the contrary, and for all their good intentions, what they suggest is mostly staccato, irregular and discordant.

In the Fall of 1992, Sterba Farm Market, a small pick-your-own farm in Markham operating on land that already had been approved for residential development, closed down for its penultimate season. The owner put up a hand-made sign for his customers that captured many of the current problems with landscapes of time and heritage (Figure 6.2). It said simply: 'Closed Temporality.' The sign, the building and the farm have long since disappeared.

The Paradox of the Closure of Temporality

In his book *Landscape of Events* Paul Virilio (2000, xi) claims enigmatically that 'history has crashed into the wall of time.' By this I think he means that somehow the past has got in the way of the future, perhaps because of failed promises of progress and the gradual evaporation of hopeful expectations, perhaps because of the surge of enthusiasm for preservation of the past. Cultural landscapes that are being made at the present time are filled with time edges, deceptions, eradications, restorations, and renewals. They demonstrate an intense and simultaneous concern for heritage and a blithe disregard for temporality. This is an inherently unstable situation and a particular manifestation of a contradiction that lies deep within modernity. The French philosopher Paul Ricoeur (1965, 276-77) wrote that 'in order to take part in modern civilization it is necessary at the same time to take part in scientific, technical and political rationality, something which very often requires the pure and simple abandonment of a whole cultural past ... There is the paradox: how to become modern and return to sources; how to revive an old, dormant civilization and take part in universal civilization.' If it is to be more than a token gesture, sustainable development has to resolve this paradox.

Temporality and Sustainability

'Progress' and 'heritage' may point in different directions, but they both externalize the experience of time in ways that are strained. William James in *The Varieties of Religious Experience* (1961, 380) wrote that 'Knowledge about life is one thing; effective occupation of a place in life, with its dynamic currents passing through your being, is another.' Of these currents, time is perhaps the most important. From the perspective of detached knowledge about life, to which James referred, time might be understood as a river on which we are floating; its flow constitutes progress, with heritage and history upstream and the future downstream. But from the phenomenological perspective of the effective occupation of a place in life, time lies within us and infuses everything we feel and do. Care, concern, anxiety, guilt, recollection, anticipation, responsibility, hope, planning, building, and all the other experiences that constitute our lives, are saturated with time (Barrett, 1962, 227-228). This dynamic, lived experience of time is what I understand by 'temporality.'

The phenomenon of temporality has been thoroughly explicated by Martin Heidegger (1962) in his phenomenological study of *Being and Time*. In this treatise he discloses temporality as a central phenomenon of human existence. Although we necessarily spend much of our everyday lives engaged in 'idle chatter' or uncritical 'curiosity' about what we see and perceive, Heidegger argues that there are moments when we become reflexively aware of the deep responsibilities that are implicit in the very fact of our existence, including responsibilities to others and to things simply because they also exist. These can be understood as moments of 'authentic care' and Heidegger proposed that 'Temporality reveals itself as the meaning of authentic care' (1962, 374). Temporality is always present both in the experience of those responsibilities we have acquired because of what has happened in our lives, and in our obligations toward the future. In temporality there is responsibility, in responsibility there is continuity, and continuity is the key to sustainability.

In a later work, *The Question Concerning Technology*, Heidegger (1977) pursued these notions of responsibility from a different angle. In this essay he interprets the essence of technology as a way of revealing truth, in which truth is to be understood not as a confirmed fact but as clarity or insight. *Techne* in its classical Greek meaning referred as much to art and poetry (*poiesis*) as to engineering (1977, 13), because both building and art were then understood as ways of revealing truth. *Techne* embraced any activity that brought clarity to the meaning of existence and the responsibilities that were associated with it. Modern technology, however, has the character of challenging and confronting whatever it works with. Instead of bringing truth and clarity, it treats nature as a 'standing reserve' to be set aside and unlocked for human use whenever required. At the core of modern technology seems to be the expectation that nature reports itself to human beings in some way or other that is identifiable through processes of calculation and accounting (1977, 23). 'In this way the impression comes to prevail that everything man encounters exists only insofar as it is his own construct' (1977, 27). This, Heidegger suggested, brings us 'to the brink of a precipitous fall,' the point where we ourselves will be taken as standing-reserve.

Heritage preservation is commendable as a strategy of last resort and a rearguard action against the destruction of the past. On the other hand, it takes history as a standing reserve, to be set aside and fenced off as an attraction, or as a commodity to be acquired and moved around.

This is the attitude that lies behind what Hobsbawm described as problems 'for which nobody has or even claims to have solutions,' the problems of global climate change, the depletion of renewable resources and failures to distribute the benefits of progress in a just and equitable way. This is also the attitude manifest in landscapes that display a self-conscious, confused and fractured sense of time. For all their good intentions, Heritage Estates, Heritage Town, Cornell and countless other preservation and development sites demonstrate the use of heritage as a standing-reserve of history, a commodity to be exploited and moved around at will. In other words these problems of heritage and of environments have to be regarded not just as separate technical matters, but as different expressions of an underlying and shared ontology that is rapidly ceasing to be appropriate.

Sustainability and temporality, when regarded as solutions to these problems, also have their technical aspects. There are ecologically and geographically sound practices of environmental management that have to be designed, implemented and maintained; similarly there are strategies for achieving townscapes that display a coexistence of times. Within these there may well be some sort of time edges, but these will have form and character that are equivalent to the breaks between movements in a symphony. They will reveal lines of change that are nevertheless part of a whole. These various practices are necessary but insufficient. True sustainability requires a change in outlook that grasps technology as 'safekeeping and coming to presence of truth' (Heidegger 1977, 33). Although what I have written here is mostly in the context of urban sustainability, safekeeping is no less important as an underlying principle for ecological and social sustainability. From the perspective of safekeeping we find ourselves aware that we dwell in the world and have responsibility for sustaining its well-being. This is not a responsibility that humanity has chosen; nor is it a responsibility that can be evaded. It simply is.

Conclusion

'Remember' wrote Alfred Lord Tennyson in his poem *Locksley Hall Revisited*, 'how the course of time will swerve, /Crook and turn upon itself in many a backward streaming curve.' He had first visited Locksley Hall sixty years earlier and in the enthusiasm of youth coined one of the great anthems of progress: 'Forward, forward let us range/Let the great world spin forever down the ringing grooves of change.' When he revisited sixty years later, after six decades of industrial and urban progress, he dramatically reversed his advice: 'Let us hush this cry of "Forward" till ten thousand years have gone.' Much of what made Tennyson alter his view of progress was that it had done so little to resolve the poverty and injustice that it had largely caused. The fact that a century later both Eric Hobsbawm and Amartya Sen could still point to issues of deep poverty and inequality, and then add to these a litany of environmental concerns, suggests that progress is more successful at creating problems than at solving them.

Sustainable development offers a possible way out of this paradox. I have argued from the context of urban landscapes that sustainability, if it is to be more than a thin gesture towards the resolution of urban, ecological, and social problems, requires a radical departure from modern habits of thought associated with economic growth and technological challenges to nature. It has to be rooted in composure toward time, technology and nature. Sustainability stands little chance of success as long as our culture is adrift in temporal confusion, so it somehow has to grasp the temporality that links past, present and future and translate this into practice. This will require the unfamiliar and difficult ways of thinking associated with phenomenology, though it can draw on precedents in poetry and philosophy, and in the accomplishments of those who have looked with care at the world around them and have had the freedom to act on their insights. There is no programmatic way either to learn it or to teach this way of thinking and seeing, and it seems that even when learned it has a tendency to slip away. There is, however, probably no better place to begin than by becoming attuned to the resonance, temporal rhythms, and modulations of life and landscape.

Acknowledgements

I want to acknowledge several graduate students: Hugh Sloan, whose remarkable 1976 thesis lingered in my thinking for 25 years (it was not published); Jennifer Hall, who studied Heritage Estates in 1999; as well as Andrew Blum, Stephanie Heidenreich, Nik Luka, and Zack Taylor.

References

Barrett, W. (1962) *Irrational man: a study in existential philosophy.* Anchor Books, New York.

Buttimer, A. (2001) 'Sustainable development: issues of scale and appropriateness' in Buttimer, A. (ed.) *Sustainable landscapes and lifeways.* Cork University Press, Cork, pp. 7-34.

Hall, J. (1999) Consuming the past at a suburban heritage site: Markham Heritage Estates, Ontario Canadian urban landscape examples 20, *The Canadian Geographer* 43, pp. 433-438.

Heidegger, M. (1962) *Being and time* (trans. J. Macquarrie). Harper Row, New York.

Heidegger, M. (1977) *The question concerning technology and other essays* (trans. W. Lovitt). Harper Torchbooks, New York.

Hobsbawm, E. (1994) *The age of extremes: a history of the world 1914-91.* Pantheon, New York.

Hobsbawm, E. (2000) *On the edge of the new century.* New Press, New York.

Homer-Dixon, T. (2000) *The ingenuity gap: how can we solve the problems of the future?* Alfred Knopf, Toronto.

James, W. (1961) *The varieties of religious experience.* Macmillan, New York.

Lynch, K. (1972) *What time is this place?* MIT Press, Boston.

Ricoeur, P. (1965) *History and truth.* Northwestern University Press, Evanston.

Sen, A. (1999) *Development as freedom.* Anchor Books, New York.

Sloan, H. (1976) Aspects of temporality and townscape with a perspective on Ontario towns, MA Research paper Department of Geography, University of Toronto.

Virilio, P. (2000) *Landscape of events.* MIT Press, Cambridge Mass.

Chapter 7

Grasping the Dynamism of Urban Place: Contributions from the Work of Christopher Alexander, Bill Hillier, and Daniel Kemmis

David Seamon

In the 1970s, Anne Buttimer taught at Clark University where I was a graduate student writing a doctoral dissertation under her direction. At Clark, one of Buttimer's central research interests was the relationship between physical and human worlds and how the particular lived dynamics of that relationship played a role in facilitating place-making and human community, particularly in cities. One question she and her graduate students explored was how the everyday time-space routines of individuals and groups helped to transform space into place and how personal and group identification with place could facilitate a sense of locality and urban neighborhood (Buttimer 1969, 1971b, 1972, 1976, 1980; Buttimer and Seamon 1980; Rowles 1978; Seamon 1979, 1993a, 1994, 2000, 2002; Seamon and Nordin 1980; Seamon and Mugerauer 1985).

At the time, social-scientific research examining these topics was dominated by positivist theory and quantitative methods. Through the Continental philosophical traditions of phenomenology and existentialism coupled with her keen interest in French social geography, Buttimer sought a more accurate, comprehensive and empathetic way for examining individual and groups' day-to-day relationships with their physical, spatial, and social environments (Buttimer 1971a, 1974, 1976, 1980, 1987; Seamon 1987b). In her research on urban social space for low-income housing projects in Glasgow, for example, she emphasized that the everyday world of urban residents is multi-layered existentially. She conceptualized this lived complexity in terms of activity spaces, environmental images, and place identification and attachment (Buttimer 1972). Later, she considered peoples' experienced links with place and environment through the phenomenological notion of *lifeworld* – the taken-for-granted dynamic of everyday experience that largely happens automatically without conscious attention or deliberate plan (Buttimer 1976, 277; Seamon 1979, 1994, 2002). She argued that environmental aspects of the lifeworld – e.g., sense of place, social space, time-space rhythms, and the lived dialectic between home and horizon – offer a uniquely geographical contribution to phenomenological research (Buttimer 1976, 1980).

Early on, Buttimer recognized that both geographical and phenomenological thinking on the environmental and spatial nature of the lifeworld was incomplete: 'phenomenological descriptions remain opaque to the functional dynamism of spatial systems, just as geographical descriptions of space have neglected many facets of human experience' (Buttimer 1976, 277). In her writings, Buttimer contributed much toward integrating these partial perspectives conceptually and practically.

In this chapter, I examine how the ideas of three current researchers – architect Christopher Alexander, architectural theorist Bill Hillier, and political philosopher Daniel Kemmis – provide important new insights for understanding the urban lifeworld and for making more vibrant places. I argue that these thinkers' conceptions of place, though considerably different in some ways, can be drawn together to offer a powerful understanding of how physical-spatial and human worlds might mutually sustain each other by bringing human beings together informally and thereby generating a sense of togetherness, particularly in cities. In turn, this possibility of spontaneous geographical gathering can support a liveliness of place and one kind of implicit environmental belonging.

Place-Making as Wholes Healing Themselves

I begin with the ideas of Christopher Alexander and Daniel Kemmis because their work illustrates a practical movement afoot in public policy and environmental design that attempts to understand useful societal change from the viewpoint of wholes healing themselves. Alexander has been the most visible proponent of this perspective (Alexander 1975, 1977, 1979, 1981, 1984, 1987, 1993, 1995, 2003).[1] He deeply believes that built environments of the past – for example, a city like Venice or Oxford, or a building like Chartres Cathedral or a Japanese farmhouse – regularly held a palpable sense of unity and harmony. In his writings and designs, he seeks a way to return a sense of wholeness, vitality, and grace to the buildings and places in which we live. A crux of his approach is what Alexander calls 'pattern language,' a practical method whereby the layperson or designer can identify and create the underlying elements and relationships in a built environment that give it a sense of order, vividness, and life (Alexander 1975, 1977, 1979, 1985; Coates and Seamon 1993; Dovey 1989; Gelernter 2000; Seamon 1987a, 1990b).

Alexander understands successful urban place-making as a collaborative process of healing whereby the city becomes more alive and healthy through an incremental growth of parts that, over time and synergistically, enriches the whole. Key aspects of this healthy city include small blocks, mixed uses, lively streets, physical and human diversity, distinctive neighborhoods, and human sociability – especially informal interactions in streets, sidewalks, and other public spaces (Adams 2001; Jacobs 1961; Lofland 1998). Crucially important is a design and decision-making process whereby new parts of the city arise in such a way that they strengthen the existing urban fabric and make it more identifiable and coherent. 'Design,' Alexander writes, 'must be premised on a process that has the creation of wholeness as its overriding purpose, and in which every increment of construction, no matter how small, is devoted to this purpose' (Alexander, 1987, 16).

In his *The Good City and the Good Life*, Daniel Kemmis (1995) explores the idea of urban wholeness and healing as it might have meaning for urban politics and citizenship: '[T]he refocusing of human energy around the organic wholeness of cities promises a profound rehumanizing of the shape and condition of our lives' (ibid. 151). Kemmis describes a way of urban life that involves individual citizens' feeling a part of the city because it provides a place for them to belong. Individuals, he argues, 'cannot be fully healthy, physically and mentally, in isolation, but only as meaningful players in a meaningful community ... [T]he healing (making more whole) of cities is serving to ... reknit ... the often frayed and sometimes severed strands of our humanity' (ibid. 152).

In this sense, Kemmis says much about the lived-process of making community happen, especially through striking vignettes drawn from his own political experiences as former mayor of Missoula, Montana, and former Speaker of the Montana legislature. Perhaps the most valuable dimension of his work is its speaking to how, in terms of a communal and political process, Alexander's theory for healing the city – a theory that Kemmis draws on directly and regularly – needs to happen. In other words, Kemmis attempts to answer the difficult question of how the practical steps of urban change are to be decided by the various parties involved. For Kemmis, this decision-making process is through and through *political*, whereby he means the realization of the city's possibilities through a civility among different citizens' views.

Though Alexander and Kemmis agree that healing the city is a central need, their understanding of how this healing happens is considerably different. Kemmis – as a politician and political philosopher – sees urban healing fostered largely through civil discourse among citizens and politicians. In contrast, Alexander – as an architect – insists that, before any such discourse can begin, there must first be a basic understanding as to what environmental wholeness is and how it can be strengthened or stymied by qualities of the physical-spatial environment and urban and architectural design. I argue here that, ultimately, *both* aspects of the healing process – physical and human, material and communal – must be considered and carried out, though I concur with Alexander that a knowledge of how the *physical* city grounds and stimulates the healing process must inform civil discourse.

Wholeness Begetting Wholeness

Before examining Alexander's conception of the physical city, it is useful to specify Kemmis's ideas on political process more exactly because they say much about how individuals come to care for the place where they live and thereby help to make it better. Early on in *The Good City*, Kemmis discusses what he calls the 'good life,' which 'makes it possible for humans to be fully present – to themselves, to one another and to their surroundings. Such presence is precisely opposite of the distractedness – the being beside – that is so prevalent in our political culture' (Kemmis 1995, 22).

For Kemmis, the city is organic in the sense that it 'organizes in its own terms a certain portion of the world' (ibid. 177). For this reason, Kemmis argues that the city dweller does not always initiate an active interest in his or her city; more often, the city, in its liveliness and attraction, activates the attention and concern of the dweller, who then contributes to the city. In this sense, place wholeness begets human wholeness and vice versa. This mutual interplay of part and whole, person and world, dweller and city is, for Kemmis, the foundation of civilization: 'This fundamental connection between human wholeness and livability and the wholeness and life of the city are all contained in ... the word "civilized"' (ibid. 12).

What Kemmis discusses here, implicitly, is the same phenomenological principle emphasized by Buttimer in her lifeworld writings: that people-are-immersed-in-world-as-world-is-immersed-in-people (Buttimer 1976; Seamon 1993a). This relationship is elusive and difficult to give grounded significance because it is a synergistic structure that becomes other than itself when conceptually broken apart (Seamon 1990a). One of Kemmis's strengths is his ability to explore and describe this person-place intimacy through his own political experience. For example, he discusses Missoula's lively farmers' market, which provides a place for the city to work on its citizens by bringing them together and providing economic and social exchange. This 'gathering role ... enables people to come away from the market more whole than when they arrived' (Kemmis 1995, 11).

Kemmis's central question is how citizens' sense of responsibility for their place can facilitate a civilized politics. All politics, Kemmis emphasizes, is about power, but the politician who can make the good city happen must always remember that his or her power 'is only a form of stewardship on behalf of those whose power it really is' (ibid. 153). Conventionally (at least in American government), power has been regulated by a system of procedural and legal checks and balances, but this system too often interferes with politicians' and citizens' exercising the personal responsibility of working out solutions together, 'which alone can make democracy work' (ibid. 154).[2]

Kemmis argues that good politicians remember they are stewards of power, which they barter to make a better city, through listening to what citizens say but also listening to the city itself. The need is to meet with many different people, to get them to talk to each other, and – when the moment seems right – to make the best decision possible on behalf of the city. In the end, says Kemmis, the mark of the good politician is 'knowing when to let the world work, and when to work on the world' (ibid. 177).

If this phrase sounds like the language of German phenomenological philosopher Martin Heidegger, it is. Throughout *The Good City*, Kemmis refers to Heidegger's essay, 'Building Dwelling Thinking' (Heidegger 1971), which argues that human beings can only design and make policy if they dwell in place and belong (Mugerauer 1994; Stefanovic 2000). For Kemmis, the good politician is open to the needs of both people and place so that, as the right moment arises, he or she can use power to make the next step toward healing the city:

> [I]f the city is constantly responding to what it has already created and to what fortune brings forward, then the next act of creation must always be some paradoxical blend of will and acceptance ... This blend is precisely the defining characteristic of the good politician (Kemmis 1995, 178-179).

On the other hand, this kind of practical openness to what the city might become cannot happen if ordinary citizens do not partake in the political process. Unfortunately today, community involvement too often becomes special-interest groups fighting for power. The need, says Kemmis, is to draw into the process people who can be civil and take responsibility for mediating extremes and finding a middle point of possibility. To be a citizen involves 'the ability to teach or encourage one another to speak so that you can actually be heard by others who do not already share your view' (ibid. 192).

Though this process of motivating citizens and healing place is not necessarily easy, immediate, or guaranteed, Kemmis believes that, if citizens really look at and listen to their place, then it can 'say' what it needs. For example, he discusses the development of a riverfront park system along the Clark Fork River just south of downtown Missoula, which in the 1980s had been largely destroyed economically by retail competition from outlying suburban malls. To stimulate an economic renaissance, Missoula's leaders implemented tax-increment financing, which leveraged private investments in downtown improvements and store renovations. This early success spawned additional downtown investments, which in turn stimulated the creation of a downtown riverfront park that would eventually spur on both sides of the river the creation of additional parks connected by walking trails.

Kemmis emphasizes that, at the start, neither citizens nor politicians could have imagined the chronological order or physical shape that Missoula's downtown revitalization would take. The success of the original risk that Missoula's leaders took in implementing the tax-incentive program led to private entrepreneurs and public and private agencies taking additional risks, many of which also proved successful. For Kemmis, the incremental, serendipitous piecing together of Missoula's downtown and riverfront is what he visualizes when he ponders Christopher Alexander's definition of environmental healing – that 'every new act of construction ... must create a continuous structure of wholes around it' (Alexander 1987, 22).

Toward the end of his account of downtown Missoula's redevelopment, Kemmis tells us that the riverfront parks and trail segments are still not connected in one complete walking loop allowing users to experience Missoula's river district as a whole. Yet he trusts, as Alexander would, that there are enough environmental parts in place so that now Missoulians feel the lack of an environmental whole. He predicts that eventually this feeling will generate the public will to complete the riverwalk system. He explains:

> The riverfront trail literally reaches out to join itself, putting pressure on the intervening parcels that prevent it from doing so, and then whenever it succeeds in filling in one of those gaps, it encourages whatever touches it to a greater wholeness as well (Kemmis 1995, 171).

Alexander's Healing of the City

As the example of downtown Missoula indicates, Kemmis's understanding of the city is much obliged to Alexander's vision of urban healing and wholeness (Alexander 1987), to which Kemmis refers regularly in *Good City*. He also argues, however, that Alexander, as an architect, gives most attention to *physical* healing but that, 'as important as the physical body of the city is, it alone cannot make the city healthy' (Kemmis 1995, 14).

Kemmis's effort to move beyond the physical healing of the city is both *Good City's* strength and weakness. On the one hand, he gives an invaluable picture of the process of human give-and-take that must underlie and motivate actual building and policy decisions. On the other hand, he seems to suppose that civilized mediation among participants will somehow lead to the right decisions as to how the city will constructively change without necessarily the need for any precise understanding or specific expertise as to what the material city is and how it works. Alexander's work suggests otherwise by laying out a process of conceiving and designing that goes far in providing a practical method for envisioning the physical revitalization of urban districts and the city as a whole.

Table 7.1 Christopher Alexander's 'Seven Rules for Urban Design'

1. *Piecemeal growth*
   The grain of development must be small enough, so that there is room, and time, for wholeness to develop. Building increments must not be too large, and there must be a reasonable mixture of large, medium, and small projects.
2. *The growth of larger wholes*
   Every building increment must help to form at least one greater whole in the city, which is both larger and more significant than itself. Everyone managing a project must clearly identify which of the larger emerging wholes this project is trying to help, and how it will help to generate them.
3. *Visions*
   Every project must first be experienced and then expressed as a vision of what is needed to heal the existing structure, not from an intellectually formed concept.
4. *Positive outdoor space*
   Every building must create coherent and well-shaped public space next to it. It is crucial that 'buildings surround space' rather than 'space surrounds buildings.'
5. *Layout of large buildings*
   The entrances, the main circulation, the main division of the building into parts, its interior open spaces, its daylight, and the movement within the building must all be coherent and consistent with the position of the building in the street and in the neighborhood.
6. *Construction*
   The structure of every building must generate smaller wholes in the physical fabric of the building, in its structural bays, columns, walls, windows, building base – in short, in its entire physical construction and appearance.
7. *Formation of centers*
   Every whole must be a 'center' in itself, and must also produce a system of centers around it.

Source: Adapted from Alexander (1987, chapter 3). Copyright 1987 by Christopher Alexander and used with permission.

In *New Theory*, Alexander (1987) seeks to develop a design process organized around seven *rules* – see Table 7.1 – that he believes can provide a healing action and lead to a renewed sense of urban place. In studying these rules, one notes that they attempt to guide the design process by fostering a good fit between new construction and the existing urban environment.[3] For example, rule one – 'piecemeal growth' – says that the best construction increments are small, thus there should be an even mix of small, medium, and large construction projects. Building on rule one, rule two – 'the growth of larger wholes' – directs how specific design projects can be seen to belong together and therefore requires that 'every building increment must be chosen, placed, planned, formed, and given its details in such a way as to increase the number of wholes that exist in space' (Alexander, 1987, p 248).

Of these seven rules, the most pivotal is the last – *formation of centers* – which, in Alexander's work beginning with *New Theory*, becomes the primary conceptual and practical means for clarifying and extending his earlier 'pattern language' ideas (Alexander 1993, 1995, 2003). Most simply, a center is any sort of spatial concentration or organized focus or place of more intense pattern or activity – for example, a handsomely designed window, a well placed kiosk, an elegant arcade, a welcoming building, a lively plaza full of people enjoying themselves, or an entire city neighborhood that is well liked and cared for (see especially Alexander 2003, ch. 3). Whatever its particular nature and scale, a center is a region of more intense physical and experiential order that provides for the relatedness of things, people, situations, and events. In this sense, the strongest centers gather the parts in a relationship of *belonging*, including city dwellers.[4] Further, where one finds life and wholeness in the city, centers are never alone but mutually implicated at many levels of scale: 'The wholeness of any portion of the world is the system of larger and smaller centers, in their connections and overlap' (ibid. 90-91).

A Waterfront Design for San Francisco

In *New Theory*, Alexander illustrates the use of centers and the seven rules through a simulation experiment conducted with architecture graduate students at the University of California at Berkeley in a design studio taught by him and faculty colleagues Ingrid King and Howard Davis (Alexander 1987). The nineteen students in this studio focused on thirty acres of the San Francisco waterfront just north of the Bay Bridge and, at the time, destined for future development. The major task the students faced was to transform these thirty acres, for the most part empty, into a district of buildings, streets, plazas and parks that would all contribute to a sense of place, vibrancy, and wholeness. Eventually, the students converted the waterfront site into an a set of places that included such elements as a pedestrian mall, a main square, a waterfront park, and a market and fishing pier (see Figures 7.1 and 7.2).

Procedurally, the students were asked to imagine themselves as developers and representatives of community groups. On the other hand, Alexander, King, and Davis took the role of an evaluation committee responsible for guiding the growth process. No student's design idea (the 'vision' of rule three) could be finalized until the committee had evaluated the idea and considered strengths and weaknesses. All faculty and students were involved in all discussions about every project, so there was much mutual understanding as to the project's progress and ultimate aim.

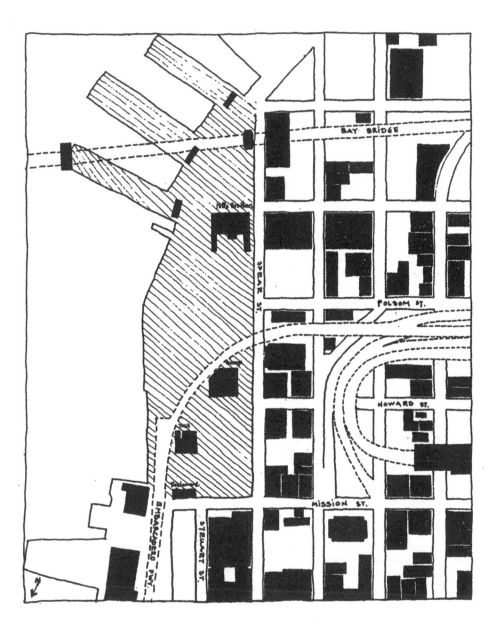

Figure 7.1 San Francisco waterfront site to be developed
Source: Adapted from Alexander (1987, 105). Copyright 1987 by Christopher Alexander and used with permission.

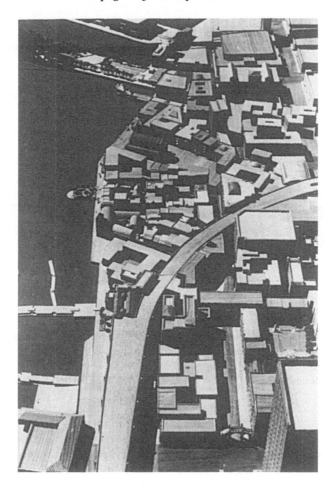

Figure 7.2 Photograph of a wooden model of the San Francisco waterfront
        project after 89 projects
Source: Adapted from Alexander (1987, 107). Copyright 1987 by Christopher Alexander and
used with permission.

To illustrate the way the simulation worked, it is useful to review the first phase of the
project, which involved development of the northern part of the site as shown in Figure
7.3. To begin, students and faculty visited the site and decided that the northern portion
(lower area of Figure 7.3) seemed the right place to start development, since the natural
entrance to the site – Mission Street – was there. This decision was then formally
announced by the committee, who invited projects that would 'enhance the entrance, and
create it strongly and dramatically' (ibid. 115). One student had the vision of a high,
narrow arching gate that would serve as a distinctive entry marker. This idea was
approved by the committee as the starting point for the site's development.

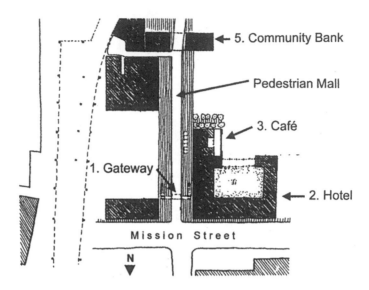

Figure 7.3 The San Francisco waterfront site after completion of five projects.
Note how the first project (the entry gate) suggests a larger whole –
a street and pedestrian mall directed south toward the center of the
site

Source: Adapted from Alexander (1987, 41). Copyright 1987 by Christopher Alexander and used with permission.

This entrance gate was important because it generated a sense of passage that started beneath the arch and continued south. In this way, the gate hinted at a larger whole – a street and pedestrian mall going south into the heart of the site. This pedestrian street was then defined more exactly by the next two projects: a hotel and a café, which fixed the street's west side and width (an existing building on the east fixed the street's east side). Soon after, project five – a community bank – established the far end of the street, which was then completed by a series of increments that included an apartment house, an office building, and various smaller-scale projects like a gravel walk and low wall.

In terms of centers, each project defining the pedestrian street did three things at once: first, it helped to complete one major center already defined; second, it helped to pin down some other, less clearly defined center; third, it hinted at some entirely new center that could emerge later. One example is the hotel (project two), which wrapped around a garden courtyard. First, in conjunction with the gate, this building helped to complete the southern edge of the site; second, it helped to establish the larger center of the pedestrian street by fixing its western edge; third, in shaping itself around an outdoor courtyard, the hotel hinted at a new center that in later increments would become a large public garden running south from the hotel and shaped by a series of apartment buildings.

Incorporating Physical and Human Worlds

After 89 projects, Alexander's team ended their experiment, though it is clear that additions and refinements were still possible and would be needed if the design were ever actually built. If one studies the model for the project, shown in Figure 7.2, one notes that the final design organized itself around three main areas: first, the area north of the expressway and composed of the pedestrian street and office and apartment buildings; second, the middle section of the site, centered around a main square and grid of streets heavily residential in character; third, the southern part of the site, which became largely a commercial area of factories, warehouses, and nautical facilities.

As this studio simulation suggests, Alexander developed his rule-based approach to urban design because he believes that there must be some sort of reasoned procedure for the actualization of wholeness, whereby decision-makers gain understanding and the city gains realization. In contrast, Kemmis, in his approach to the city, has much less interest in such practical understanding and clear-cut procedure. Rather, he seems to believe that, if citizens and politicians begin to put the welfare of their city *first*, an understanding of what the city needs to become will automatically arise through civil discussion, mediation, and compromise: 'As citizens become more practiced at working together with the city's best interests at heart, it is precisely such structures of wholeness that recommend themselves to their attention' (Kemmis 1995, 194).

Alexander might not disagree with this perspective, provided the participants had some degree of informed awareness of what the wholeness of place is and some set of guidelines to keep this wholeness in mind. On the other hand, Alexander says little about how his rule-based approach, through citizen involvement, might actually go forth into building. How, in other words, can his reasoned procedure – the seven rules – be given direction and actualized through real-world participants?[5]

In the studio experiment, the role-playing of students and faculty was obviously artificial and arbitrary. Ultimately, students had to agree with the judgments of the faculty committee and to work in relation to the rules whether they personally agreed with them or not.[6] At the same time, the resulting designs were completed only as paper plans and a scale model that never had to face the real-world evaluation of the residents, developers, city officials, politicians, and other constituencies ultimately providing approval, funding, and public support.

In regard to applied direction, this is where Kemmis's ideas are such an important complement to Alexander's approach: Kemmis provides an extended picture of what is necessary, in terms of getting different parties to discuss and compromise, if the urban wholeness and healing that Alexander provides a design means for is to be supported and carried out by the citizenry. On the other hand, Kemmis seems less aware of how a city works physically and spatially. Again, we return to the basic phenomenological principle that people are immersed in their worlds, which first of all are physical and spatial. In this sense, Alexander's design vision is an essential complement to Kemmis's hopeful politics of place, though I next want to argue that this vision is incomplete and needs supplementing by Bill Hillier's innovative theory of how qualities of physical space play an inescapable role in making lively urban places.

Hillier's Theory of Space Syntax

Since the mid 1970s, researchers at the Bartlett School of Architecture and Planning, University College, London, have developed convincing conceptual and empirical evidence that the physical-spatial environment plays an integral part in sustaining active streets and an urban sense of place. Largely conceptualized by the architectural researchers Bill Hillier and Julienne Hanson, this research examines the relationship between physical space and social life, or, more precisely, 'the social content of spatial patterning and the spatial content of social patterning' (Hillier and Hanson, 1984, x-xi). Most often, this work has come to be called Bill Hillier's 'theory of space syntax,' the label used here (Hanson 1998; Hillier 1989, 1993, 1996, 1999; Hillier and Hanson 1984, 1989; Hillier et al. 1982, 1987; Murrain 1993; Peponis 1989, 1993).

Throughout his writings, Hillier asks if there is some 'deep structure of the city itself' that contributes to urban life (Hillier 1989, 5). He finds this deep structure in the relationship between spatial configuration and natural co-presence – that is, the way the spatial layout of pathways can informally and automatically bring people together in urban space or keep them apart: 'By its power to generate movement, spatial design creates a fundamental pattern of co-presence and co-awareness, and therefore potential encounter amongst people that is the most rudimentary form of our awareness of others' (Hillier 1996, 213). Hillier argues that, through a particular kind of spatial configuration – what he calls the *deformed grid* – cities have historically 'exploited movement constructively to create dense, but variable, encounter zones to become what made them useful: mechanisms for contact' (ibid. 174).

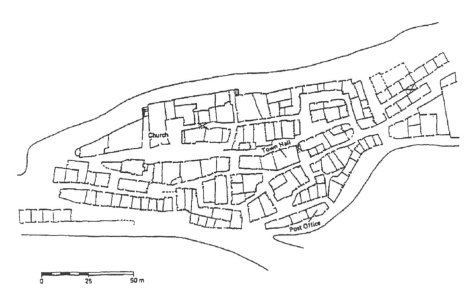

Figure 7.4 The French village of Gassin
Source: Adapted from Hillier and Hanson (1984, 90). Copyright 1984 by Cambridge University Press and used with permission.

One means that Hillier uses to demonstrate the relationship between spatial configuration and pedestrian movement is a careful examination of the street and open-space fabric of many different settlements throughout the world (Hillier and Hanson 1984; Hillier et al. 1982). Many of these places – for example, the French village of Gassin illustrated in Figure 7.4 – regularly incorporate the following topological characteristics that together create what Hillier calls the *beady-ring structure*:

1. All building entrances face directly onto the village open spaces, thus there are no intervening boundaries between building access and public space.
2. The village open spaces are continuous but irregular in their shapes; they narrow and widen, like beads on a string.
3. The outdoor spaces join back on themselves to form a set of irregularly shaped rings.
4. This ring structure, coupled with direct building entry, gives each village a high degree of permeability and access in that there are at least two paths (and, typically, many more) from one building to any other building.

The next question Hillier asks is whether this beady-ring structure can be described and measured more precisely. At the start, one faces a difficult recording problem: in terms of everyday function, a settlement's open space is one continuous fabric but, formalistically and spatially, this fabric is composed of many different parts – streets, alleys, squares, plazas, walls, buildings, and the like. How can this unwieldy network of spaces and things be defined and measured without destroying the seamless nature of the settlement's open spaces?

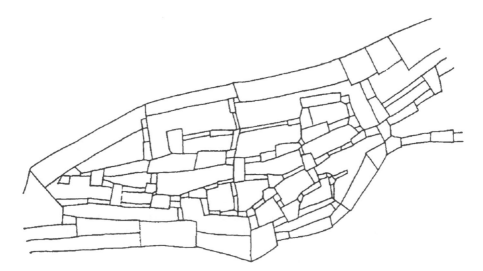

Figure 7.5 Gassin's convex map
Source: Adapted from Hillier and Hanson (1984, 92). Copyright 1984 by Cambridge University Press and used with permission.

To address this conceptual difficulty, Hillier suggests that any open space can be considered in terms of its *convex* or *axial* qualities. A *convex space* refers to the *two*-dimensional nature of open space and is best exemplified by plazas, squares, and parks. In that they can have considerable breadth in relation to width, convex spaces relate to the *beadiness* of the beady-ring structure. In terms of environmental experience, convex spaces typically become local places – e.g., the site of a weekly market, an open space where children regularly play kickball, or a place where older people gather on sunny afternoons. By identifying the least number of convex spaces accounting for all streets, plazas and other outdoor space, one can construct a *convex map* as shown for Gassin in Figure 7.5.

In contrast to convex spaces are what Hillier calls *axial spaces*, which depict the *one*-dimensional qualities of space and therefore relate to human movement through the settlement and to the *stringiness* of the beady-ring structure. Axial spaces are best illustrated by long narrow streets and can be represented geometrically by the maximum straight line that can be drawn through an open space before it strikes a building, wall, or some other material object (see the axial map of Gassin in Figure 7.6).

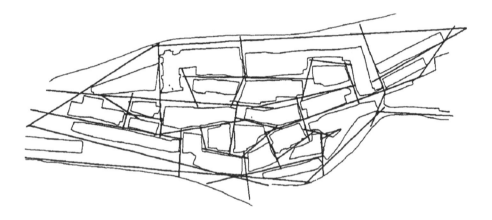

Figure 7.6 Gassin's axial map
Source: Adapted from Hillier and Hanson (1984, 91). Copyright 1984 by Cambridge University Press and used with permission.

Whereas convex spaces speak mostly to the *local* qualities of a space, axial spaces are significant for understanding a settlement's *global* pattern – that is, the way the particular spatial configuration of the pathway fabric lays out a potential movement field that draws people together or keeps them apart. *Natural movement* is the term Hillier uses to describe the potential power of the pathway layout to automatically stymie or facilitate movement and such related environmental events as co-presence, co-awareness, informal interpersonal encounters, and lively local places and street activity (Hillier 1996, 161).

Integration and the Deformed Grid

To establish precisely the amount of natural movement that a particular pathway configuration potentially generates, Hillier introduces the concepts of *integrated* and *segregated* pathways. The former is a pathway that makes itself readily accessible to many other pathways and therefore is *shallow* in relation to them. In other words, many other pathways and the users on these pathways feed into this pathway, thus it is well integrated in relation to the surrounding grid structure and more than likely a well-used route along which many people travel. In contrast, few or no other routes feed into *segregated* pathways, which are poorly accessible and *deep* in relation to the surrounding grid. Segregated pathways typically are dead ends or elements in treelike grids; one thinks, for example, of the 'cul-de-sac and loop' pattern of low-density, automobile-dependent suburbs, or the hierarchical circulation layouts of modernist housing estates.

To measure and map the relative integration of all pathways in a particular pathway system, Hillier develops a quantitative procedure that he calls *measure of integration* (Hillier and Hanson 1984, 108-09). One product of this procedure is an integration map like the one for Gassin in Figure 7.7, which summarizes the integration values for all pathways in the village. The streets marked by solid lines depict the village's *integration core* – those streets that have many other streets feeding into them. These streets have the most chance for being alive with street activity, public life, and commerce. In contrast, the hatched lines identify Gassin's *segregation* core – the streets that deflect activity away from themselves and therefore indicate pockets of quiet and seclusion that are typically residential in character.[7]

Hillier next asks if these lines of greater and lesser integration indicate some deeper topological structure underlying the settlement *as a whole*. In fact, after studying the integration and segregation cores of many settlements, both Western and non-Western, Hillier concludes that such a larger global structure exists, which he calls the *deformed wheel*. The rim, spokes and hub of this wheel are the pathways with high integration values (in Figure 7.7, the solid lines). Typically, these streets are the most used by a settlement's residents and are also the main entry routes and therefore used by strangers – for example, a farmer bringing his produce to weekly market or tourists exploring the settlement. Also, most of the largest convex spaces and location-dependent uses, like shops, are on the most integrated streets of the deformed wheel, since these streets are the places of greatest movement.

In the interstices between the most active streets are the most segregated, less used pathways (in Figure 7.7, the hatched lines). Hillier concludes that, for many traditional settlements, the most active areas abut the quietest areas: the places of street life, publicness, and strangers' mixing with residents are a short distance from the more private areas used mostly by residents only. Movement and rest, activity and place, journey and dwelling, difference and locality, publicness and home, lie apart yet together! Hillier explains:

> By linking the interior of the settlement to the periphery in several directions – and always in the direction of the main entrances to the settlement and the neighboring towns – the effect of the integrated lines is to access the central areas of the town from outside, while at the same time keeping the core lines close to the segregated areas, in effect linking them together. Since the core lines are those that are most used by people, and also those on which most space-

dependent facilities like shops are located, and the segregated areas are primarily residential, the effect of the core is to structure the path of strangers through the settlement, while at the same time keeping them in a close interface with inhabitants moving about inside the town. The structure of the core not only accesses strangers into the interior of the town, but also ensures that they are in a constant *probabilistic* interface with moving inhabitants (Hillier 1989, 11).

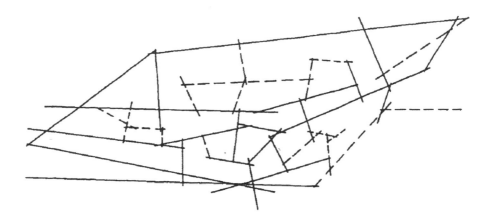

Figure 7.7 Gassin's 'deformed wheel.' The streets of potentially greatest movement are marked by the solid lines, whose shape roughly suggests a wheel with hub and spokes. In contrast, the hatched lines indicate streets of potentially less movement; overall, they are located in between the more active streets

Source: Adapted from Hillier and Hanson (1984, 117). Copyright 1984 by Cambridge University Press and used with permission.

In regard to cities, Hillier demonstrates that most urban pathway systems have traditionally been an integrated fabric of smaller deformed wheels (usually associated with designated neighborhoods and districts – for example, London's Soho, Bloomsbury, or City), whose most integrated pathways join together to shape a much larger *deformed grid* that founds the dynamic of natural movement for the city as a whole (Hillier 1996, ch. 4). '[E]ach local area,' explains Hillier (ibid. 171), 'has its heart linked to the supergrid lines that surround it by strong integrators. These form an edge-to-centre structure in all directions, and the less-integrated areas are within the interstices formed by the structure.' Hillier is highly critical of most twentieth-century urban design and planning because it often eviscerated this relationship between local and global integration by replacing integrated pathway configurations with treelike systems of segregated pathways. The long-term result is that these 'spatial designs create serious lacunas in natural movement,' which in turn undermines the informal sociability of streets and neighborhoods and may in time attract 'anti-social uses and behaviours' – for example, unsafe streets and higher crime rates, particularly in the mazelike pathway systems of many public housing projects (ibid. 178).

Implications for Alexander's Theory of Urban Design

In his theory of space syntax, Hillier appears to provide incontrovertible evidence that pathway structure plays a major role in establishing the kind of place that a settlement becomes. Specifically, he identifies the particular kind of pathway configuration required to foster lively streets and the informal public encounters and sociability that make cities such effective places of contact and exchange (Hillier 1996, 174).[8]

If Hillier's conclusion is true, it points toward two revolutionary possibilities for understanding and designing cities: first, that urban designers must deal with space before they deal with form; second, that in dealing with space, urban designers must understand the city's global configuration as it is and can be rendered through the configuration of axial spaces and variations in integration. Hillier emphasizes that if urban designers ignore the city's spatial configuration and resulting natural movement, they risk 'eliminating all the properties of density, good spatial scale, controlled juxtaposition of uses, continuity, and integration of the urban grid on which the well-ordering and well-functioning of the city depends' (ibid. 179).

From this perspective, Hillier would be highly critical of Alexander's approach to urban design. He would claim that, ironically, Alexander's urban theory is piecemeal. Because it ignores the integrative power of pathways, the global whole is reduced to its local parts.[9] For sure, Alexander envisions the city as an organic whole of rightly placed, interrelated parts: 'Every increment of construction must be made in such a way as to heal the city' (Alexander 1987, 22). In practice, however, Alexander's seven rules unintentionally reduce the interconnectedness of urban place to *buildable parts* – in the San Francisco project, for example, the entry gateway, the pedestrian mall, the hotel, or main square. Alexander's urban vision holds no inkling of the deformed grid and its potential to energize the city's parts through a permeable, integrated pathway system. Without this awareness, Alexander does not have available the primary engine for the organic wholeness he so sincerely hopes to create.

This lack of global interconnectedness is seen in the San Francisco waterfront design. Though there is much about the project to praise, its major failing is a poorly envisioned street grid that inhibits local interconnections and movement and provides no clear pathway commingling with the street fabric of the larger city. In his evaluation of the project at the end of *New Theory*, Alexander is aware of this weakness: '[T]he large-scale structure is not as profound as we wanted it to be. Although the general disposition of the main square, mall, small grid, and so on, is quite nice, and is suitably informal, it does not yet have the profound unity of a place like Amsterdam or Venice' (1987, 234–35).[10]

The larger problem here may well be Alexander's concept of centers, which, by their very nature, involve focused intensity and are thus much more local than global in their conception and effects. No doubt, the weblike structure of the deformed grid is a kind of center that contains within its mesh smaller, interconnected centers that identify a city's functioning neighborhoods and districts. The key point is that the deformed grid is global, citywide, and thus *whole* in its manifestation and results. If Hillier is correct, any theory of urban design must begin at the city's global scale, carefully studying pathway configuration. This understanding then becomes the starting point for determining how a particular district, through new and existing pathway connections, might gain, through natural movement, a vital relationship with the integration fabric of the larger urban whole. As it stands, Alexander's urban theory does not have the means to identify or actualize the underlying, integrative power of the deformed grid.[11]

Integrating Alexander, Hillier and Kemmis

Does its global weakness mean that Alexander's theory of urban design is a failure? In fact, Alexander admits in the introduction to *New Theory* that his approach is 'full of holes and incomplete' (Alexander 1987, 2). He also recognizes, however, that process rather than form is the heart of wholeness and that 'if we create a suitable process there is some hope that the city might become whole once again. If we do not change the process there is not hope at all' (ibid. 3).

Alexander believes that, by regularly having to consider how a particular design part connects to the larger urban fabric, participants will be able to envision projects that both contribute to and speak for the whole. Within the limits imposed by the abilities of participants, each of the design parts presents the whole just as, simultaneously, the whole is present in each of the parts. This approach is extremely uncertain, however, and its central danger has been called the 'hazard of emergence':

> A part is only a part according to the emergence of the whole that it serves; otherwise it is mere noise. At the same time, the whole does not dominate, for the whole cannot emerge without the parts. The hazard of emergence is such that the whole depends on the parts to be able to come forth, and the parts depend on the coming forth of the whole to be significant instead of superficial. The recognition of a part is possible only through the 'coming to presence' of the whole (Bortoft 1985, 287; also see Bortoft 1996).

Alexander's great contribution, not only to urban design but to environmental thinking in general, is to seek ways to bring to presence a built world that will enable and ennoble human worlds. On one hand, there can be no real architectural and environmental whole if the vision of that whole dominates, and the parts themselves are forced into place. On the other hand, the parts are only valuable if they all have their proper place and can be together in a way that is real and non-arbitrary. There is a possibility of *genuine belonging*.

This idea of genuine belonging is the region of confluence between Alexander's ideas and those of Hillier and Kemmis. Like Alexander, Hillier and Kemmis interpret the city as its parts belong and have a place. Alexander holds the most utopian hope in the sense that he believes, through correct understanding and practice, we might intentionally recreate the lively urban places and dynamic neighborhood street life that unfolded in earlier cities mostly with no directed effort at all. In spite of its deficiencies, Alexander's theory is powerful exactly because it attempts to incorporate *all* aspects of the process of making vital places. Through his vision of wholeness and healing, he hopes to transform failing urban districts, using as a means his seven rules and a constructive dialogue among all parties involved.

As Hillier's discoveries demonstrate, Alexander's process of urban design must incorporate a more accurate understanding of how existing 'good cities' work, particularly their less obvious global structure. Hillier helps us to see that the city is often fractured today exactly because the wholeness of its pathway fabric has been disrupted. Alexander's approach, particularly through a deeper formulation of centers and a revised set of rules incorporating the idea of pathway configuration, should be able to accommodate Hillier's insights. Without this revision, Alexander's approach will continue to fail because, in its current format, it assumes a localist approach to place in which the relationship to larger global structure is incompletely understood.[12]

In turn, Kemmis's work is important to Alexander's vision because he unravels a practical middle way whereby citizens, putting their place first, might set aside differences and become involved in the much broader intentional process that Alexander works to construct. In this sense, Kemmis marks the start of a phenomenology of the way by which individuals and groups come together to envision and make the 'good city.' Such interpersonal and inter-group dialogue is crucial if Alexander's approach is to generate a working relationship between participants in the urban-design process and any rules that guide that process.

Ultimately, Alexander, Hillier, and Kemmis all aim to facilitate an environmental and place wholeness, whereby the city is an interconnected mesh of vital neighborhoods and districts. Hillier gets us to see that the heart of urban liveliness is an integrated pathway fabric, while Kemmis gets us to see that urban liveliness can ignite citizen involvement and a virtuous circle of more and more wholeness. In its comprehensiveness, Alexander's vision holds a place for the ideas of both Hillier and Kemmis and points toward an integrated theory and practice that might transform urban placelessness into lived places of exuberance (Adams 2001; Murrain 1993; Relph 1976, 2001).

At Clark University in the 1970s, we only dimly understood how the physical and spatial aspects of the urban lifeworld contribute to its liveliness, community, and a sense of place. We knew, phenomenologically, that the habitual body's expression of routines in time and space contributed to the possibility of place regularity (Buttimer 1976, 1980; Seamon 1979, 2002; Seamon and Nordin 1980), but we had no clear understanding of the ways the intimate dialogue between physical and human worlds could facilitate or inhibit lived place. Nor did we give attention to how citizens and policy makers might come together to decide fruitful planning and design for their place.

In this sense, the works of Alexander, Hillier, and Kemmis are pivotal for remaking place in the city and for extending in unimagined but crucial ways the work on urban lifeworlds begun by Buttimer and her students at Clark in the 1970s. When coupled together, the ideas of Alexander, Hillier, and Kemmis offer fresh insights into Buttimer's Clark legacy and provide a remarkable vision for recreating, through intentional design and generous participation, robust urban places.

Notes

1   Several of Alexander's works include other contributors. Because Alexander is the principal author, I use his name as the citation, but all contributors are listed in the reference entry.

2   Kemmis examines the political basis for this argument in his earlier *Community and the Politics of Place*, which argues for 'a politics which rests upon a mutual recognition by diverse interests

that they are bound to each other by their common attachment to a place' (Kemmis, 1993, 123; also see Kemmis 2001).

3   In studying the seven rules carefully, one notes that they have two related functions: first, rules one, two and seven help the designer to recognize and understand environmental wholes; second, rules three, four, five, and six help to create new parts in the whole that should lead to healing and a stronger environmental order.

4   On the theme of belonging, see Bortoft 1985, 1996; Seamon 1990a, 1993a.

5   This is not to imply that Alexander is oblivious to the many deeper structural changes required socially, economically, and politically for his design approach to actually work in the real world. In fact, he has discussed this need for major societal changes in a number of writings – e.g. see Alexander 1984, ch. 3 and 4; and Alexander 1987, 239-242.

6   Interestingly, Alexander points out that this unspoken agreement between students and committee became stronger as the students had more experience with the rules: 'in the last stages of development, the students were able to function almost entirely without guidance from the committee, since the rules had been completely absorbed and understood' (Alexander 1987, 110).

7   Eventually, the question must be asked if in fact the pathways indicated by the integration measure as being potentially most used, actually are. To answer this question, Hillier and his research team counted pedestrian volume on pathways and compared the rate with the pathways' integration values. These studies have shown that spatial integration is a good predictor of movement, in the sense that highly integrated spaces are typically more used for movement than spaces that are weakly integrated (see Hillier 1996, 161; Hillier et al. 1993, 29-66; Marcus 2000, 109-111; Read 1999, 251-264).

8   Hillier provides an effective argument for claiming that 'movement in the urban grid is, other things being equal, generated by the configuration of the rid itself' (Hillier 1996, 5 and 167-170). He convincingly demonstrates that the grid presupposes other elements like placement of functions, land uses, and densities: 'the structuring of movement by the grid leads, through multiplier effects, to dense patterns of mixed use encounter that characterize the spatially successful city' (ibid. 6).

9   In his most recent work, Alexander (2003, 417) briefly discusses Hillier's ideas, explaining how they demonstrate that 'it is not really possible to keep function and space separate.' Unfortunately, Alexander seems to miss Hillier's much more important point that global structure presupposes and animates local parts.

10   In his most recent work, Alexander (2003, ch. 5) identifies fifteen properties of centers and attempts to move beyond the piecemeal criticism by incorporating more integrative qualities like 'levels of scale,' 'deep interlock and ambiguity,' and 'gradients.' Even so, one still worries that he continues to emphasize the parts of the whole at the expense of its underlying degree of global interconnectedness (Hillier's relative integration). As a means to understand art works, decorative objects, and the formal properties of architecture, Alexander's emphasis on local qualities of wholeness provides helpful insights because these things are more or less independent physical entities that do not house human lifeworlds. On the other hand, the fifteen properties cast an incomplete understanding when one applies them to the larger-scale environmental fabric within which the lifeworlds of real human beings actually unfold.

11   An exceptional effort to develop a simple but effective way for urban designers to incorporate pathway integration and permeability is illustrated in Bentley et al., 1985. This work draws indirectly on Alexander's 'pattern language' format and therefore also offers a useful indication of how Alexander and Hillier's vision of the city might be integrated.

12   Because of their localist emphasis, Hillier is strongly critical of most current scholarly and design efforts emphasizing 'place.' He levels his harshest criticism again Oscar Newman's theory of 'defensible space' (Newman 1973, 1980) but no doubt he would also criticize phenomenological research as well as other place-grounded designs like architect Andres Duany's 'new urbanism.' Hillier writes: 'The current preoccupation with 'place' seems no more

than the most recent version of the urban designer's preference for the local and apparently tractable at the expense of the global and intractable in cities … Places are not local things. They are moments in large-scale things, the large-scale things we call cities. Places do not make cities. It is cities that make places. The distinction is vital. We cannot make places without understanding cities' (Hillier 1996, 151).

References

Adams, P.C. (2001) 'Peripatetic imagery and peripatetic sense of place' in Adams, P.C., Hoelscher, S. and Till, K.E. (eds.) *Textures of place.* University of Minnesota Press, Minneapolis, 186-206.

Alexander, C. (1979) *The timeless way of building.* Oxford University Press, New York.

Alexander, C. (1981) *The Linz Café.* Oxford University Press, New York.

Alexander, C. (1993) *A foreshadowing of 21st century art: the color and geometry of very early Turkish carpets.* Oxford University Press, New York.

Alexander, C. (2003) *The nature of order, vol. 1: the phenomenon of Life.* Center for Environmental Structure, Berkeley.

Alexander, C., Anninou, A., King, I. and Neis, H. (1987) *A new theory of urban design.* Oxford University Press, New York.

Alexander, C., Black, G. and Tsutsui, M. (1995) *The Mary Rose Museum.* Oxford University Press, New York.

Alexander, C., Davis, H., Martinez, J. and Corner, D. (1984) *The production of houses.* Oxford University Press, New York.

Alexander, C., Ishikawa, S. and Silverstein, M. (1977) *A pattern language.* Oxford University Press, New York.

Alexander, C., Silverstein, M., Angel, S., Ishikawa, S. and Abrams, D. (1975) *The Oregon Experiment.* Oxford University Press, New York.

Bentley, I., Alcock, A., Murrain, P., McGlynn, S. and Smith, G. (1985) *Responsive environments: a manual for designers.* The Architectural Press, London.

Bortoft, H. (1985) 'Counterfeit and authentic wholes: finding a means for dwelling in nature' in Seamon, D. and Mugerauer, R. (eds.) *Dwelling, place and environment: toward a phenomenology of person and world.* Columbia University Press, New York, pp. 281-302.

Bortoft, H. (1996) *The wholeness of nature.* Lindesfarne Press, Hudson, Hudson, NY.

Buttimer, A. (1969) Social space in interdisciplinary perspective, *Geographical Review* 59, pp. 417-426.

Buttimer, A. (1971a) *Society and milieu in the French geographic tradition.* Association of American Geographers Monograph 6, Rand McNally, New York.

Buttimer, A. (1971b) Sociology and planning, *Town Planning Review* 42, pp. 145-180.

Buttimer, A. (1972) Social space and the planning of residential areas, *Environment and Behavior* 4, pp. 279-318.

Buttimer, A. (1974) *Values in Geography.* Resource Paper 24 Association of American Geographers, Washington, D.C.

Buttimer, A. (1976) Grasping the dynamism of lifeworld, *Annals of the Association of American Geographers* 66, pp. 277-292.

Buttimer, A. (1980) 'Home, reach, and the sense of place' in Buttimer, A. and Seamon, D. (eds.) *The human experience of space and place.* Croom Helm, London, pp. 166-187.

Buttimer, A. (1987) A social topology of home and horizon, *Journal of Environmental Psychology* 7, pp. 307-319.

Buttimer, A. and Seamon, D. (eds.) (1980) *The human experience of space and place.* Croom Helm, London.

Coates, G.J. and Seamon, D. (1993) 'Promoting a foundational ecology practically through Christopher Alexander's pattern language: the example of Meadowcreek' in Seamon, D. (ed.) *Dwelling, seeing, and designing: toward a phenomenological ecology.* State University of New York Press, Albany, New York , pp. 331-354.

Dovey, K. (1989) The pattern language and its enemies, *Design Studies* 11, pp. 3-9.

Gelernter, M. (2000) Sun-filled windows, *Architectural Research Quarterly* 4, pp. 190-193.

Hanson, J. (1998) *Decoding homes and houses.* Cambridge University Press, Cambridge.

Hanson, J. and Hillier, B. (1987) The architecture of community, *Architecture and Behavior* 3, pp. 233-250.

Heidegger, M. (1971) 'Building, dwelling, thinking' in Heidegger, M. *Poetry, language, thought.* Harper and Row, New York, pp. 145-161.

Hillier, B. (1985) The contingent and the necessary in spatial form in architecture, *Geoforum* 16, pp. 163-178.

Hillier, B. (1989) The architecture of the urban object, *Ekistics* 56, pp. 5-21.

Hillier, B. (1993) Specifically architectural knowledge, *Nordisk Arkitekturforskning / Nordic Journal of Architectural Research* 6, pp. 11-37.

Hillier, B. (1996) *Space is the machine.* Cambridge University Press, Cambridge.

Hillier, B. (1999) The hidden geometry of deformed grids; or why space syntax works, when it looks as though it shouldn't, *Environment and Planning B: Planning and Design* 26, pp. 169-191.

Hillier, B. and Hanson, J. (1984) *The social logic of space.* Cambridge University Press, Cambridge.

Hillier, B. and Hanson, J. (1989) Introduction, *Architecture and Behavior* 3, pp. 197-199.

Hillier, B., Hanson, J., Leponis, J., Hudson, J. and Burdett, R. (1982) Space syntax: a different urban perspective, *Architect's Journal* 30, pp. 47-63.

Hillier, B., Hanson, J. and Peponis, J. (1987) Syntactic analysis of settlements, *Architecture and Behavior* 3, pp. 217-232.

Jacobs, J. (1961) *The death and life of great American cities.* Vintage, New York.

Kemmis, D. (1993) *Community and the politics of place.* University of Oklahoma Press, Norman, Oklahoma.

Kemmis, D. (1995) *The good city and the good life.* Houghton Mifflin, New York.

Kemmis, D. (2001) *This sovereign land: a new vision for governing the West.* Island Press, San Francisco.

Lofland, L.H. (1998) *The public realm.* Aldine, New York.

Marcus, L. (2000) *Architectural knowledge and urban form.* Kungliga Tekniska Högskolan, Stockholm.

Mugerauer, R. (1994) *Interpretations on behalf of place.* State University of New York Press, Albany, New York.

Murrain, P. (1993) 'Urban expansion: look back and learn' in Hayward, R. and McGlynn, S. (eds.) *Making better places: urban design now.* Butterworth, London, pp. 83-94.

Newman, O. (1973) *Defensible space.* MacMillan, New York.

Newman, O. (1980) *Community of interest.* Doubleday, New York.

Peponis, J. (1989) Space, culture, and urban design in late Modernism and after, *Ekistics* 5, pp. 93-108.

Peponis, J. (1993) Evaluation and formulation in design: the implication of morphological theories of function, *Nordisk Arkitekturforskning / Nordic Journal of Architectural Research* 6, pp. 53-62.

Read, S. (1999) Space syntax and the Dutch city, *Environment and Planning B: Planning and Design* 26, pp. 251-264.

Relph, E. (1976) *Place and placelessness.* Pion, London.

Relph, E. (2001) 'The critical description of confused geographies' in Adams, P. C., Hoelscher, S. and Till, K.E. (eds.) *Textures of place.* University of Minnesota Press, Minneapolis, pp. 150-166.

Rowles, G.D. (1978) *Prisoners of space? Exploring the geographical experience of older people.* Westview Press, Boulder, Colorado.

Seamon, D. (1979) *A geography of the lifeworld.* St Martin's, New York.

Seamon, D. (1987a) 'Phenomenology and environment-behavior research' in Moore, G.T. and Zube, E. (eds.) *Advances in environment, behavior and design* (volume one). Plenum, New York, pp. 3-27.

Seamon, D. (1987b) Phenomenology and the Clark experience, *Journal of Environmental Psychology* 7, pp. 367-377.

Seamon, D. (1990a) 'Awareness and reunion: a phenomenology of the person-environment relationship as portrayed in the New York photographs of André Kertész' in Zonn, L. (ed.) *Place images in the media.* Roman and Littlefield, Totowa, New Jersey, pp. 87-107.

Seamon, D. (1990b) 'Using pattern language to identify sense of place: American landscape painter Frederic Church's *Olana* as a test case' in Selby, R. (ed.) *Coming of age: proceedings, Environmental Design Research Association.* Environmental Design Research Association, Oklahoma City, Oklahoma, pp. 171-179.

Seamon, D. (1991) 'Toward a phenomenology of the architectural lifeworld' in Hancock, J. and Miller, W. (eds.) Architecture: back ... to ... Life, Proceedings of the 79th Annual Meeting of the Association of Collegiate Schools of Architecture, ACSA Press, Washington, D C, pp. 3-7.

Seamon, D. (1993a) 'Different worlds coming together: a phenomenology of relationship as portrayed in Doris Lessing's *Diaries of Jane Somers*' in Seamon, D. (ed.) *Dwelling, seeing, and designing: toward a phenomenological ecology.* State University of New York Press, Albany, New York, pp. 219-246.

Seamon, D. (ed.) (1993b) *Dwelling, seeing, and designing: toward a phenomenological ecology.* State University of New York Press, Albany, New York.

Seamon, D. (1994) The life of the place: a phenomenological reading of Bill Hiller's space syntax, *Nordisk Arkitekturforskning / Nordic Journal of Architectural Research* 7, pp. 35-48.

Seamon, D. (2000) 'A way of seeing people and place: phenomenology in environment-behavior research' in Wapner, S., Demick, J., Yamamoto, T. and Minami, H. (eds.) *Theoretical perspectives in environment-behavior research.* Plenum, New York, pp. 157-178.

Seamon, D. (2002) Physical comminglings: body, habit, and space transformed into place, *The Occupational Therapy Journal of Research* 22, pp. 42S-51S.

Seamon, D. and Mugerauer, R. (eds.) (1985) *Dwelling, place and environment: towards a phenomenology of person and world.* Columbia University Press, New York.

Seamon, D. and Nordin, C. (1980) Market place as place ballet: a Swedish example, *Landscape* 24, pp. 35-41.

Stefanovic, I. (2000) *Safeguarding our common future: rethinking sustainable development.* State University of New York Press, Albany, New York.

Chapter 8

Rent Rhythm in the Flamenco of Urban Change

Eric Clark

Rhythms: music of the City, a picture which listens to itself, image in the present of a continuous sum. Rhythms perceived from the invisible window, pierced in the wall of the façade ... but beside the other windows, it too is also within a rhythm which escapes it. *Henri Lefebvre*

Rhythm, it seems, marks all manifestations of life. There are all the obvious rhythms of daily life, the familiar 'place ballets', often hidden under the cloak of habit and conformity. There are rhythms at vast varieties of geographical and temporal scales, and of distinct yet emergent nature: geophysical, biological, social, cultural.[1] As a universal that stretches over scales and boundaries, physically given and socially constructed, time-space provides a measure we readily use to grasp reality, leaping between and connecting parts of complex wholes. Havelock Ellis (1929, 35) argued that 'Dancing and building are the two primary and essential arts', all other arts flowing from them. That the two connect, and that building is not without its own sets of rhythms, is captured in Goethe's felicitous metaphor of architecture as 'petrified music' (Eckermann 1998, 303). At greater time scales, this music is not at all petrified, but in movement, punctuated by the rhythmic flux and flow[2] of capital investment and disinvestment in the built environment, a movement conducted by the rhythms of land rents.

As a regulator of flows of labor and capital applied to land, in agriculture and in building, land rent has been a central concept of political economy from its beginnings over three centuries ago. Intellectual endeavors since Sir William Petty's (1662) effort 'to explain the mysterious nature of [Rents]' have probed the forces at play. Debates on how best to conceptualize the origin, anatomy and consequences of land rent nevertheless continue to flourish. The literature is vast, generated by numerous issues that continue to divide perspectives. The position assumed here views land rent as a mechanism connecting societal rhythms at various time scales, from daily rhythms geographically orchestrated and manifested in land-use patterns, to the long-term cycles of urban development and redevelopment.[3]

In earlier studies, I have together with Anders Gullberg analyzed how land rents relate to rhythms of capital investment in the built environment (primarily construction activity and transportation infrastructure) and urban development regimes in the specific case of Stockholm City's post-war redevelopment (Clark and

Gullberg 1991; 1997; Clark 1995a). Gullberg's (2001) monumental *City: drömmen om ett nytt hjärta* (*City: the Dream of a New Heart*) provides an impressively broad and detailed history of these relations. What interests me in the following analysis is a very specific issue, one peculiarly neglected in the literature. The issue is: how does the interplay between differential rent I and differential rent II, initially analyzed by Marx and much elaborated upon in the 1970s and 1980s, stand in relation to rent gaps?[4] Use of these esoteric terms is unfortunately necessary to formulate the issue briefly. The concepts will be explained in the following effort to show how they relate – to each other and to societal rhythms of the built environment.

A common distinction in rent theory dating back at least to Thünen's *Isolated State* (1966 [1826, 1842]) is that between composite rent (sometimes called estate rent or building rent, which includes payment for fixed capital) and the 'pure' rent of land. Pure rent is an abstraction still commonly used among economists. Messy material and monetary flows do not fall simply or neatly into our theoretical distinctions, not even very broad ones like land, capital and labor. Thünen noted 'how difficult it is to work out the land rent of any given farm' and that 'nearly every such attempt has miserably failed in practice' (1966, 212). The same could be said today. Operationalizations are tricky and necessarily laden with inadequacies.[5]

I mention this as a caveat. Flaubert (1982 [1856], 218) observed that human speech is 'like a cracked kettle on which we tap crude rhythms for bears to dance to, while we long to make music that will melt the stars.' From the bustling bear dances of city life, rent theory seeks to spin pure harmonies.[6]

To illustrate the crude rhythms of land rents I present an historical example, unique yet not without similarities to the histories of urban blocks everywhere. The interplay of differential rents and the rent gap will be presented separately, before being compared. In the end, I aim to show how they are complementary in the sense of marking distinct yet imbricate rhythms of the same slow dance of urban transformation.

Kvarteret Stralsund in Malmö

On the east side of the Old Town in Malmö lies an unassuming seven-storey red brick building covering all but a corner of the block of Stralsund.[7] The land underneath it was marshland outside the town wall through the 1700s. Demolition of fortifications to the south and east and construction of new canals during the first decade of the 1800s brought the 'New Town' into seamless unity with the Old Town, nearly doubling the area of the town. (A century later, the surgery forgotten, 'Old Town' came to designate the entire area within the canals, including what had previously been called the New Town.) The Stralsund block was originally developed in 1813 for commercial gardening, supplying the town with fresh produce for over four decades. Rapid urbanization during the mid 1800s, with Malmö's population nearly doubling between 1840 and 1860, gave rise to speculative development 'under forms which inevitably bring to mind similar phenomena in the New World during the worst gold fever' (Bager 1954, 35). Gold fever reached Stralsund, then one of the last white spots on the map of the town within the bridges, in 1857. The land was subdivided into 43 properties in two blocks, Rostock and Stralsund, separated by a new street, Little Garden Street. During the 1860s and 1870s the area was developed for a combination of residential, workshop and retail uses. Shoemakers, tailors, painters, carpenters, blacksmiths and petty dealers lived and worked here in small one- or two-storey buildings (see Figure 8.1).

Figure 8.1 Lilla Trädgårdsgatan 1961. Looking east, with Stralsund on the right and Rostock on the left
Source: Adapted from Clark 1987b.

Land use was predominantly residential, the residents predominantly newcomers to Malmö with manual occupations. One or two rooms per dwelling was the rule. Streetfront stores and workshops in the back were important elements of land use. The residential profile and land use mix changed little over the years, though occupations changed with technological and economic development, as did commodities produced and traded in the shops and stores. This was a typical 'småfolksstråk,' similar to many others in rapidly developed outskirts of nineteenth century Swedish towns, where simple ordinary people lived and worked.

Malmö continued to experience rapid urbanization, the blocks of Rostock and Stralsund slowly turning from peripheral 'småfolksstråk' to inner city low-rent area housing small households (typically pensioners and young singles) and small businesses. When property capital began systematically purchasing properties in the blocks during the 1950s with an eye towards redevelopment, the population of Malmö had increased more than tenfold since their development in the 1850s and 1860s. Properties changed hands frequently between 1954 and 1967, as the number of property owners fell from 32 to two. In the early 1970s, the two blocks again became one – Stralsund – and Little Garden Street disappeared under a seven storey apartment building with cellar parking and ground floor office and retail space (see

Figure 8.2; note that 8.1 and 8.2 were taken from nearly identical perspectives). Similar redevelopments took place during the 1960s and 70s in many neighboring blocks on the east side of the Old Town, and across the canal in Östra Förstaden (East Suburb).

Figure 8.2 Lilla Trädgårdsgatan 1987. View from same position as Figure 8.1
Source: Author. Adapted from Clark 1987b.

Interplay Between Differential Land Rents

Marx distinguished between four main types of land rent: monopoly rent, absolute rent, and two forms of differential rent. The bases of differential rents are differences in relative fertility, situation, or intensity of investment. Central to Marx's analysis of differential rents is the idea that there is a systematic interplay between two types of differential rent such that 'while the first is the basis of the second', they 'at the same time place limits on one another' (Marx 1981, 871).[8]

Assume that the application of capital to land is equal across an isolated region. The differential land rent accruing to different sites will be due to relative advantages in fertility or location. Marx called this differential rent I (DR I). It is simply based on the application of equal capital to unequal lands. Turn this around and assume unequal application of capital to lands of equal relative advantage in terms of fertility or favorableness of location (i.e. lands of equal DR I). Investments in new technology or increased intensity of capital per square meter enables greater yield (surplus value) per square meter. Part of this surplus value can be captured by landowners. The resulting differences in land rent will be due to these differences in yields associated with differences in capital investment. Marx called this differential rent II (DR II). It is simply based on the application of unequal capital to equal lands. While DR I is based on differences in the land which are relatively stable over the long run, DR II is based on differences in capital investment which may be highly transient.

In reality of course we find neither equal capital applied to unequal lands nor unequal capital applied to equal lands, but rather the application of unequal capital to unequal lands, and various concrete expressions of a single undivided land rent. Nevertheless, we can analytically distinguish between the two forms of rent, and their significance for urban space economies is not diminished by difficulties of measurement.[9]

How do these two forms of differential rent form the basis of and set limits on one another? A key to answering this question, and to understanding ties to rent gap theory, is the notion of 'normal' capital investment. For any given site at any given time there is a 'norm' regarding how much capital is commonly invested. This is not a norm in the sense of social mores. Normal in this context is regulated by given levels of technological development and societal pressures to maximize profit. For instance, at one site, given the surrounding uses and surrounding investments, a normal capital investment may be, say, fifty thousand dollars in a single family home. At another site this would be an absurd investment because given its surrounding uses and built environment, existing technologies and a capitalist economy, 'normal' capital investment is rather fifty million dollars in a tall office building. 'The concept of "normal" capital becomes as variegated as the variegated fertilities [and variegated locations in relational space] to which that capital is applied' (Harvey 1982, 356).

The interplay between DR I and DR II reflects and regulates the flow of capital onto land. DR I is the basis for DR II in the sense that DR II arises through deviation from normal capital investments which historically establish DR I (even at the tricky margin, where DR I = 0). By venturing to invest in new technology or above normal capital, greater yields may be gained, some of which landowners can lay claim to in the form of DR II.

DR I sets limits on DR II in the sense that the additional above normal capital employed at one site, the basis of DR II, will if successful gradually become the norm for land of similar fertility/location. In a process generated by profit-maximizing behavior and competition, transient DR II turns into DR I. DR II sets limits on DR I in the sense that above-normal investments of capital result in higher yields and lower prices, throwing marginal land out of use and thereby diminishing DR I. To the extent this occurs, DR II is appropriated not in addition to but at the cost of DR I.

Let us return to the Stralsund block in Malmö and relate it to other sites in Malmö. First, let us compare with a more peripheral site, farther out in Östra Förstaden. Let us assume that equal capital is applied to these sites during the short phase of development in the 1860s. Any difference in land rent will be due to the difference in locational characteristics of the sites. This is DR I. The relatively central Stralsund will have a DR I higher than the site in Östra Förstaden, but lower than a site adjacent to Stortorget (Central Square). It is however not very likely that capital investment in buildings at Stortorget will be equal to that in Stralsund or in Östra Förstaden. 'Normal' capital investment on central sites is generally much greater than on peripheral sites.

Now let us switch to compare Stralsund with another site with 'equal' locational characteristics. Of course, no two sites are identical, so by 'equal' is meant equal DR I. With equal DR I, it can be assumed that 'normal' capital investment for the sites will be equal. However, under pressure to maximize profits and utilize new technologies, some developers will invest more capital than is normal in similar conditions. Now, let us assume that investment in Stralsund exceeds 'normal' capital investment for these and similar sites. The resulting difference in land rent between Stralsund and other sites with equal DR I is due to the additional intensity of capital investment on the land. This is DR II. If successful, above normal investment will jack up the 'normal' capital investment for similar sites.

Rent Gaps

The term rent gap denotes a disparity between actual land rent and potential land rent (Smith 1979; see Figure 8.3). The potential land rent of a site is determined solely by the site's 'highest and best' use, while the actual land rent of a site is a function of also the site's current intensity and type of land use.[10] Development on the site will involve an intensity of fixed capital investment designed to accommodate the site's 'highest and best' use, i.e. will be appropriate for the procurement of potential land rent at that point in time. Thus, actual land rent will equal potential land rent as 'the full resources of the site' are developed (Marshall 1961, 797). In the course of time, surrounding conditions may change, allowing for a higher and better possible use of the site, while the existing building fixed to the site constitutes an element of inertia for adaptation to a higher and better use. In this way, the building may come to 'no longer correspond to the changed circumstances' (Engels 1975, 20), as urban growth pushes up the site's potential land rent to a level corresponding to a greater intensity of capital investment and/or a 'higher' type of use. A rent gap arises, the expansion of which acts as an incentive for the property owner to disinvest. Consequential neglect of repair and maintenance contributes to the pace of building depreciation, which in turn influences actual land rent negatively as use of the existing structure shifts to 'lower uses' (e.g. filtration in the housing market). If the rent gap continues to expand, as the actual land rent associated with current use becomes increasingly removed from the potential land rent associated with the site's changing 'highest and best' use, the property will eventually be considered an interesting object of redevelopment by property capital agents. Speculation on future land rent income sets in and there occurs an upswing in capitalized land rent (and in its corresponding imputed contribution to annual land rent) during the years prior to redevelopment. Redevelopment may take the form of demolition and new construction or renovation and improvement of the existing building.

In a nutshell, the rent gap constitutes initially an economic pressure to disinvest in the fixed capital on a site, which consequently becomes increasingly inappropriate to the site's 'highest and best' use, and eventually an economic pressure to redevelop the site in order to accommodate higher intensity and type of use.

Fundamental to the rent gap is the condition that investments in the built environment involve a spatial 'fix' (Harvey 1982, 1985). In urban contexts, capital investments on land are generally several times the value of the land. Though adaptations in the form of additional building investment (e.g. adding on floors or annexes) or upward shifts in use (e.g. conversion to office space) are not uncommon, the original investment *tends* to lock the site into a given range of intensity and type of use for the duration of the economic life of the building. It is due to the sheer size of building investments, the durability of buildings, and primarily the interest of financers to harvest returns on investment that we do not experience instantaneous and continuous adaptation in the urban space economy to every small change in potential land rents. 'Sunk costs' call for return, and this is a powerful force of inertia in the built environment (Clark and Wrigley 1995, 1997). The dream world of constant equilibrium in the space economy through instantaneous responses to every small shift away from equilibrium corresponds to a nightmare of chaos in the built environment resembling what Harvey aptly describes as a 'frenetic game of musical chairs' (1982, 393). In the real world, the game of musical chairs in the built environment is much slower and less continuous, with local bursts of sweeping change rather than constant smooth progressions of marginal change.

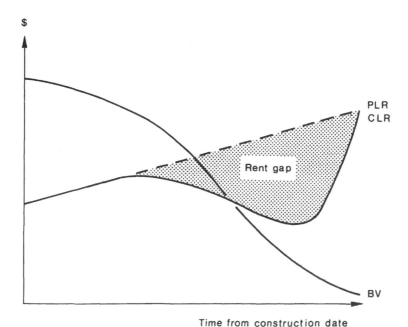

Figure 8.3 Modified conceptual outline of a development cycle, given increasing potential land rent (PLR)
Source: Adapted from Clark 1987b.

The closest real cities come to approximating the hypothetical extreme of near-instantaneous adjustments is during periods of very rapid urbanization. In Chicago, for example, during a period of exceptionally rapid growth, Hoyt could note that 'thirteen-storey skyscrapers with a structural life of a century or more have been torn down to give room for twenty-two- or forty-four-storey tower buildings' (1933, 335). Rapid urban growth, like rapid technological change, means that profit-maximizing behavior requires the destruction of recently invested capital. It also underlies rapidly increasing potential rents. Nevertheless, the need to recover sunk costs in the form of building investments entails inertia, however diminished by rapid growth, and this spatial fix constitutes a real basis for the identification of actual land rent as distinguished from potential land rent.

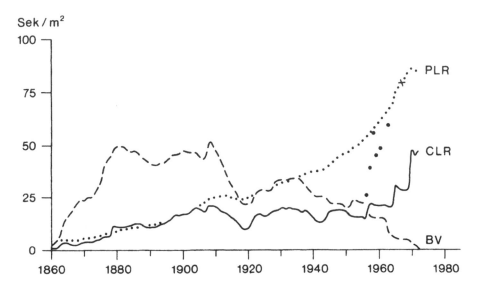

Figure 8.4 Building value (BV), capitalized land rent (CLR), and potential land rent (PLR) of Stralsund 33. The X on the potential land rent curve marks the value given by sale prices, used as target in generating the curve. The thick dots mark the values of [(sale prices-building values)/sq m sold], for the years 1956-1962

Source: Adapted from Clark 1987b.

The history of the Stralsund block in Malmö fits well the rent gap view of urban change. Compare the empirical data in Figure 8.4 with the simplified schema of Figure 8.3.[11] Development of the site in the 1860s from agricultural use to a mix of residential and business uses involved capital investments which put the land to its (then) 'highest and best use.' As Malmö grew – in population, geographically and economically – the 'highest and best use' shifted along with the site's potential land rent, while the sunk costs of fixed capital on the site locked it into a lower intensity and type of use, such that the actually capitalized land rent remained at low levels. Additional investments in the first decades adapted the site to its changing social and economic surroundings, but roughly fifty years after development, a rent gap appeared and expanded over the following decades. As the rent gap grew it became rational for property owners to disinvest in the existing building stock. Large property companies began purchasing properties in the block during the 1950s, when speculation pushed capitalized land rents upwards again. Consolidation of ownership enabled a large-scale and 'rational' redevelopment of the site in the early 1970s. Capitalized land rent and potential land rent were again at par and a new cycle began.

A Long Slow Flamenco

Flamenco is a rhythmically complex and compound art form. There are many types and styles which vary in mode, flavor and complexity, but they all build upon multiple layers of intricate rhythm. The compás (basic rhythms) and contratiempos (counter rhythms) of cante (song) and toque (guitar), palmas (hand clapping) and pitos (finger snaps), cajon (rhythm box, percussion instrument) and palo (stick, beat on the ground), jaleos (shouts of encouragement) and castanets, and of course baile (dance), an essential part of which is zapateado (footwork producing elaborate rhythms), weave together distinct rhythms into a whole performance. I think flamenco provides an apt metaphor for the ties between rent gaps and the interplay of differential rents, and between them and other urban rhythms at various temporal scales, because elements of each layer of rhythm are simultaneously constitutive of other layers of rhythm.

There is a pulse in the interplay between DR I and DR II, and a cadence in the playing out of rent gaps, and elements of both rhythms constitute beats in the other. Both focus on the flux and flow of capital onto land, so it is not surprising to find this to be the connecting strand. The relevance of the interplay between DR I and DR II hinges on the distinction between 'normal' capital investment on 'equal' sites and above normal investment of capital. The relevance of rent gap theory hinges on the distinction between capital investment appropriate for a site's 'highest and best use' and historically dated investments which have become below normal investment for 'equal' sites. These are different distinctions, but distinctions which resonate.

The two notions emphasize different comparisons indicative of different rhythms. The interplay between DR I and DR II emphasizes a *synchronic comparison across space of differences in capital investment*, especially in terms of normal and above normal. In so doing, it allows us to see how above normal investments on some sites influence not only the land rent of those sites, but also the land rents of other sites, either close by or otherwise 'equal' sites in the urban region. It also brings our attention to the tendency of above normal capital investments to become normal, and eventually below normal. The notion of rent gap emphasizes a *syntopic comparison across time, of differences in actual and potential land rent* which correspond to different types and volumes of capital investment. In so doing, it allows us to see a tension which constitutes a pressure to change capital fixed to a specific site.

Of course, we cannot conceive of a rent gap without there being comparisons across space with neighboring land and fixed capital. Indeed, potential land rent rests upon just such comparisons in relational space. Nevertheless, the very notion of rent gap emphasizes the view from a single property over time. Likewise, we cannot conceive of interplay between DR I and DR II without developments over time. And yet, the very notion of differential rent is based on the view of synchronic comparison between properties in terms of relative fertility and location and in terms of level of capital investment.

Let us return to Stralsund one last time. The history of Stralsund can be seen as a very long and slow polyrhythmic flamenco dance, a couple of the numerous intertwining rhythms being the interplay of differential rents and the development and closure of rent gaps. At the time of development in the 1860s, capital investments were at least 'normal' for sites of this relative location. The 'full resources' of the site were developed securing actual rents on par with potential rents. Some properties were developed more intensively than others (note the three-storey building on the left in Figure 8.1). Perhaps the work of craftsmen in the area employed new technologies, resulting in higher yields. These above normal investments are the basis for landowners in Stralsund to capture increased rents in the form of DR II. In the following decades, the city of Malmö grew, and what was in the 1860s above normal capital investment on such sites became rather the norm. The DR II appropriated by Stralsund landowners turns into DR I. Continued urban growth eventually leads to a situation where the capital fixed to the land, originally above normal, has not only become normal, but is now below normal. This influences actual land rents negatively, and a rent gap arises. Experience from other sites of similar locational characteristics indicates that the potential land rent of Stralsund is considerably higher, but would require entirely different capital investments. Property agents begin to buy in to the area, with an eye toward speculative gains in the near future in association with redevelopment. When the land is finally redeveloped in the early 1970s, the intensity of capital investment (now seven-storey apartment building with cellar parking and ground floor office and retail space) exceeds the norm for 'equal' sites, and again, DR II can, temporarily, be appropriated ... until this level of capital investment – under the incessant pressure of the capitalist space economy to secure profit – becomes the new norm ... and the dance continues.

Coda

So what? Understanding these rhythms is all good and well, but what if we would rather do another dance altogether? These rhythms lead to violent conflict in many cities (Smith 1996). Can we as researchers just turn aside and avoid issues of values in geography (Buttimer 1974)? Maybe we need 'to participate far more courageously in the dynamic' (Buttimer 1985, 315). I believe a comparative analysis aimed at understanding why this long slow flamenco dance turns into tumult in some places and not in others would find two key factors to be degree of social polarization and practices surrounding property rights. In places characterized by a high degree of social polarization, short on legally practiced recognition of the rights of users of place and long on legally practiced recognition of the rights of owners of space, the conflict inherent in these rhythms becomes inflammatory (cf. Blomley 1997, 2002). Not so in places characterized by relative equality and legally practiced recognition of the rights of users of place. If so, this indicates a direction for courageous engagement. Conflicts arise between interests associated with linear rhythms of 'consecutiveness and reproduction of the same phenomena' (the users of place, seeking continuity in place) and interests associated with cyclical 'rhythms of new beginnings' (owners of space, seeking new profit), as rents flow through the circuit of built environment. The 'essential and determinant factor is money' argued Lefebvre, and concluded: 'When relations of power take over relations of alliance, when the rhythms of "the other" make impossible the rhythms of "the self," then a total crisis explodes, with the deregulation of all compromises, arhythmy, implosion-explosion of the city' (Lefebvre 1996, 231, 225, and 239).

Visiting Malmö, Neil Smith asked me to show him the battlefields of gentrification. Perplexed, I did not know how to explain that there were processes of gentrification in Malmö, but no battlefields. Conflicting interests, displacement, personal tragedies, yes, but not the desperation behind battlefields. The cumulative outcome of political and legal battles in Sweden during the twentieth century set the stage for less violent ways of dealing with inherently conflictual processes of urban change. I believe it is fair and accurate to say this is changing, with increasing polarization and decreasing concern for the rights of users of place. Perhaps there will in the foreseeable future be gentrification battlefields also in Sweden. That depends on our willingness to face up to the 'faces of oppression' (Young 1990; cf. Harvey 1993), to develop relations of alliance between the interests of linear and cyclical rhythms. It depends on our willingness to participate far more courageously in the dance of building cities.

Notes

Cecilia Clark, Ewa Flycht, Anders Gullberg and Neil Smith provided necessary inspiration. Trevor Barnes, Tom Clark, Dan Hammel, Anders Lund Hansen, Yahia Mahmoud, Tom Mels, Lars-Olof Olander, and Margareta Rämgård-Hirdman gave valuable critique of various drafts. Some of that critique may still be relevant, but I gratefully acknowledge their showing me where more work needed to be done.

1   On societal rhythms, see Young and Schuller (1988). On place ballets, see Seamon (1979). On hidden rhythms, see Zerubavel (1981). There is a large literature on social rhythm disorders and therapy, much of which links them to biological circadian rhythms, which are in turn based on geophysical conditions (e.g. Frank, Swartz and Kupfer 2000). On a different time scale (18-19 year cycles), Berry (2000) links rhythms of inflation and economic growth with geophysical rhythms. At yet a greater time scale, Devezas and Corredine (2001) link long waves (Kondratieff waves) with biologically determined conditions of 'generational learning.'

2   The etymological root of rhythm, the Indo-European sreu-, means to flow (*American Heritage Dictionary* 1976, 1542).

3   See Clark (1987b) for a review of the history of land rent theory. The quote is from Petty (1662), chapter 4 paragraph 12.

4   I have suggested and alluded to the connection earlier, but the only analysis going beyond the mention of connection is that of Sheppard and Barnes (1990, 259), in which the rent gap is viewed from the perspective of intra-class conflict over differential rent.

5   That we need operationalizations connecting the abstractions of theory with the concrete terms of observation bears witness to the ontological depth of reality, which it is the purpose of abstract analyses to penetrate. It is easy to fall for the empiricist fallacy of equating an abstract concept with a naively observable event, often imbedded in a plea to demystify: abstractions are after all so complicated; let's stick to bare facts! But efforts to work with a 'flat ontology', for instance by reducing rent to observational contract rent, fail to explain differences in rent or how rent influences other phenomena. If rent is nothing but payment to a landowner, Thünen's statement makes no sense. See Clark 1987a and 1995b for critiques of recent examples of empiricist notions of rent.

6   Given the context of this book, I cannot help but let the non-Swedish speaking readers in on a playful twist here: in the language Anne Buttimer speaks at home, 'rent' means clean or pure. Etymologically, the Swedish and English words are not related, though they appear to become 'false friends' in the context of theory on 'pure rent.'

7   Malmö is Sweden's third largest city, located across the Sound from Copenhagen. Greater Malmö has a population of ca 500,000, roughly half of whom live in the city of Malmö. City blocks are commonly given names in Sweden. In public records, properties are referred to primarily by the name of the block followed by a number, and only secondarily by street address.

8   Pivotal literature is Marx (1968, 1981), Ball (1977), Fine (1979), Harvey (1982) and Sheppard and Barnes (1990). Hall (1966) identifies a similar interplay between Thünen's concepts of extensity rent and intensity rent.

9   The importance of differential land rent in the space economy is highlighted by historical instances where it has been more or less eliminated politically and judicially. In the Soviet Union, for instance, the role of differential land rent as regulatory and distributive mechanism was replaced by a bureaucratic apparatus, which provided for the exaggerated 'needs' of various administrations and ministries for land, without any pressure to use or manage the land. The consequences were urban sprawl, extensive stretches of unused land centrally located amidst dense built environments, the world's widest railroad corridors, and other manifestations of what one critic called 'a disgraceful attitude toward land' (A *Kommunist* Roundtable 1990, 22). While this flattened the geography of exchange value for land, it did little to change the use values of locational advantage. Thus, differential rents may have become marginal in terms of monetary flow, but they did not so much disappear as turn into consumer and producer surpluses, occasionally valorized through black markets.

10   In the following I will use actual land rent and capitalized land rent interchangeably since the analytical distinction between them has little bearing on the connections I want to draw. Actual land rent refers to the value of use of a site for a given period (for instance, annual) given the current intensity and type of use. Capitalized land rent refers to the present value of all future annual land rents given the limits on intensity and type of use imposed by the current building, but allowing for speculation on redevelopment, involving change in intensity and type of use (Table 8.1, below).

11   Operationalizations and research design underlying Figure 8.4 are summarized in Clark 1988. For more detail, see Clark 1987b.

Table 8.1 Four abstract concepts of land rent distinguished in terms of time and (intensity and type of) use, and two concrete categories of cash flow distinguished in terms of time

	Periodic	Infinite
Abstract concepts		
Current use	Actual land rent	Capitalized land rent
'Highest best' use	Potential land rent (p)	Potential land rent (i)
Concrete concepts		
Naively observable	Contract rent	Land price

Source: Adapted from Clark 1995b.

References

A *Kommunist* Roundtable (1990) You can't outsmart life: a discussion of the proposed law on property in the USSR, *Problems of Economics* 33, pp. 6-30.

Bager, E. (1954) 'Nya Staden' och 'Gamla Staden', *Malmö Fornminnesförenings Årsskrift* 22, pp. 33-54.

Ball, M. (1977) Differential rent and the role of landed property, *International Journal of Urban and Regional Research* 1, pp. 380-403.

Berry, B. (2000) A pacemaker for the long wave, *Technological Forecasting and Social Change* 63, pp. 1-23.

Blomley, N. (1997) The properties of space: history, geography, and gentrification, *Urban Geography* 18 286-295

Blomley, N. (2002) Mud for the land *Public Culture* 14, pp. 557-582.

Buttimer, A. (1974) Values in geography, *Commission of College Geography Resource Paper* 24, Association of American Geographers, Washington, DC.

Buttimer, A. (1985) Farmers, fishermen, gypsies, guests: who identifies?, *Pacific Viewpoint* 26, pp. 280-315.

Clark, E. (1987a) A critical note on Ball's reformulation of the role of urban land rent, *Environment and Planning A* 19, pp. 263-267.

Clark, E. (1987b) *The rent gap and urban change: case studies in Malmö 1860-1985.* Lund University Press, Lund.

Clark, E. (1988) The rent gap and transformation of the built environment: case studies in Malmö 1860-1985, *Geografiska Annaler B* 70, pp. 241-254.

Clark, E. (1995a) An arrow-cycle interpretation of the redevelopment of Stockholm's central business district, *Danish Journal of Geography* 95, pp. 83-91.

Clark, E. (1995b) The rent gap re-examined, *Urban Studies* 32, pp. 1489-1503.

Clark, E. and Gullberg, A. (1991) Long swings, rent gaps and structures of building provision: the postwar transformation of Stockholm's inner city, *International Journal of Urban and Regional Research* 15, pp. 492-504.

Clark, E. and Gullberg, A. (1997) 'Power struggles in the making and taking of rent gaps: the transformation of Stockholm City' in Källtorp, O., et al (eds.) *Cities in transformation: transformation in cities.* Avebury, Aldershot, pp. 248-265.

Clark, G. and Wrigley, N. (1995) Sunk costs – a framework for economic geography, *Transactions of the Institute of British Geographers* 20, pp. 204-223.

Clark, G. and Wrigley. N. (1997) Exit, the firm and sunk costs: reconceptualizing the corporate geography of disinvestment and plant closure, *Progress in Human Geography* 21, pp. 338-358.

Devezas, T. and Corredine, J. (2001) The biological determinants of long-wave behavior in socioeconomic growth and development, *Technological Forecasting and Social Change* 68, pp. 1-57.

Eckermann, J.P. (1998 [1836, 1930]) *Conversations of Goethe.* Da Capo, Cambridge MA.

Ellis, H. (1929 [1923]) *The dance of life.* Random House, New York.

Engels, F. (1975) *The housing question.* Progress, Moscow.

Fine, B. (1979) On Marx's theory of agricultural rent, *Economy and Society* 8, pp. 241-278.

Flaubert, G. (1982 [1856]) *Madame Bovary.* Random House, New York.

Frank, E., Swartz, H. and Kupfer, D. (2000) Interpersonal and social rhythm therapy: managing the chaos of bipolar disorder, *Biological Psychiatry* 48, pp. 593-604.

Gullberg, A. (2001) *City: drömmen om ett hjärta. Moderniseringen av det centrala Stockholm 1951-1979.* Monografier utgivna av Stockholms Stad 113 (två band), Stockholmia förlag, Stockholm.

Hall, P. (1966) 'Introduction' in Thünen, J.H. *The isolated state and its relation to agriculture and national economy.* Pergamon, Oxford, xi-liv.

Harvey, D. (1982) *The limits to capital.* Blackwell, Oxford.

Harvey, D. (1985) *The urbanization of capital.* Blackwell, Oxford.

Harvey, D. (1993) Social justice, postmodernism and the city, *International Journal of Urban and Regional Research* 16, pp. 588-601.

Hoyt, H. (1933) *One hundred years of land values in Chicago.* University of Chicago Press, Chicago.

Lefebvre, H. (1996) 'Seen from the window' and 'Rhythmanalysis of Mediterranean cities' in Kofman, E. and Lebas, E. (eds.) *Writings on cities.* Blackwell, Oxford, pp. 219-240.

Marshall, A. (1961) *Principles of economics* (ninth variorum edition, volume one). Macmillan, London.

Marx, K. (1968) *Theories of surplus value* (part two). Progress, Moscow.

Marx, K. (1981) *Capital* (volume three). Penguin, Harmondsworth.

Petty, W. (1662) *A treatise of taxes and contributions.* London.

Seamon, D. (1979) *A geography of the life-world.* Croom Helm, London.

Sheppard, E. and Barnes, T. (1990) *The capitalist space economy: geographical analysis after Ricardo, Marx and Sraffa.* Unwin Hyman, London.

Smith, N. (1979) Toward a theory of gentrification: a back to the city movement by capital not people, *Journal of the American Planning Association* 45, pp. 538-548.

Smith, N. (1996) *The new urban frontier: gentrification and the revanchist city.* Routledge, London.

Thünen, J.H. (1966 [1826, 1842]) *The isolated state and its relation to agriculture and national economy.* Pergamon, Oxford.

Young, I.M. (1990) *Justice and the politics of difference.* Princeton University Press, Princeton.

Young, M. and Schuller, T. (eds.) (1988) *The rhythms of society.* Routledge, London.

Zerubavel, E. (1981) *Hidden rhythms: schedules and calenders in social life.* University of Chicago Press, Chicago.

Chapter 9

Time-Space Rhythms and Everyday Urban Life

Ann-Cathrine Åquist

Economic Geography and the Gender Division of Labor

The introductory chapter of the textbook *Introducing Human Geographies* (Cloke, Crang, and Goodwin 1999, x) includes two photographs, one showing men working at a conveyer-belt in an automobile factory and the other showing a woman vacuuming a home. The text under the photographs says 'Which of these photographs of work looks more like it should be in a Human Geography textbook to you? Why?' The authors argue that human geography has largely ignored women and that economic geography has paid little attention to women's unpaid housework. Feminist geographers first raised these issues in the 1980s. Unfortunately, feminist studies within human geography still tend to be pursued in isolation, with little effect on the discipline as a whole, and especially not on economic geography.[1] When women are included in economic geographic studies today, the focus is generally on labor market issues and paid work.

It is hardly surprising that economic geography does not deal with unpaid housework. The gender division of labor existing today goes far back in history. We tend to take this division of labor for granted; it seems 'natural.' Intertwined with the gender division of labor is the separation of the private and public spheres. The home, where unpaid housework is performed, belongs to the private sphere, while paid work, organized through the labor market, belongs to the public sphere. And economic geography, as well as economic theory in general, deals with work organized through the labor market and taking place in the public sphere.

The gender division of labor and the separation of the private and public spheres has a bearing on the production (and reproduction) of knowledge. The knowledge that we use (that is, reproduce) and produce is connected to the everyday world that we live in. This holds true for all kinds of knowledge, both everyday and scientific. The social world in which we live, the work we do in our everyday life, and our experiences affect our thinking and our use and production of knowledge. Our position in society gives us a certain perspective from which we know the world. These ideas are not new; they have been researched for a long time (Åquist 1995). In Donna Haraway's (1996) elaboration of this theme, she speaks of 'situated knowledges.' Male researchers tend to pose questions and pursue inquiries that have a bearing on men's experiences and men's everyday life. Men dominate research in economic geography. That is why it is not surprising that economic geography ignores women's unpaid work in the home.

At first sight, Richard Florida's book *The Rise of the Creative Class* seems to be an exception. The cover of the book claims that the creative class 'is transforming work, leisure, community and everyday life.' Florida thus widens the perspective of economic geography beyond paid work to encompass all the different spheres of an individual's everyday life. However, creativity is the dominant element in his thinking. Creativity is one of the forces driving economic and societal change, and Florida argues that its importance is increasing:

> The Creative Class has shaped and will continue to shape deep and profound shifts in the ways we work, in our values and desires, and the very fabric of our everyday lives (Florida 2002a, ix).

The aim of this chapter is to discuss one aspect of Florida's interesting and thought-provoking theory – his representation of everyday life. While it is interesting that Florida deals with this topic at all, I find his understanding of everyday life wanting. Here, I will present a different understanding of everyday life, taking into consideration a gender perspective and the importance of time-space rhythms.

The Meaning of Creativity

It is not easy to define creativity. Florida quotes a dictionary, defining it as 'the ability to create meaningful new forms' (Florida 2002a, 5). Creativity involves the ability to synthesize, to browse through large amounts of material and come up with new and useful combinations. It also involves the willingness to take risks and requires self-assurance (ibid. 31). Creativity is not restricted to a few geniuses; it draws on the ordinary abilities of human beings. It usually evolves in a four-step process:

1. Preparation, which means conscious study of a task.
2. Incubation, 'the mystical step, is one in which both the conscious mind and the subconscious mull over the problem in hard-to-define ways' (ibid. 33).
3. Illumination, in which a solution suddenly appears.
4. Verification and revision, which means ordering and working though the idea and its consequences.

The creative class is defined as consisting of people whose own creativity is the most important tool in their work. Artists form a well-known group. Florida also includes people working in science, engineering, architecture, design, education, health care, law, and business. Even though factory workers are not included in the creative class, many factory jobs require some creative ability since many companies now have continuous-improvement programs on the factory floor (ibid. 10). Florida argues that currently as much as 30 percent of the workforce in the US belongs to the creative class. This marks a change from the first half of the twentieth century, when the working class was the biggest and most important class. It was replaced in the 1960s and the 1970s by the service class, researched by Daniel Bell and many others. Lately the creative class has risen to be the most important class in the contemporary economy. Florida argues that creativity is more characteristic of contemporary society than the often-cited notions of knowledge and information. In his view 'knowledge and information are the tools and materials of creativity' (ibid. 44).

The Importance of Place

Florida writes that he began to understand these changes as he studied regional economic development in order to find out why certain regions and cities grow and prosper while others lose jobs and migrants. He found that the old pattern of people moving to jobs was beginning to change. Now companies are moving to and developing in places where there is an abundance of skilled people. The next building block in his theory came when Florida learnt about the research of Gary Gates, who studied the location of gay people in the US. The location pattern of companies employing skilled people, especially high-tech industries, coincided to a large extent with places with the highest concentrations of gay people. Florida constructed a Bohemian index, measuring the concentrations of artists, writers, and performers in regions, and found that even the Bohemian index showed a strong correlation with high-tech industries. His conclusion was that

> economic growth was occurring in places that were tolerant, diverse and open to creativity – because these were places where creative people of *all* types wanted to live (Florida 2002a, x).

> I often tell business and political leaders that places need a people climate – or a creative climate – as well as a business climate (ibid. 7).

An atmosphere characterized by openness, tolerance, and diversity is important for the creative class and consequently for the economy of a city. Florida mentions the following cities as having this particular atmosphere: San Francisco, Boston, Seattle, Austin (Texas), Boulder (Colorado), Santa Fe (New Mexico), and Dublin in Ireland. Not every city or area with famous universities is attractive to the creative class; for example, the well-known North Carolina Research Triangle 'lacks the "hip" urban lifestyle,' and even Silicon Valley is losing its attraction (ibid. 284–285).

Transformations of Everyday Life

Along with these changes come transformations of everyday life. Florida explains these changes by taking his readers on a time-trip. He asks us to imagine 'a typical man on the street' (Florida 2002a, 1) from the year 1900 and take him to the year 1950. Then imagine another man, typical of the 1950s, and move him to the present day. Which would recognize himself most in the new environment? Florida argues that the man from 1900 would experience more conspicuous changes such as cars *en masse*, airplanes, the appearance of cities, supermarkets and the use of electric appliances such as radios, television sets, washing machines, and so on. But he would find everyday life much the same:

> Someone from the early 1900s would find the social world of the 1950s remarkably similar to his own. If he worked in a factory, he might find much the same division of labor, the same hierarchical systems of control. If he worked in an office, he would be immersed in the same bureaucracy, the same climb up the corporate ladder. He would come to work at eight or 9 each morning and leave promptly at 5, his life neatly segmented into compartments of home and work. He would wear a suit and tie ... He would seldom see women in the workplace, except as secretaries, and almost never interact professionally with someone of another race. He would marry young, have children ... probably work for the same company for the rest of his life (Florida 2002a, 2-3).

This is the image of the 'company man,' whose life is structured by an organization. It was much the same in 1950 as in 1900. The man from 1950 would find far more profound changes in everyday life if he were to visit the year 2000. At work much would be different: individuality and self-expression are preferred instead of conformity; more relaxed dress codes have replaced suits and ties; women and nonwhites can be managers; and people seemingly come and go as they please without scheduled working hours. There will be mixed-race couples and same-sex couples. According to Florida, these changes are caused by the increasing importance of creativity in the economy. Creativity comes in many forms and will thrive in environments characterized by diversity and tolerance. It cannot be turned on and off at will. That is why 'people would seem to be always working and yet never working when they are supposed to' (ibid. 3). People with creative jobs, like artists, have always chosen their own working hours; now this way of organizing time is spreading to new groups in the labor market.

> Job security is traded for autonomy. People favor opportunities to learn and grow at work, to be able to influence the content of work, and to control their own schedules. Companies have had to adapt to the changes and must try to create workplaces that are attractive to creative people (Florida 2002a, 13).

In an article with the same title as his book (*The Rise of the Creative Class*), but with the subtitle 'Why cities without gays and rock bands are losing the economic development race,' Florida reports a conversation with a young student at a university in Pittsburgh:

> I asked the young man with the spiked hair why he was going to a smaller city in the middle of Texas, a place with a small airport and no professional sports teams, without a major symphony, ballet, opera, or art museum comparable to Pittsburgh's. The company is excellent, he told me. There are also terrific people and the work is challenging. But the clincher, he said, is that, 'It's Austin!' There are lots of young people, he went on to explain, and a tremendous amount to do: a thriving music scene, ethnic and cultural diversity, fabulous outdoor recreation, and a great nightlife (Florida 2002b, 2).

To this young man it is not a problem to move down in the urban hierarchy to a smaller city in Texas as long as it has specific attractions for young people in creative jobs: other young people, a particular music scene and nightlife, diversity and outdoor recreation facilities.

Florida gives us a glimpse of his own life:

> I have a nice house with a nice kitchen but it's often mostly a fantasy kitchen – I eat out a lot, with 'servants' preparing my food and waiting on me. My house is clean, but I don't clean it, a housekeeper does. I also have a gardener and a pool service; and (when I take a taxi) a chauffeur. I have, in short, just about all the servants of an English lord except that they're not mine full-time and they don't live below stairs; they are part-time and distributed in the local area. . . . The woman who cleans my house is a gem: I trust her not only to clean but to rearrange and suggest ideas for redecorating (Florida 2002a, 76–77).

The form of everyday life represented in Florida's theory is the everyday life of wealthy bachelors. However, not everybody lives a bachelor's life. Society would fall apart if we all did. Florida mentions in his book that families live differently, but this point is not thematized and has no bearing on his theory. In what follows I will present a more elaborated understanding of everyday life.

Housework

Many of the tasks associated with housework are done in the home, in the private sphere. They are usually unpaid, unless done by servants or 'out-sourced.' As long as work is unpaid, it usually attracts no attention from research in economics. However, housework is to a large degree necessary work. Housework is *labor for physical survival* on a day-to-day basis, as well as from a generational perspective, relating to the natural time-space rhythms that find expression in our biological needs for such things as food and drink. Housework is also *labor for our well-being and comfort*. Further, it is *labor for cultural reproduction*, for example, in the upbringing and socialization of children, that takes place in the home.

In Florida's representation of the everyday life of the creative class, which he argues comprises about 30 per cent of the American workforce, little attention is paid to housework. This lack of attention can be compared to figures on the amount of time spent in housework. I know of no contemporary studies of this from the USA, and will instead rely on studies of how people in Sweden spend their time. The Swedish bureau of statistics has made several such surveys, the latest in 1990-1991 (Statistics Sweden 2000). Even if this survey is now some ten years old, it probably represents the current situation reasonably well.[2] The survey shows that women and men on average work equal number of hours per week, but that women spend more hours in unpaid work[3] and men more hours in paid work. On average, women spend a little more than 33 hours per week in unpaid work (and 27 hours in paid work); the figure for men is a little more than 20 hours in unpaid work (and 41 hours in paid work). The amount of time spent in unpaid work varies over the life cycle and with family situation. In families with young children, unpaid tasks take a lot of time. For example, among couples with children younger than seven years, women spend an average of 50 hours per week in unpaid work, while men in the same situation spend about 25 hours on this. (It is interesting to note that single mothers with children younger than seven years spend an average of 40 hours per week in unpaid work – about ten hours less than cohabiting women.) Children are not the only ones who depend on others for care; so do many elderly people. One question in the survey concerned whether the respondents had one or more persons living outside their own household who were dependent on their care. Among the women 14 per cent answered yes, and among the men nine per cent.

The unpaid work is important since it is labor for biological survival, for peoples' well-being and comfort, and for cultural reproduction. And it takes a lot of time. The survey of the use of time showed that people on average spend about 26 hours per week in unpaid work, not considering gender differences – as compared with about 34 hours in paid work. When Florida announces that the lifestyle of the creative class will bring a profound transformation in everyday life, he does not seriously take into account the unpaid work that is also an important part of everyday life.

Organization of Housework

Household tasks can be organized in various ways. In Sweden today housework is generally done by the members of the household, mainly by women, as unpaid work. In the case of Richard Florida, the work is done by paid 'servants,' spending a part of their working day in his home and other parts of their working day in his neighbor's homes. A negative aspect of this form of organization is that it requires a society with high differences in wages. In order to be able to pay servants a decent wage, you must have a rather high wage yourself. An interesting question is whether people working as servants earn enough to pay servants to do their housework. In *Nickel and Dimed: Undercover in Low-wage America* Barbara Ehrenreich (2002) reports on her investigation of the living conditions of those who live on very low wages. As part of her study, she took low-waged jobs herself for several months. In Portland, Maine, she was offered a cleaning job by Merry Maids, a company that hires people to clean other peoples' homes, at $5 to $6 per hour. She turned the offer down because it was impossible for her to support herself on that wage. She writes that according to the National Coalition for the Homeless, in 1998 it took, on average nationwide, an hourly wage of $8.89 to afford a one-bedroom apartment (Ehrenreich 2002, 3).

In her book *The Grand Domestic Revolution*, Dolores Hayden (1981) presents her research on 'the material feminists.' These were women in the USA in the latter half of the nineteenth century who wanted to get rid of their housework in order to get time for other activities. They had several ideas about how to organize housework other than as unpaid work in the home. One of them, Melusina Fay Peirce, argued for 'cooperative housekeeping.' Her idea was to bring housework out of the private sphere in the home into separate buildings where it could be performed by many women together, in combination with hired servants. That would make the work more efficient by giving advantages of scale and enabling efficient use of machines. She actually started such a cooperative in 1896 in Cambridge, Massachusetts, but had to close down after some time, according to Hayden because of resistance from husbands (Hayden 1981, 80). But the idea of cooperative housekeeping survived. Ebenezer Howard, who is famous for the idea of the garden city, became a spokesman for cooperative housekeeping. In Letchworth, the first garden city, Howard lived in a kitchenless apartment that was part of a complex in which housework was organized cooperatively (ibid. 231). This part of the story of the garden city is seldom told.

A Perspective on Everyday Life

Though largely ignored in economic geography, everyday life has long been a theme in philosophy and the social sciences. The German philosopher Edmund Husserl, who founded phenomenology, developed the notion of the lifeworld, which can be understood as emerging from an everyday life perspective. Every individual is the center of his or her own lifeworld, which consists of our immediate experiences and actions in the world, in a pre-conceptual sense. Anne Buttimer has elaborated the connections between phenomenology and human geography. In her essay 'Grasping the Dynamism of Lifeworld,' she writes about the lifeworld as the 'cultural defined spatiotemporal setting or horizon of everyday life' (1976, 277) and relates it to the human geographical concepts of a sense of place, social space, and time-space rhythms.

Another set of perspectives on everyday life derives from the women's movement in the 1960s and 1970s. This movement criticized the gender-blind mainstream social sciences for obscuring the reality of women's lives and experiences. 'The private is political' became a slogan. This theme was taken up in the emerging field of women's studies in the early 1970s. In the early days, feminist researchers often asserted that the aim of women's studies was to present women's reality and experiences. This theme has been elaborated by the Canadian sociologist Dorothy Smith (see below).

The Characteristics of Everyday Life

Everyday life pertains to the mundane and concrete activities that make up the daily time-space rhythms. The concept of everyday life has several dimensions and can be defined in various ways. Charlotte Bloch has defined it as the activities through which we recreate ourselves at the same time as we contribute to the recreation of society (Bloch 1991, 31).

One dimension of everyday life, mentioned by several researchers, is that we take our everyday lives for granted. Birte Bech-Jørgensen (1997) argues that our everyday activities to a large extent pass unnoticed. We do not usually direct our attention toward everyday life, as we tend to carry through everyday activities without giving them much thought. But we can easily transfer our attention to everyday life, making it the focus of reflection and discussion.

Another characteristic of everyday life is its routinization, as expressed in time-space rhythms. Buttimer argues for an understanding of human life based on phenomenological thinking and the concept of lifeworld. Time-space rhythms are an important aspect of the lifeworld, and they are based in both nature and culture. Anne Buttimer writes (see also the introduction to this volume):

Lifeworld experience could be described as the orchestration of various time-space rhythms: those of physiological and cultural dimensions of life, those of different work styles, and those of our physical and functional environments (Buttimer 1976, 289).

Some time-space rhythms emerge from nature, others from culture. Those emerging from nature have to do with biological body rhythms as well as physical environments. There are interesting connections between the two types of time-space rhythms. For example, research conducted by Marianne Frankenhauser shows that the level of stress hormones in men's bodies decreases in the late afternoon as they leave their workplaces in order to go home. By contrast, in women's bodies, the level of stress hormones remains high in the late afternoon as women leave their paid workplace and go home to their housework (quoted in Friberg 2002).

The time-space rhythms emerging from our culture are to a large extent set by the hours of paid work, which often dominate our daily program. Even those who are not engaged in paid work often have their daily programs set by working hours; for example, children have to adjust to their parents' schedules. Usually, we do much the same the things every day from Monday through Friday. Friday night often marks the start of another pattern, which dominates weekends. The routines of weekends are usually different from those of weekdays. Those who work irregular hours do not have the same patterns for their daily programs as those who work nine-to-five, Monday through Friday. Besides paid work, another factor that shapes everyday life is unpaid work such as cooking, cleaning, and doing laundry, as well as caring for children and others.

The time-space rhythms of daily routines give everyday life a certain time-organization, though these routines vary with the different phases of life. There are differences between the routines of small children, children going to school, teenagers, young adults living single lives, and so on. Everyday life also has a spatial organization. It is situated in space, in a specific place. Various activities are often localized to different places. Paid work is usually done somewhere other than our homes. Most children spend parts of their days in day-care centers or schools. The housing segregation of urban areas is another aspect of the spatial organization of everyday life. We live our daily lives in concrete spaces/places created in the interaction of the physical and the social.

An Illustration of the Time-Space Rhythms of Everyday Life

The time-space rhythms of everyday life can be illustrated by a study of housework that I undertook several years ago (Åquist 1984). The parents in twenty households with at least one child of pre-school age kept time-diaries for a week and were interviewed afterwards. My study confirmed the gender division of labor that is documented in official statistics. Women performed more unpaid work than men, although there were variations. In one household the woman was a full-time housewife and did all of the housework. At the other extreme was a household where the parents were sharing the period of parental leave, so that they both performed 50 per cent of the full-time paid work. That household had the most equal sharing of unpaid work, including childcare, in my study. Table 9.1 exemplifies the time-diaries for one day of each of the parents in one of the households. They have three children: one ten years old, another eight and the third two.

Table 9.1 Monday time-diaries of parents in a Swedish household with three young children. Mother working part-time in paid work (75 per cent), father fulltime.

Mother	7.30 a.m.	make breakfast, eat
	7.45 a.m.	load the washing machine
	8.30 a.m.	clear the table, load the dishwasher
	9.10 a.m.	take the youngest child to the day-care center
	9.40 a.m.	go to work
	3.00 p.m.	come home from work
	3.20 p.m.	drive the two oldest children to a piano lesson
	3.30 p.m.	go shopping for groceries
	4.10 p.m.	pick the children up after their piano lesson
	4.15 p.m.	drive to the day-care center and pick up the youngest child
	4.50 p.m.	come home
	4.55 p.m.	unload the washing machine, hang the laundry, reload the washing machine
	5.10 p.m.	prepare dinner
	5.20 p.m.	read to the youngest child
	6.00 p.m.	dinner
	6.30 p.m.	clear the table, load the dishwasher, make coffee
	8.00 p.m.	empty washing machine, hang the laundry
	8.30 p.m.	put the two oldest children to bed
	8.45 p.m.	work
	11.30 p.m.	pick things up and straighten things out
Father	7.15 a.m.	wake the children up, lay the table
	7.25 a.m.	shower
	7.45 a.m.	breakfast
	8.05 a.m.	go to work
	5.45 p.m.	come home from work
	5.55 p.m.	lay the table
	6.05 p.m.	dinner
	6.30 p.m.	clear the table
	6.35 p.m.	have coffee
	7.30 p.m.	put the youngest child to bed
	8.00 p.m.	work

Source: Adapted from Åquist 1984.

These time-diaries illustrate several issues in the time-space rhythms of everyday life. First, the frame imposed by the biological need for sleep, food, and drink is clearly shown. Secondly, the gender division of labor is illustrated. However, in general, this particular couple shares a little more equally than comes through on this particular day. They take turns taking the youngest child to the day-care center and on Saturdays both of them put in several hours of housework. The time-diaries also illustrate the pace of working hours. Both the woman and the man in this example have the option of bringing some work home. Both of them use that option so that they can come home earlier in the afternoon in order to spend more time with their children, illustrating the importance of the late afternoon and early evening hours for family life. Although not marked in this example, another factor influencing time-space rhythms is the hours at which institutions such as day-care centers, food stores, libraries, and dentists are open.

The Everyday World as Problematic

Dorothy Smith (1987) has formulated an interesting theoretical perspective on everyday life in her book *The Everyday World as Problematic*. Her work is inspired by phenomenology and Marxist theory. She argues for beginning inquiry outside and prior to textual discourses. One site for such a beginning is the everyday world. Smith argues that our world can be understood as containing two layers. One layer is the world of ruling, which is primarily a world of texts and discourses. It is generalized and without a specific localization. Mainstream social science, the mass media, and state and parliamentary politics belong to this layer. The other layer contains the everyday world. This world is particular and local, with every person at the center of her or his own everyday world. It is a world of concrete social practices and a basis for the world of ruling. It is like a 'service department' where human beings are reproduced. Smith gives an example of this divided world:

> When I went into the university or did my academic work at home, I entered a world organized textually . . . and organized to create a world of activity independent of the local and particular . . . But I went home or put down my books and papers to enter a different mode of being. I cleaned up after, fed, bedded down, played with, enjoyed, and got mad at two small children. I inhabited a local and particular world – the parks we would go to, the friends they had, my neighbors (ibid. 6).

Smith argues that modes of understanding are created in the world of ruling in the form of discourses and ideologies that become tools for understanding both layers of the world. Smith's project is to develop a way of creating knowledge from a different point of view, from the everyday world. Here, the world is known otherwise, it is directly felt, sensed, and responded to, outside discourses. Her approach is part of her feminist position. She argues that the world of ruling is dominated by men and men's perspectives. Women's work and activities are, to a greater extent than men's, contained within the everyday world. In order to make women's experiences and realities visible, the point of departure must be the everyday world. The aim is not merely to develop social scientific inquiries from women's perspectives but to address society and social relations from a standpoint outside the world of ruling. That is the standpoint where questions should originate, although it is not necessarily the level on which the answers will be found. In the world of ruling, the social and economic processes that constitute the conditions for the everyday world are controlled.

Conclusions

Richard Florida's theory of the increasing importance of creativity in economic change is interesting. One reason why it seems plausible is the increasing economic importance of the design of products, of brand names, and of logos. Florida argues that the importance of creativity has resulted in the rise of a new class with creativity as its main tool. This new creative class 'has shaped and will continue to shape deep and profound shifts in the ways we work, in our values and desires, and in the very fabric of our everyday lives' (Florida 2002a, ix).

However, I find his superficial understanding of everyday life problematic. His representation of everyday life relates only to paid work and to 'hanging out' with colleagues and friends at sports arenas and 'hip nightspots.' The members of the creative class are said to work long and irregular hours (which may eventually result in health problems since such irregularity interferes with biological rhythms). As the example of Florida's own everyday life shows, he seems to assume that housework can be 'out-sourced' to 'servants.' Taking into account how much time people currently spend on housework, cleaning, cooking, laundry, tending the home, caring for children and others, Florida's view is not realistic. Another unrealistic aspect of his reasoning is the idea that everyday life is structured only by the conditions and experiences of working life. Other structuring factors, such as the time-space rhythms emerging from biological needs and from responsibility for the well-being of children and others depending on our care, are left out.

There are other theories of everyday life that are much more elaborated and convincing, including Dorothy Smith's theory with its analytical distinction between the world of ruling and the everyday world. Adopting Smith's perspective, Florida's theory is contained within the world of ruling. It would be interesting to address the economic and societal changes caused by the increasing importance of creativity from the perspective of the everyday world, using the approach of Dorothy Smith.

Notes

1   There are some signs of change. Economic geography seems to be widening its focus to include issues like consumption and housework, as indicated in the reference to *Introducing Human Geographies* (Cloke, Crang, and Goodwin 1999).
2   In 1992 the Swedish economy went into a recession that lasted for most of the 1990s, and it has not yet fully recovered. Rates of unemployment were unusually high for several years in the 1990s. The percentage of women and men in the labor force did not increase during the 1990s. In 1999, 87 per cent of women aged 35 to 54 and 90 per cent of the men in the same age group were in the labor force.

3   Unpaid work includes tasks like cooking, washing dishes, doing laundry, ironing, cleaning, but also caring for children and other people in need of care, repairing the home and gardening.

References

Åquist, A-C. (1984) On changes in housework (unpublished manuscript).

Åquist, A-C. (1995) Kön, praktik och kunskapsproduktion (Gender, social practices and the production of knowledge), *Nordisk samhällsvetenskaplig tidskrift* 21, pp.41–49.

Bech-Jørgensen, B. (1997) 'Symbolsk orden og hverdagskultur' (Symbolic order and everyday culture). in Christensen, A-D., Ravn, A-B. and Rittenhofer, I. (eds.) *Det kønnede samfund. Forståelser af køn og social forandring.* Aalborg Universitetsforlag, Aalborg, pp. 109–135.

Bloch, C. (1991) I lust och nöd: om vardagsliv och känslor (For better or for worse: on everyday life and emotions), *Kvinnovetenskaplig tidskrift* 2, pp. 31–42.

Buttimer, A. (1976) Grasping the dynamism of lifeworld, *Annals of the Association of American Geographers* 66, pp. 277–292.

Cloke, P., Crang, P. and Goodwin, M. (eds.) (1999) *Introducing human geographies.* Arnold, London.

Ehrenreich, B. (2002) *Nickel and dimed: undercover in low-wage America.* Granta Books, London.

Florida, R. (2002a) *The rise of the creative class.* Basic Books, New York.

Florida, R. (2002b) The rise of the creative class: why cities without gays and rock bands are losing the economic development race, *The Washington Monthly* (http://www. washingtonmonthly.com/features/2001/0205.florida.html).

Friberg, T. (2002) 'Om konsten att foga samman: kvinnors förflyttningsprojekt i tid och rum' (On the art of bringing together: women's projects of movements in time and space) in Schough, K. (ed.) *Svensk kulturgeografi och feminism – rötter och rörelser i en rumslig disciplin.* Karlstad University Studies 2002 3, pp. 67–77.

Haraway, D. (1996) 'Situated knowledges: the science question in feminism and the privilege of partial perspective' in Agnew, J., Livingstone, D.N. and Rogers, A. (eds.) *Human geography: an essential anthology.* Blackwell, Oxford.

Hayden, D. (1981) *The grand domestic revolution.* MIT, Cambridge, Massachusetts.

Smith, D. (1987) *The everyday world as problematic: a feminist sociology.* Northeastern University Press, Boston.

Statistics Sweden (2000) *Women and men in Sweden 2000.*

PART IV
REANIMATING EMBODIED
LANDSCAPES

Chapter 10

Place and Identity: The Life of Marie de l'Incarnation (1599-1672)

Anne Godlewska

Introduction

This paper argues the importance of place in understanding the human experience. It contends, together with Malpas, Buttimer, Entrikin and a whole host of geographers, that place is integral to identity (Buttimer and Seamon 1980; Buttimer 1983; Entrikin 1991). Who you are, who I am, has everything to do with the places we have inhabited and inhabit now. In other places we would behave differently, we would think differently and we would probably conceive of ourselves differently. But place is no simple concept. It is not a matter of here or there or of Africa or Canada. If place is integral to identity so is identity integral to place. Place is a complex network of subjective experiences, objective projections, embodied limitations, social expectations, opportunities and forces, and physical forces. As Foucault, Malpas and many others have pointed out, the dichotomy between the world wholly within me and the world wholly outside me does not exist (Foucault 1978; 1979; Malpas 1999). Place and identity, then, are inseparable. If we wish to understand identity, then we must struggle to understand that complex network that is place. If we wish to understand place, then we must struggle to understand identity in all its complexity.

One of the dimensions of the complexity of place is its nested character, or as Merleau-Ponty and Foucault, following him, describes it: the folded character of place. I prefer to use Malpas' term, the nested nature of place. Perhaps others read Foucault differently but I find in Foucault, Deleuze's final explanations not withstanding, a hopelessness about human agency (Deleuze 1988, 94-123). For Foucault there seems to be no place within us that is personal, that is individual and that can resist power (Gutman 1988, 103; Hutton 1988, 135). Knowledge, thought, behavior, and subjectivity are all shaped by forces external to the individual. On one level I accept this. Much of our understanding, thinking, modes of behavior, etc. have been imposed on us. But we do struggle with the impositions, try to make sense of them, select and block influences, and constantly compose autobiographies around our struggles, both through the impositions and across them. And then we impose on others. Or join others in imposing on others. There is resistance and complicity in our identity, though it is often not clear to us when we are resisting and when we are complicitous. Our view of this at any given time depends on some sort of combination of recent experiences, what we have been reading or seeing, the state of our bodies, pressures imposed by society, whether it is raining or not and the current state of our autobiographical musings. Perhaps it is a matter of emphasis. Foucault believed that, as a result of some sort of a will to power and control, humans have consistently created forms that imprison human creativity. Yet he also believed that in the way we live and through the informed choices we make we are capable of creating meanings that can reconstitute human nature.

It is this battle of resistance and complicity, this striving for both creativity and power and control that is such a fascinating and inscrutable aspect of human identity. While, for many, geography is consigned to the realm of maps, mountains and measurements, the true realm of the geographer is place – in all its subjective, representational, corporeal, social and physical complexity. This paper explores the relationship between identity and place in the life of Marie Guyart (who became Marie de l'Incarnation upon becoming an Ursuline nun). It explores her strategies to create meaning and freedom for herself within the constraints of seventeenth-century Catholicism. These involved manipulating one of the most intimate dimensions of place, the body, in order to transform identity and through that transformation to create for herself a vastly expanded series of places in the world, which in turn shaped her identity.

The concept of nested place is critical to understanding Marie de l'Incarnation and the power she wielded in her own time. The nested places of her life, and the places upon which she acted include her body, the convent, the Church, the town of Quebec, French North America, the realm of French Catholicism, and the spaces of God. Each of these places can be seen as containing and hence bounded. But as we will see, to act upon the body, or the convent or the town, for example, with the right conjunction of networks, contacts and meanings, could be to act on many if not all of these nested places. So these containing places, through the actions of powerful and knowing individuals, turned or folded outwards. A woman with a strong sense of her identity and a deep understanding of her own time, although enclosed within a convent, could and did reach out and shape places well beyond the confines of her convent.

Why Marie de l'Incarnation?

Saskia Sassen, writing about internet technologies, may be right that the twenty-first century will be characterized by a transition from a long-established preoccupation with controlling bodies to controlling flows of information (Sassen 2002). In this transition lies a new recognition that flows of information may be more important in shaping place than the disposition and movement of bodies. The world of seventeenth-century religious women was one in which the body was strictly controlled. In the Christian mystical tradition, of which Marie de l'Incarnation was a part, the body was to be overcome to allow the soul to achieve a higher plain of communication with God. But the rejection of the body can be seen in another light. Seventeenth-century French society was one in which the woman was subordinate: subordinate to father, to husband and ultimately to son. Most women more or less accepted this state of affairs most of the time. Some found ways to escape some of these constraints some of the time. A violent rejection of the body and its gendered identity effectively freed Marie Guyart from the traditional control of father, husband

and son and even provided her with considerable freedom and importance within the Church. By sacrificing the body, indeed by embracing amongst the most severe disciplining of the body, Marie was able to slip the bonds of gender, class and poverty and to find one of the rare spaces of freedom for seventeenth-century French women, the places created by networks and flows of information. Even in the seventeenth century, flows of information shaped place and space and enhanced power and social position. Dwelling for thirty-three years in a convent of no more than 2808 sq ft (261 sq m), the places of Marie de l'Incarnation's life extended beyond the confines of the convent and its gardens. While her body was enclosed, at least in the Ursuline convent in New France, information flowed freely, giving Marie de l'Incarnation access to and influence on places beyond the convent walls. The frequent and often exalted visitors to the convent informed her of events and conditions in the colony and beyond. Those visitors also allowed her to act and influence beyond the convent. Her voluminous correspondence, her pragmatic writings and her spiritual writings also allowed her to know and act upon places beyond the reach of visitors to the convent. Two of the most important relationship that defined the extent and nature of her place in the world were those with her 'family' members, especially with her son, and those with the Jesuit fathers both in France and in Quebec.

Marie de l'Incarnation was a very important and widely known religious mystic (in the tradition of Teresa of Avila) just before mysticism fell out of favor with the Catholic Church. From an old aristocratic family on her mother's side but from the artisan class on her father's side, Marie was born into a pious family. She was married to a silk merchant at 17, according to the wishes of her family, gave birth to a son, experienced the bankruptcy of her husband's business and was a widow by the age of 19. She then moved in with her father for a period of a year and went to work in her sister's household for a subsequent nine years. She is worthy of our attention on a number of grounds. She was an intelligent and resourceful woman living in and through Catholicism in the period of tight social control associated with the Counter-Reformation. She was the foundress of the Ursuline Order in Quebec City, which became one of the institutional foundations of the new social order of New France. She was one of the first female missionaries in northern North America with considerable contact with the indigenous peoples of the St. Lawrence Valley and well beyond. She also recorded her thoughts in letters written through the course of her life and in spiritual writings, both of which were published posthumously.

As a woman and as a widow in debt and without means, in the 1620s and 1630s her prospects were probably not very bright. Remarriage was possible and appears to have been what was expected of her. But she had not found marriage a positive experience. She described her husband's bankruptcy as a difficulty 'larger than a person of my gender, my age and my ability could carry' (Sp 2: 58; See also Sp 1:24; 1: 173).[1] But her dissatisfaction went beyond financial concerns. She described her marriage as captivity and marriage in general as 'a state of victim-hood' (Sp 1: 463) and, while discussing the inexplicable nature of temptations that assailed her in the first few years of her convent life, she described sex as something that had never given her much pleasure (Sp 1: 146 and 320). Although the relationship with her son evolved as he attained adulthood, when he was a child, her bond to him seems to have been more limited than would appear to have been the norm in seventeenth-century France (Ariès 1960). In her spiritual writings, the great preponderance of references to her son arises because he represented an obligation or a bother (Sp 1: 262, 328-329; 2: 160; 166-168; 183; 184-185). At times she saw him principally as a challenge to her vocation (Sp 1: 278-279). She admitted to having been very neglectful of his needs (Sp 2: 74). Even when he was relatively young, her contact with him seems to have been limited. He lived with a wet nurse for at least the first two years of his life, and he recounts that his mother never kissed him and certainly did not kiss or hug him good-bye when she left him for the convent (Sp 2: 82, 85; 1: 279, 283; 2: 167). Marriage and maternal life does not seem to have held any attraction for Marie Guyart. While servitude of a more or less exalted sort in her sister's home was a possible alternative, this too brought her little satisfaction as she described her brother-in-law's home as one she was happy to leave (Sp 2: 177).

The Church represented a significant alternative for Marie Guyart. She describes in her *Relation* of 1633 that since her childhood she had had a high comfort level with the things of the Church. Even when she could not understand them, she professed admiration for the preachers she must have encountered in her parish (Sp 2: 54). She was drawn to and impressed by all Church ceremonies and processions and she particularly liked Church music and the sung mass (Sp 2: 279; see also [653]). She envied the priests their belonging to a special corps and she had, before becoming an Ursuline, a desire to be a part of that corps which, once she was an Ursuline, became the 'happiness of being in the house of God and a portion of his heritage' (Sp 2: 176). Today it is easy to underestimate the very apparent power of the Institution of the Church in seventeenth-century France. It was the authority on matters spiritual and on many issues considered secular in much of the modern West. It was the source of most writing, music, and art of the period. Its architecture dominated French, Italian, and Spanish cities, towns and countryside. One has but to wander under the vaulted ceilings and arches of the Cathedral of Toledo, past its gold encrusted statuary, its fine carved pews and priceless works of art to understand the permanency of power assumed by contemporaries. The Church, then, had more to offer Marie Guyart than lieutenancy in her brother-in-law's military supply business (Sp 1: 248 and 165-166).

This observation could be read as constructing Marie Guyart as a calculating opportunist. This is not its intention. Marie Guyart was a deeply religious woman whose faith in God and the Church, although sorely tried, grew through the course of her demanding and interesting life. Nonetheless she also had a sense of her own potential and importance and was very aware of what entering the Church and embracing a life of mysticism had given her. She is said to have told her son that thanks to her entry into the convent and his consequent education by the Jesuits: 'you have been raised in a manner that considerably surpasses your condition' (Sp 2: 251). She was in no sense a cynic. Somewhere, somehow she had learned or devised a strategy that both freed her from many of the bonds of her gender, her class and her poverty, and gave her life rich spiritual meaning. That strategy, which was echoed by the narratives that structured many Catholic lives, evolved into a way of life, which no matter how alien today, had considerable integrity and internal consistency.

Breaking (the Bonds of) the Body

Rejecting the Body

Marie Guyart's strategy revolved around the body and denial of its most basic desires. From the vantage point of Western consumer culture of the early twenty-first century, deep asceticism is difficult to understand. Even those who criticize what appears to be mindless and irresponsible consumption would expect, for example, to find pleasure in the contemplation of nature or simply in good companionship. But asceticism has a long and venerable history and is far from dead in the modern world. Behind it lies a view of the individual as both a physical and a spiritual being. The spiritual is considered the greatest of these aspects and it is the more purely physical – or animal – impulses that must be denied to allow the spiritual to thrive and realize its potential. For Catholics, the realm of the spiritual is indistinguishable from God. For Marie Guyart the body may have had particularly negative connotations. It was the body and everything associated with it, including marriage, family obligations and affective bonds that not only kept her from God, in an abstract sense, but prevented her from living the kind of life she felt she could and should live. Ironically, but by no means without precedence in the Christian tradition (Armstrong 1986, 8-9, 159, 184-5), it was also through her body that she found the means to change her state from that of an impoverished middle class woman to a privileged choir nun (a position generally reserved for the upper class and wealthy elite), to a heroic missionary and one of the most respected women of her century.

According to her letters and spiritual writings, negative feelings about her own body date from the period of her marriage and were most intense prior to her entry into the convent. The Ursuline Rule to some extent imposed a more balanced approach to the body's basic needs. Nevertheless, throughout her life she most consistently characterized her body as prison (Sp 1: 231; 1: 235; 1: 389; 1: 391; 2: 143, 149 and 379). Arguably, the body did function as a prison for Marie Guyart throughout her life. Prior to her entry into the convent, the expectations of being a wife, mother, daughter and sister kept her in occupations that were relatively meaningless for her. As an Ursuline she still felt constrained by the rules of gender: she could not join the Jesuit missionaries in their travels throughout the countryside ([224, 356, 507, 555, 890]); she did not feel free to write and publish ([515, 521, 642, and about another woman's published writings: 868]); and she lamented the weakness of her woman's body and her female mind ([26-27, 156, 395, 485] and Sp 2: 205, 246, 316). These feelings were expressed in often extreme language and action. She likened the body to a dead dog, maggoty and rotting on the roadside (Sp 1: 191 and 490). She described it as her worst enemy and as 'miserable' ([849]; Sp 2: 109). It was something so horrible she could not fathom it (Sp 2: 100). At the very best, it was a fragile earthen vessel tending strongly to evil and breakage but that nevertheless held the treasure of the soul (Sp 1: 129, 444). Everything attached to it was suspect, from the mind and understanding, to the very self. Even that part of the soul most closely attached to the body was to be mistrusted (Sp 2: 344-345, 293, and 308). As far as she was concerned, the body was 'a terrible place to live' (Sp 1: 320).

Self-mortification is first mentioned in her writings in association with her brother-in-law's home. There she began by sleeping on kindling and then moved to hard boards. She wore full-body hair shirts to bed and beat herself with handfuls of nettles, thistles and chains (Sp 1: 176). She exposed herself to cold (Sp 2: 109). She flavored her meat with absinthe to render it so bitter as to be totally unappetizing (Sp 1: 176). While trying to rid herself of the negative emotions and sentiments that she considered natural to her body, she bled herself to the point of weakness (Sp 2: 294). She wore belts of steel with protrusions (Sp 1: 216). She deprived herself of sleep, approached rotting carcasses and open wounds (Sp 2: 97) and often deprived herself of the sight of the beautiful in the world by either closing her eyes or keeping them downcast (Sp 1: 180).

The Body as Social

These punishments were described by Marie de l'Incarnation as a private matter between herself and God. But of course, they were very far from private and carried significant social meaning. Once she came under the care of a spiritual director, self-mortification was a privilege accorded, or denied, by the confessor. She did not then, and not as an Ursuline, have the right to inflict self-punishments without hierarchical permission. The very involvement of the religious hierarchy in her self-punishment made them public acts. Even before she had secured a confessor, it is unlikely that her self-punishment went unnoticed. It was known and commented on in her brother-in-law's home (Sp 2: 150). Beyond the home her extreme religiosity was noticed. According to her son, in her pre-convent days in Tours, 'when she walked in the streets it was with downcast eyes, a controlled step, and with a humble gravity that delighted everyone' (Sp 1: 180). She had a strong desire to humiliate herself publicly and while her confessor refused to publish her sins, her manner of dressing was that of a penitent (Sp 2: 81). She described frequently walking around in a state of religious ecstasy, which was recognized as such by at least one priest she had not previously known (Sp 1: 225). Her son described her grace of union with God as visible in that it lent her 'such a majesty' (Sp 1: 347). Her entry into the Ursuline convent was something of a mini-procession and her son remembered a street audience for that procession (Sp 1: 284). Her vision of the Holy Trinity in the Feuillant Church in Tours lasted the duration of several masses, which, of course, would have been noticed in a relatively small community (Sp 2: 122). By the time she entered the convent, she was known beyond Tours for her religiosity. Before Marie had chosen the order into which she would seek entry, the Bishop of Dol, then visiting Tours, invited her to found a Visitation Monastery in his town (Sp 1: 271). After she had been accepted but before she entered the Ursulines, the Bishop of Tours requested a personal meeting with her (Sp 2: 161). Later, her inclination to the Canadian missions was also widely known. Within the convent, it was the major topic of conversation and she had the entire convent in Tours praying for the souls of North American indigenous children (Sp 2: 215). Her discussion of the conjunction of her will with God's (more about that later) attracted the attention of the French ecclesiastic and author, Bernières-Louvigny (1602-1659), who became a life-long correspondent (Sp 2: 222). And later in her life, when she had achieved notoriety for her work in New France, the famous Father General of the Jesuits in Quebec, Jérôme Lalemant, described her as 'known and honoured in all of France' (Sp 2: 305).

Appearances Matter

There is subtle evidence that Marie was aware of the impact of her appearance on others, and that it mattered to her. In her relations with both God and authority there is a repeating motif around the words look, looking and looks. This is so much the case that her description of these encounters calls to mind sixteenth- and seventeenth-century portraits of saintly individuals:

> I contemplated all of these truths with a gentle and simple look, in the union that I had with God, and I was delighted with ease by that he is all and I am nothing, with that all is from him, and I have nothing of me (Sp 1: 475).

Or again, in referring to God:

> He interrogated me, I would say, not by words as he did with Saint Peter, but by his looks, which so filled my heart, that it could not contain his fire (Sp 1: 482-483).

She recognized Mme de la Peltrie, her patroness, not by her features but by the 'innocence and mildness of her face' (Sp 2: 234). She described her encounter with the Virgin Mary (in her dream calling her to Canada) and her meeting with the Queen of France, Anne of Austria, prior to embarking for Canada, in very similar terms: 'Madame the countess of Brienne took us to Saint Germain where her Majesty was. She, by her goodness and piety, looked at us with a love that was altogether particular' (Sp 2: 239). Marie de l'Incarnation remembered that when during the Atlantic crossing they encountered an iceberg and annihilation looked likely: 'I arranged my clothing so that, when the wreckage took place, I could only be seen with decency' (Sp 2: 244). Later in New France, she described with fascination and awe the Jesuit fathers who had been tortured and bore the disfiguring scars of that torture. She commented: 'Their saintliness is so visible to everyone that each person is delighted with it' (Sp 2: 319). Looks, saintly looks, mattered. They were a source of a particular kind of power.

The Issue of Will

Breaking the will and denying the self was part of breaking the body. It was possible and important to annihilate oneself in work, and to be annihilated in God (Sp 1: 450, 462). The mind and rationality was particularly dangerous and while study might be appropriate for men, only love and submission was appropriate for women (Sp 2: 169). Intellectual curiosity, or 'the desire to know by one's own labours' was equally dangerous for men and women (Sp 2:169). Thus, when Marie de l'Incarnation found herself playing what she considered a male instructional role, such as when she was preaching to the novices in Tours, she felt more comfortable attributing her intellectuality to a force beyond herself:

> I carried in my soul a grace of knowledge that sometimes made me say what I would neither have wanted, nor have dared, to say without this abundance of spirit (Sp 2: 194).

Part of what breaking the will entailed was total abjection and humiliation before authority. There are some passages in her writings which suggest this stance vis-à-vis her confessors (Sp 1: 168). However will and obedience were very far from simple concepts in the life and mind of a woman as active and determined as Marie de l'Incarnation. She described herself as obedient to her spiritual director, to her brother-in-law, to her father and to her sister (Sp 2: 170). But she defied all three when she felt it was the will of God to do so. When one of her first confessors, Dom Raymond resisted her joining the Ursulines, she convinced him that it was the will of God (Sp 2: 156-157). When she left her father, the very symbol of masculine authority in seventeenth-century France, to join the convent, she described him as 'crying out in lament' (Sp 2: 161). Yet she went. She was even prepared to argue with God over the missions as she considered them of paramount importance to her celestial spouse, Christ (Sp 2: 199-200). When her declared vocation for Canada was mocked by what she considered a mediocre confessor, she lapsed into silence and went around him, according to the instructions of the Divine Majesty who 'pressed me strongly interiorly to set aside all my fears and to go above all humans' (Sp 2: 214-216). In spite of all the discussion of the will of God and the will of the community, it was Marie de l'Incarnation who chose the Ursuline colleague who was to accompany her to Paris and to New France (Sp 2: 235). She even referred to the Bishop of Paris' decision to withdraw his offer of a particular Paris Ursuline nun for the mission to Canada as having 'withdrawn his obedience' (Sp 2: 239). She declared her unwillingness to accept the Bishop of Quebec, Laval's, austere constitution of 1661, effectively forcing him to either compromise or to employ open force. Compromise became the order of the day (Lapointe 1974, viii). She seems to have found a partial solution to the problem of obedience and will in an encounter with God, that she described in almost sexual terms. She wanted to go to Canada. She felt that God wanted her to go to Canada. But where did her will and His will meet? And how could she justify breaking all the social taboos that she would need to break in order to get to Canada? Her encounter with God would give her that power:

> Even as this more-than-adorable Majesty looked upon me, which signified to me that I wanted to ravish his will, but that through love he wished to triumph over mine... Oh! My love! Oh my great God! I want nothing, I cannot want anything. You have ravished my will! How could I want, as you have ravished my will and rendered me incapable of will (Sp 2: 212-213).

Her son's interpretation of this event gives us some insight into how she and her contemporaries saw this encounter:

> But one day she underwent an extraordinary event. As while forcing herself to undertake the divine will, to never leave it and to bend it to the establishment of his son's realm over all nations, our Lord then took hers. Since, she has not had a will of her own. Only the will of God has been her will as it is impossible for her to want anything that is not the will of God (Sp 2: 223).

To have one's will indiscernible from that of God would make one a victim, in the sense in which Marie understood the victim-hood of marriage (Sp 1: 463). But in another sense, to have one's will indistinguishable from that of God would make one supremely socially powerful.

This encounter or insight did not resolve all issues around will and obedience for Marie de l'Incarnation. The problem was too complex and contradictory for resolution as is clear from her reaction to the 1650 fire that destroyed the Ursuline convent in Quebec. In spite of knowing precisely which human action had led to the fire, she saw this as an act of God particularly directed at her and her willfulness: 'His design was accomplished by our annihilation and especially as concerns me, because I had that house built and I worked hard to succeed in putting it into the condition in which it was. And I suffered great contradictions [to her will]' (Sp 2: 325-326). Her words of acceptance, although apparently absolute, have a bitter ring to them.

> You did that, my chaste husband. Bless you! You have done well. Oh! How well you have done all that you have done! It is my satisfaction that you be satisfied in what you have done! (Sp 2: 325).

The matter of will was an ongoing issue for Marie de l'Incarnation both in relation to God and in relation to her contemporaries. Public rejection of the most intimate place of the body won Marie de l'Incarnation notoriety. Her paradoxical struggle both to relinquish will and yet to find a place to preserve, cultivate and protect it allowed her to move into the public realm of an apostolic (if enclosed) missionary with considerable authority and personal and institutional power.

To Find Place and Space in the World

The Openness of Enclosed Space

To a very large extent, the world that Marie de l'Incarnation made for herself was the realm of the Ursuline convent in New France. Based on the incidental references in her correspondence, it is clear that the Ursuline convent was a focal point of communication in the world of newly Christianized natives. The Ursuline's contacts with the local Algonquin, Huron and ultimately Iroquois tribes were extensive. In the early 1640s, she recorded taking in 41 seminarists (female Indian students) although, as Marcel Trudel has pointed out, very few stayed in the convent for more than a season or two (Trudel 1999, 60, 79, 103). She recorded 700 visits from indigenous men or women in 1640 and 800 in 1641 ([132, 144]). In the 1640s, the Ursulines fed visiting groups of between 60 and 80 natives daily over the entire winter. After the Iroquois destruction of the Huron homeland in 1649, and as she was then in charge of the convent's supplies, it was she who handed the food out to the destitute Huron families that gathered in Quebec City (Sp 2: 320). Marie de l'Incarnation's experience was that her convent existed at the crossroads of the Montagnez, Algonquins, Abenaquis peoples and the people of the Saguenay who congregated in the Ursuline parlor for at least the first five to six years of the Ursuline presence in Quebec ([122]; Trudel 1999, 60; Sp 2: 258). Indigenous leaders came to the Ursuline parlor to seek her advice ([222]). No doubt, the point for many of the indigenous visitors in the 1640s was survival in a world made suddenly extremely dangerous and hostile by the imbalance in power created by the arrival of the Europeans and the sustained resistance of the Iroquois. But for Marie de l'Incarnation and her colleagues, the provision of food was the pretext for the saving of souls through education, conversion and ongoing interaction to support that conversion. Indigenous converts seem to have traveled significant distances to speak to the Ursulines and Marie de l'Incarnation recounts stories told by them to her of the conditions of their lives and relations between indigenous groups, in addition to accounts of their own proselytizing. She may have learned, especially from the stories and accounts of indigenous women, details and aspects of indigenous life that were not recounted to her colleagues just next door, the Jesuits ([181ff, 329, 329-330, 330, 331, 563, 604-605, 787]). The places created for Marie de l'Incarnation by her interactions with indigenous peoples were real and important to her (and no doubt to the other Ursulines) as she sought to share these places with her European correspondents.

Marie de l'Incarnation's convent was also at the crossroads of French and sometimes European communication. Certainly she had regular discussions with the Jesuit fathers, but more about that later. She also received visits from Bishop Laval, the fort commander, Monsieur Dupuis, the widow of Louis d'Aillebout from Montreal (who actually came to live with the Ursulines for a time), Monsieur d'Argençon, the Governor of Quebec who visited frequently, a merchant who came to speak to her of his commercial affairs, Le Chevalier de Chaumont who recounted his experience of fighting the Iroquois, l'abbé Fenelon, the first Intendant, Jean Talon, and Médart Chouart Groseilliers with whom she had numerous warm meetings prior to his departure for service with the English ([583, 667, 673, 678, 772, 841, 882, and 874]).

As a result of these sorts of exchanges Marie de l'Incarnation demonstrated considerable awareness of daily events in the Colony. She was aware of the limits of French settlement. She was also aware of some of the activities of enemies of France or rivals on the edges of the French territory ([665]). Thus she heard about and reported it to her correspondents on the English taking of Acadia [(807]). Similarly, she reported on the trading activities of the Dutch and the Iroquois ([648]), the disputes between the Dutch and the English ([742]), the climate, cash crops and the shortage of food crops in the West Indies and the way in which the missions in those islands were administered ([759]), and the movement of Monsieur de Monts into Plaisance Newfoundland, the state of the cod fishery and the mapping that was carried out there ([683]). Closer to home, she informed herself about the movement of the Sulpicians into Montreal ([614]), the work of the Hospitaliers in dealing with epidemics carried by ships from France in 1664 and again in 1665([724 and 754-755]), the financial problems of the Hospitaliers and how they were dealt with ([633]), the state of the crops after harsh winters or when the colony was under threat from the Iroquois ([627-628, 631]), the conditions of the poor in Quebec and some of the structural sources of their poverty ([684, 759]), and the impact on the populace of the earthquake of January 1663 ([686, 687]). Indeed, she seems to have joined the Jesuits in trying to determine just how far afield the earthquake was felt ([686, 692, 699]). Although she was more immediately interested in the missionary activities of fellow religious, she was not uninterested or unaware of the commercial activities taking place in the colony and the importance of commerce to her own mission. She understood that without commerce the religious enterprise could not succeed in

Quebec ([398, 631, 632, 637]). She was aware of the impact of the wars with the Iroquois on the fur trade in Montreal and indeed on the trade between the Dutch and the Iroquois ([479, 484, 648, 808]). And she understood the nature and importance of the triangular trade between Canada, the West Indies and France ([873]). These were the themes and events on which she kept informed through interactions with residents or visitors to the colony. And she passed the news she considered important on in her correspondence with colleagues, friends, and prospective donors back in France.

The Places and Spaces of Writing

It was her correspondence that gave maximum extension to the places of Marie de l'Incarnation's life. Letters kept her informed of events in France including the disputes in the Royal House of the King of England, the troubles between France and Italy in 1646 and 1663, famine in France and particularly along the Loire River, the progress of the Jansenist heresy, and at least some of the incidents of possession in convents in France ([339, 294-295, 677, 344, 588, 610, 384]). She was well informed about ecclesiastical matters directly in (or against) the interests of the Ursulines of Quebec. There must have been some considerable correspondence between the Ursuline convents of Tours, Paris and Quebec over the contractual arrangement made by Mme de la Peltrie (the patron of the Quebec Ursulines) on the possible return of the Quebec Ursulines to a new mother house in France ([426-427]). She followed with considerable interest the dispute between France and Rome over to whom the Bishop of Quebec should report ([614]). She also followed key events in a number of Ursuline convents in France, including those of Tours, Paris, Dijon, Bordeaux and Mons ([Tours: 817, 242, 517, 879, 646, 882, 647, 574, 579, 778, 762, 818, 763, 894, 497; Paris: 868, 578, 854; Dijon: 805, 312; Bordeaux: 851-852, and Mons: 882, 860]). She was aware of and sensitive to the paths through which information flowed. In one letter to the Superior of the Ursuline convent at Mons, she wrote in concern that she had learned about political matters interior to the Church and the Mons community from a resident of Mons with no formal attachment to the convent there ([718-719]). She knew the paths her letters would take and informed herself of their fate when they were diverted ([631, 408, 758]). She was even sufficiently interested in information flows in the colony to note the remarkable 'apostolic courier,' M. Boquet, who devoted his life to carrying correspondence to and from the often remote Jesuit missions ([841]). She was interested in the orders in which she had family members and followed their reforms, amalgamations, movements into foreign lands, and changes to their rules and constitutions. Her correspondence was not always that of a passive observer of events. She used her letters to influence others, both in Rome and in France, to in turn influence the Pope to create a Bull of Unification for the Quebec Ursulines ([229-230, 269, 344, 378]). She wrote to find and cultivate possible benefactors for the convent in Quebec. She wrote to secure recruits of the requisite quality. She wrote to her son to ask him to influence the behavior of her niece, perhaps by himself corresponding with her superiors and she wrote to her son to try to help him with his own spiritual voyage (Sp 1: 185). She may have sent Jérôme Lalemant to the Ursuline convent in Tours to smooth ruffled feathers over the possible establishment of a new French mother house in France, should the Ursulines be forced from New France by the Iroquois ([427]). And she used her correspondence to help the Ursuline order in Paris prepare a chronicle of the entire Ursuline order ([854]).

It is very clear that Marie de l'Incarnation appreciated the liberating power of letters and sought to preserve and protect that power. She wrote to a colleague in Tours that although the seal on letters sent from the convent was always broken, in conformity with the dictates of the rule, both she as mother superior and her successor in that post had assiduously refused to open or read the letters so as to preserve open and free communication with other members of the order ([405, 644, 765, 853]). There were other reasons to protect the confidentiality of letters. Marie de l'Incarnation wrote about spiritual matters in some of her letters and both she and her correspondents regarded these issues as personal and private (Sp 2: 216). There were times in her life, particularly during the many years that she was superior of the convent that she found the weight of her official correspondence (hundreds of letters a year written in the few months that the boats were in harbor) overwhelming. But she was aware that letters could overcome what would otherwise be insuperable bonds and bounds of place ([77]). In the years during which she spent most of her time working in Algonquin and Huron, they relieved her discomfort at frequent communication in, what seemed to her, barbarous tongues ([102]). She had a strong sense that distance could be overcome by a combination of memory and continued affection expressed in letters ([561]). And she wrote with appreciation of the power of the technology she was wielding, evidenced by her recording of the astonishment of Indians at the capacity of a small piece of paper with squiggles to convey an appreciable volume of reliable and important information over great distances ([918]).

Apart from her letters, Marie de l'Incarnation engaged in two other forms of writing: the composition of pragmatic works designed to advance the work of the missions, probably including the Quebec Ursuline Rule, and more spiritual writings. In the course of her life in Canada, having herself learnt the relevant languages, she wrote Huron and Algonquin catechisms, a *Sacred History* written in Algonquin (Davis 1995, 98-99, 106), Algonquin prayers and an Algonquin dictionary. Although she mentioned these writings rarely, they were fundamentally important to her, part of her contribution to the missionary work of the religious orders and writings she wanted to leave to posterity, or at least to her Community for use in the mission of conversion. Her efforts to learn these languages both allowed her to extend her influence into indigenous communities and also gave her more regular contact with her Jesuit instructors and fellow students in the indigenous languages. Sadly, although some dictionaries have survived and are on display at the Ursuline museum, those composed by Marie de l'Incarnation were ultimately lost, or worn out, in missions further to the north.

While she hoped to leave her pragmatic writings to posterity, in contrast, she sought to destroy her spiritual writings ([888]). In fact, she shared these writings with her son on condition that they be destroyed ([517]). It is not clear if this request was sincere. It was not seemly for a woman and a nun to seek publication and certainly it was dangerous for a mystic to write about a close personal relationship with God. In any case, her son did not destroy the letters but instead gathered them and published them together with an account of her life. Marie de l'Incarnation makes clear in both her spiritual writings and in her letters that she undertook to write about spiritual matters in Tours as a result of her Jesuit directors' insistence that she do so and she describes her 1654 *Relation* as an act of obedience to an order from her Jesuit Director, Jérôme Lalemant, to write about her inner spirituality (Sp 1: 335, 414; [426]). It seems clear that Marie de l'Incarnation was aware of the danger of writing, as she hid the fact that she was writing about her relationship with God from her colleagues and did destroy her spiritual writings when she feared they might fall into unintended hands ([515, 526, 425-426]). Writing about spiritual matters was itself an access or a way to God, a voyage that in letters, in any case, she might share with a spiritually sensitive and receptive correspondent. But she felt that any description of the depth of her spiritual engagement with God was a poor reflection of that relationship and thus nothing worth preserving in and of itself (Sp 2: 128, 130, 155, 213). Such writings might be a source of spiritual inspiration for her correspondent, but they were not meant to be preserved for posterity, as it was the act of contemplative writing that mattered, not the marks on the page that remained. It was through the act of writing that she was able to come to an understanding of the combined divinity and humanity of Christ. Writing was itself a contemplation of the 'sacré verbe incarné.' It opened Marie de l'Incarnation to a place, or more correctly a space, beyond the convent. This was the space of God.

Intimacy: Proximity in Distance

Perhaps the best example of a correspondence and a relationship that overwhelmed the bounds of the convent and yet was shaped by that place was the one that evolved with her son. As we have seen, Marie de l'Incarnation entered the Ursuline convent in Tours, at thirty-two years of age, 12 years after the death of her husband, and when her only son was about 12 years old. She did this after careful cultivation of a number of the key religious communities and figures in Tours. Again, her attitude to marriage and to motherhood was at best ambivalent and she was able to hand the financial care of her son to her sister and brother-in-law until she was ready to leave for Canada. His education and spiritual well-being was seen to by the Jesuit fathers. Her son, 'Claude Martin', initially bereft at the sudden loss of his mother, fought her entry into the convent, going so far as to present himself at the gate of the convent, pounding on the door and crying for his mother's return. Marie de l'Incarnation remained at the convent under what she took to be some disapproving stares. Her departure for Canada in 1639 when he was about 21 years of age would have represented a second overwhelming loss to him.

There are two principal phases in her correspondence-based relationship with her son: the period prior to his entry into the Benedictine order in about 1643 and the period after that. In the former period her relations with him were warm but restrained. On some level she regarded the filial tie as an obstacle to her spiritual growth and a source of possible invasion. The threat from family bonds is a theme that recurs in her writings. In her account of her Ursuline companion's life, Marie de Saint Joseph, who had to break with her family to accompany Marie de l'Incarnation to Canada and to defy them to die in Canada, Marie de l'Incarnation considered that her companion had taken the higher road, the road to God, in spite of the pain it had caused both Marie de Saint Joseph and her family ([436-466]). In correspondence with her son on the subject of her niece, who also became an Ursuline, she expressed considerable anxiety over her niece's continued relations with her family's circle of acquaintances after her entry into the convent ([216, 230, 586]). She expressed in general terms that the love of God required the painful distancing of family and friends ([106]).

Her anxiety over the filial bond to her son gradually dissipated from her correspondence after Dom Claude Martin announced his entry into the Benedictine order. The personal loss they had both suffered was a periodic subject of correspondence between them as they worked that dimension of their relationship out and arguably the wound was never entirely healed for either of them ([527, 725, 836, 898]). But a sea change appears to have come when she could share with him on a different, in a sense equally personal, but not quite so maternal, plane. She expressed this spiritual intimacy with him in very spatial terms in closing a long letter in which she had recounted some of the exploits of the Jesuit fathers.

> For you, I do not leave you [as long as we are] close to God. Let's live together in this vast ocean, and let's live there, here below, as we await the eternity that we will really see ([224]).

It was, then, in the spaces of the infinite that she could find a place of unfettered friendship and love for her son. And that place was created through their correspondence.

But the correspondence remained, and perhaps became increasingly, corporeal as Dom Claude insisted that it be so. Only her side of the correspondence survives but her answers indicate that Dom Claude frequently asked about the details of her everyday life, about the internal configuration of the convent, how her health was, how she looked, and how she had changed. She generally answered his questions with such brevity that the answers cannot have satisfied. However, in one letter she wrote to her son that she had invited 'an honest young man,' who was soon leaving for France, into the convent and had raised her veil so that when he saw Dom Claude in Séez (175 km west and slightly south of Paris) he could tell Dom Claude that he had seen her with eyes that Dom Claude could himself see ([384]). Similarly it was very important to her to see someone who had seen Dom Claude and to have her son send his sermons to her so that she could come close to hearing him ([214, 862, 187]). When she could send her letters to him by the hand of a visiting Jesuit who would willingly visit both Marie de l'Incarnation and Dom Claude, she did so ([395, 403, 809]). There is in this exchange a striving for intimacy that was spiritual. Toward the end of her life Marie de l'Incarnation told her son that, were they living within a possible proximity, she would have taken him as her confessor. On earth, for Marie de l'Incarnation, there was no closer relationship. There are no reminiscences of his childhood in her side of the surviving correspondence. But there is an intimacy that is personal and corporeal, a place of intimacy created in the spaces of their correspondence.

Their correspondence was of vital importance to both Marie de l'Incarnation and Dom Claude. Through this correspondence, she was able to keep abreast of events in the Church in France in particular, and, with the readily accorded permission of his superior, in their two orders. She spoke of his order and indeed of her own as one might of family: as a rich extension of her experience ([228-229, 797]). He was an important part of a network of contacts and she felt that she could ask him to contact a third party on her behalf. As time passed she seems to have needed his letters more as she sometimes chided him for not writing sufficiently ([569, 571, 613, 790]). Dom Claude was convinced of his mother's sanctity and openly sought to collect and preserve her letters, thus according her, as she perhaps understood, another kind of place in history.

The Ursulines and the Jesuits: Gender Barriers and Gender Bonds

It is through her relations with the Jesuits that the places of Marie de l'Incarnation's life both extended well out into the world and folded back into her to give both place and space considerable depth of meaning in her life. Marie de l'Incarnation shared at least three nested places and spaces with the Jesuit fathers: the places of their missionary work; the places of their writings, and the spaces of spirituality. These shared places and spaces are evident in Marie de l'Incarnation's correspondence. Of course, the patterns of their sharing would emerge more clearly from a comparison of the *Jesuit Relations* and Marie de l'Incarnation's writings, supplemented by archival documents. Indeed one of their most important shared nested places is substantially inaccessible in Marie de l'Incarnation's correspondence: the shared narrative forms of the city of Quebec. Nevertheless much can be gleaned about their shared places and spaces from this single source.

The Shared Places of Mission

Marie de l'Incarnation's attachment to the Jesuits began early in her life. It is clear that the inspiration for her own mission of conversion of the native peoples of Canada was closely tied to and indeed perhaps inspired by theirs. The language of her dream or vision with its 'great and vast spaces,' its 'hideous rocks,' its 'silence,' and its 'dreadful precipices' (Sp 2: 190), which dream moved her to attempt to join the mission in Canada, reflects some of the language in the early *Jesuit Relations*. Canada as 'an endless forest,' 'a solitary and unexplored country,' with 'perilous heights and narrow passes,' a land from which nothing can arise but 'cold and gloomy vapours,' 'a horrible wilderness,' protected by dangerous rocks and 'frightful and horrible precipices' (Thwaites 1846-47, 7: 254; 1: 219; 2: 41; 3: 61; 3: 33-35; 3: 209 and especially 7: 177-179). She describes her fascination with these 'labourers of the Gospel to whom I felt myself tightly tied' (Sp 2: 199). As early as 1635, the Jesuit fathers and, particularly, Father Poncet were sending her copies of the *Jesuit Relations* ([443]; Sp 2: 205; Sp 2: 259). But the Jesuits were more than an inspiration to Marie de l'Incarnation. They offered very practical assistance. The Reverend Father Paul Lejeune advocated for her to come to Canada. Other Jesuits, including the Reverend Fathers Poncet and Lalemant, worked with her and her co-conspirators against social norms to forge an alliance of laborers in the missionary fields of conversion ([64]). They actively sought partners with whom they could share the places of conversion and conquest. It was a letter in the *Jesuit Relations* and authored by Lejeune calling for a wealthy benefactress to fund a mission of teaching nuns to come to Canada that first attracted Mme de la Peltrie to the Canadian cause ([904]). It was a Jesuit who put

Mme de la Peltrie and Marie de l'Incarnation in touch with each other. It was a Jesuit who sought to provide Mme de la Peltrie with the appearance of a marriage initially to satisfy her father and preserve her right to any future inheritance and then, once her father had died, to protect her money from the claims of other potential heirs. Thus did they ensure that in spite of being an unmarried woman she could dispose of her wealth as she (and as they) wished, in the Canadian mission ([906]). It was substantially, but not exclusively, the Jesuits who provided the women with protection so that they could travel to and live in Paris as they prepared their expedition to Canada ([907, 908]). It was the Jesuits who made preparations for their departure, even renting them a ship for their Atlantic traverse ([909]). And it was with the Jesuits that Marie de l'Incarnation was able to share the humor of having duped the relatives of Mme de la Peltrie to overcome the restrictive social norms governing the behavior and status of single women ([909]). Once in Canada, it was the Jesuits who governed the Ursulines, provided their spiritual directors, composed the first *New French Ursuline Constitution* and, when the convent burned down in 1650, loaned them the funds to rebuild.

Throughout her life Marie de l'Incarnation followed the travels of the Jesuits. From her letters it is clear that she was aware of their activities in China, the Caribbean, Rome, France, Canada and well into the future United States. Of course she most closely followed their travels and efforts in continental North America. Her circle of knowledge and interest expanded in all directions, not blocked by successful indigenous resistance, as the Jesuit fathers converted peoples in a larger and larger circle around the new French settlements; beginning with Sillery, Tadoussac and Acadia and stretching beyond to Newfoundland, the Upper Saguenay, James Bay, Nippissing, Cataraqui, Manicougan, the East side of Lake Superior, New York, and Maryland. She kept track of where and what they had built, where they were, what they were doing and the conditions of their lives ([132]), for example, their activities at Sillery ([132]), Tadoussac ([123]), with the 'Peuples du côté nord' ([143]), the Iroquois and Hurons ([195]), at the Baie des Chaleurs near Miscou ([199]), with an indigenous group some 300 leagues beyond the Hurons ([224]), concerning the Iroquois war and peace ([254, 254-255]), with the Nippising ([270]), the Attikamek ([279]), the Abinakis ([279]), the Winnipeg ([284]), the Anastohé (Hurons in Virginia) ([332]), and the Ottawa (Northern Algonkians) ([379]). She followed the territorial incursions and counter incursions with the Iroqouis ([500, 510ff]), a proposed mission in Ontario ([507]), a mission in Kingston amongst the Iroquois ([544]), knowledge of the China Sea ([545]), the New York Dutch and Maryland English Catholics and the Jesuit work there ([605-606]), discovery of the entry to a great Northern sea ([699]) where there were a large number of peoples who had not heard of God ([728, 729]), a mission to the Montagnais who lived near the Manicougan R. ([729]), Father Nicolas' work with the Nez Percé ([810]), and a report on the missions for 1668 ([810]). Later she followed the missionary work of the fathers in the Iroquois territories ([840]), and the travels/mission of Father Dablon ([841]). She was aware of the departure of Dollier de Casson and Barthélemy (Galinée actually went with Dollier de Casson) for the Upper Mississippi ([841]), and mission work at the Baie des puants ([872-873]). She received news from Marquette's mission work on the east side of lake Superior ([873]), Dalois' mission to the Fox ([902]) (including news of the light, weather, vegetation, fauna, people and snow conditions of the Far North, together with information on the more than 40 fathers traveling here and there ([902-903]). There are suggestions that she respected the Jesuit fathers in part because of the distances they traveled, the isolation they suffered and the physical challenges they faced ([105, 118, 805, 841]). The theme of distance measured in suffering and overcome by spiritual drive is a common one in her writings ([118]). Her knowledge of the Jesuit activities was not restricted to what she could read in the *Jesuit Relations*. Both before their travels and after their return many of the Jesuit fathers came to see Marie de l'Incarnation and told her of what they had seen and experienced. A small amount of that information found its way into her letters, a relatively small number of which have survived. She did not consider herself a passive observer of these events. The Jesuits and the Ursulines formed a spiritual association by virtue of their common mission of conversion ([277]). This association extended even beyond the bounds of life as the Ursulines believed that the martyred fathers could be called upon as advocates in support of their continued mission of conversion ([352]).

The Shared Places of Writing and Spirituality

As a result of this association, the Ursulines and Jesuits also shared the places created by their writing. Marie de l'Incarnation's writings have been treated by most authors as ancillary to the *Jesuit Relations*. It is true that much of her knowledge of Canada was derived from the knowledge of Jesuits. But the *Jesuit Relations* represented but a small part of the knowledge and experience of the Jesuits, and indeed, as Marie de l'Incarnation herself pointed out, of the experience and activities of all the other orders in Quebec, including the Hospitaliers, the Sulpicians and the Recollets, to say nothing of the secular experience of the colony ([802-803]). She pointed out that much of the cloistered work of the Ursulines was invisible to the Jesuits, who probably had the most extensive visiting privileges to the Ursuline convent. In fact, rather than opposing the two sets of correspondence, it makes sense to see them as the Ursulines and Jesuits did: as part of a combined effort to generate support of one sort or another for the Canadian missions. Marie de l'Incarnation continued to read the *Jesuit Relations* once in Canada and seems to have seen the Canadian version prior to its editing in Paris. From time to time there is a clear sense of continuity of account as she would sometimes curtail an account of her own with the comment that the sequel could be read later in the *Jesuit Relations* ([112, 137, 349, 350, 505, 521]). The importance of the fidelity of her own letter writing relative to the activities of the Jesuits is clear from one letter in which she was describing a both subtle and distant phenomena: spirituality among the Indians in a new missionary field. In that case she passed her letter to a Reverend Father who had been to that mission to assure that her account reflected the Jesuit perception with fidelity ([918]). Arguably, their writing was a shared narrative which shaped their perceptions of place. A place- and space-sensitive comparison of the letters of Marie de l'Incarnation and the *Jesuit Relations* is clearly called for, as the spaces of their interaction were complex. The shared places of their correspondence were sometimes political, as in the case of Marie de l'Incarnation's support and advocacy for the Jesuits, even in their disputes with the French hierarchy over the planned imposition of a Bishop of Quebec (and despite her own desire to have a Bishop in Quebec) ([295, 433, 597]). They were also often tangibly physical, as when the Jesuit fathers carried her letters and words to colleagues and especially to her son in France. But, most importantly, they were spiritual.

The greatest degree of intimacy between the Jesuits and the Ursulines was achieved through the institution of spiritual direction, or the relationship between the penitent and her personal confessor. The director or confessor was both the religious confidant of the confessed and a guardian of Church power and hierarchy. In the latter third of the seventeenth century, there was a particular tension around the relationship due to the teachings of Miguel de Molinos, the founder of Quietism, who argued, amongst other things, that the relationship between the Director and the penitent fell beyond the purview of Bishops and the Church hierarchy. For the Catholic Church, the danger of this particular teaching was analogous to the threat of direct access to God, embodied by Protestantism. Further, as Molinos himself demonstrated, there was considerable risk of complicity in such a deeply personal relationship. Molinos was condemned by the Church in 1687. The relationship of a spiritual person with his or her spiritual director, then, was a very sensitive one, both deeply personal, a relationship of power (both personal and institutional) and linked to some of the most intense religious debates of the seventeenth century. Of course, when the confessed was a woman, the relationship of power was complicated by gender. Also of course, the institution was an upper class institution available to educated and privileged people only. According to her own account, Marie de l'Incarnation claimed the right to her first confessor through a very public demonstration of piety and contrition when she virtually assaulted a Feuillant father and demanded that he hear her confession (Sp 2: 70). In her writings, Marie de l'Incarnation named a succession of at least 8 confessors, but she certainly had more. From 1635 her confessors were always Jesuits and the most important of these for Marie de l'Incarnation was unquestionably Jérôme Lalemant. In a sense, it is impossible to know anything about these relationships, as they were not public. But Marie de l'Incarnation's spiritual writings (which she, herself, distinguished from her historical writings – though they are interspersed amongst the 65 chapters of her 13 states of prayer) were instigated by one of her early spiritual directors and were directed to him. Her *Relation de 1654* was a continuation of this spiritual writing but was written for her son (Sp 1: 142; Sp 2: 262, 340).

Her spiritual writings and one or two letters do provide us with some sense of the nature of the relationship between penitent and confessor. For Marie de l'Incarnation, who admitted numerous times in her writings to not being much of a reader (Sp 2: 188-189, 382), the relationship was a source of Church dogma and language (Sp 1: 164; Sp 2: 54, 77). It was through her confessors that she learned how to express herself acceptably on matters on which there would be very little tolerance of deviance: on the meaning of the Trinity, for example, or on matter of dreams and visions, or on the meaning of possession, or on the limits of reason and understanding (which was to be a source of certainty, not inquiry or critical skepticism). Father Poncet was a particularly important figure for Marie de L'Incarnation. It was he who sent her the *Jesuit Relations* in Tours to entice her to come to Canada. He was involved in the practical arrangements to get her, Mme de la Peltrie and their companions to Canada. When both she and Poncet were in Canada, they seem to have met frequently in her convent and to have shared their adventures, sometimes with humor. He may or may not have served as her confessor while he was in Canada. She did write to him about the state of her soul and after his retirement to France requested that he consider returning to Canada to assume the role of her personal confessor and the spiritual director of the Quebec Ursulines ([857]). There is a particularly interesting exchange between Father Poncet and Marie de l'Incarnation concerning the demonic possession of Mère de Saint Augustin of the Hospitaliers which reflects some of the complexity of the relationship between the educated priest and the spiritual nun. The woman in question had nursed a famous case of possession in Quebec for an extended period of time and had, unknown to most of her colleagues, herself become the victim of possession. Father Poncet, who might have written to any number of people in New France, wrote to Marie de l'Incarnation from Rome to ask what she thought of this case. She responded twice to the question. In her first response she began by declaring her lack of authority to speak of such matters ([886-888]). She described the good works and excellent behavior of Mère de Saint Augustin, and then directed Poncet to the woman's spiritual director, Father Chastelain, for further information. She commented that the woman was said to have rescued important people from these demons by herself wrestling with them in silence. She again expressed her reluctance to pronounce on these matters as 'men of science' were themselves uncertain and preferred to live in doubt than to place too much trust in these extraordinary visions. Her hesitation may have been derived, in part, from her personal experience of the possessions at Loudon (Sp 2: 179-180). She cited Father Ragueneau's more positive views on the matter. Then she commented that it was strange that Mère de Saint Augustin had never shared any of this with her superior, who was an enlightened, experienced and very virtuous woman. She concluded by saying that good works and fidelity to the rule were more important to her than visions of any sort. She concluded that letter with a discussion of the state of her own soul and the observation that she had had only one vision in her life: the one that had shown her Canada. In a second letter written more than a month later she retracted all possible criticism of Mère de Saint Augustin and seemed to accept that this Hospitalier was playing a crucial role in preserving the colony from devils ([911]). What is discernable in these two letters is considerable reticence to engage in discussion around these issues. All discussion was hedged with reference to male authority and judgment. These issues were dangerous and perhaps particularly so for women. Nevertheless, in the first letter, Marie de l'Incarnation expressed considerable skepticism about the truth of the matter – a skepticism that was entirely in keeping with her long-held views on dreams, visions or apparitions of the imagination. Between her first and her second letter she may have been influenced by the views of her spiritual director, or perhaps by a genuine concern for the reputation of Mère de Saint Augustin, or perhaps the second letter is a continued expression of anxiety about making any pronouncements on the matter.

It is impossible to state definitively what Marie de l'Incarnation thought of visions as her attitude shifted subtly with the changing religious politics of the century and in the context of the French engagement with the religious traditions of the peoples encountered in North America. Visions of one sort or another were a significant issue for Marie de l'Incarnation. Later in life she claimed to have had only one vision in her life: her vision of Canada in a dream. In her various writings, however, she mentioned a number of visions which she called either *visions imaginaires* or *visions intellectuelles*. One was of a building composed of crucified bodies (Sp 2: 237), another was of Seraphim (Sp 2: 119), another of the fires of Hell (Sp 2: 268), and she had recurring visions of the Trinity (Sp 2: 123, 129, 132). Those of the Trinity she tended to describe not as visions of the imagination, but as more of an insight ('des saintes impressions'), which she saw as akin to being visited by God. The distinction she made between imaginary visions and spiritual impressions is related to her views concerning the body and the contaminating nature of the body and the senses. Imaginary visions had to be embodied to be perceived by the senses, whereas an impression was purely spiritual and thus more noble and pure (Sp 2: 385). She expressed some anxiety and a lot of care around the issue of visions. She claimed to have been visited by the Devil, soon after hearing of the Loudon possessions. She was alarmed and frightened by the Loudon possessions and her relatively negative attitude to visions appears to have been reinforced by the excessive importance she felt the indigenous peoples of French Canada attributed to visions in dreams ([855]). In 1666 she advised her son that arriving at the Holy Spirit through faith was better than visions: it was more reliable, more meritorious, and better for one's humility ([764]). She was, however, prepared to accept dreams and visions if they came from 'persons of great virtue' such as those who claimed to have had a dream or premonition of the fire that destroyed the convent in 1650 (Sp 2: 326).

In addition to providing vocabulary and a sense of orthodoxy, her confessors also clearly played a role in limiting harmful and dangerous excesses. It was her first or second named confessor, either Dom François de Saint-Bernard, or Dom Raymond de Saint-Bernard, who prevented her from abandoning her son before the age of 12 (Sp 2:104). It was one of her early confessors who refused to post her sins publicly on the Church door (Sp 2:98). It was the Jesuit Father de la Haye, her fourth known confessor, who spent much of the first three years after her entry into the convent trying to help her cope with a deep depression, for which he prescribed writing (Sp 1: 139-140, 320-321, 334). And numerous confessors chastised her for excessive or un-prescribed self-punishment (Sp 1: 216, 390; Sp 2: 71). The latter may have been as much about directorial control as about concern for her spiritual or bodily well-being (Sp 2: 287).

It would have been natural and inevitable in such an intimate relationship that both confessor and woman would have shared something of their daily lives and the politics of their existence. For example, as a younger woman, Marie de l'Incarnation described the initiation of her relationship with her new confessor Dom Raymond de Saint-Bernard with some satisfaction:

> He interrogated me on my way of life and generally he wanted to know me in depth. He regulated me in all things and for prayer he forbade me to continue to avoid meditation and commanded me to abandon myself entirely to the conduct of the spirit of God who until then had directed my soul. I was to give him an account of all that was happening, which I did without fail all the time I was under his direction (Sp 2: 79).

It is unimaginable that a woman as connected as Marie de l'Incarnation would not have sought and exchanged information about mutual acquaintances, the Church in France and the missions both in Canada and beyond with her confessors. Inevitably, some of that information would have been both personal and pragmatically political as much as spiritual. Indeed, Marie de l'Incarnation on occasion condemned herself for using her exchanges with her confessor for the exchange of information she described as frivolous (Sp 2: 273; Sp 2: 293). But the relationship was primarily about spiritual guidance and in that realm too, in the case of women of strong conviction, the relationship was very far from simple or one sided (Sp 1:370-372, 194, 366).

Teresa of Avila (1515-1582) is a symbolic reference that recurs in Marie de l'Incarnation's life and in writings about her, beginning with her first biographer, her son Dom Claude Martin. Although most commentators have described the link between the two women as a certain strong-minded mysticism, perhaps their greatest similarity is their stance of independence and power vis-à-vis and through their confessors. It is clear that Teresa of Avila often chose her own confessors. When they did not measure up to her spiritual standards, effectively she fired them. As is clear from her correspondence with Father Poncet, Marie de l'Incarnation also sought to choose her confessors. We know from her son's biography that she had views about the quality of her confessors and must have conveyed those views to her son (Sp 1: 334, 330 n. a., 325, and Sp 2: 53). She herself chose to switch from seeking confessors among the Feuillant fathers to choosing amongst the Jesuit fathers (Sp 2: 182). Yet there are some passages in her spiritual writings – particularly but not exclusively in the Tours (or early) period of her writing which suggest a total abjection and humiliation before her confessors (Sp 1: 168). Great care must be employed in interpreting these passages as, given that humility was a feminine ideal and extreme humility an appropriate response to the presence of God (Sp 2: 152), in whose stead the confessor (and any superior) stood, this behavior may have been somewhat formulaic. Further, given the appropriateness of humility, it may also have been something of a control strategy (Sp 2: 98 but also Sp 2: 111). Extreme humility could bring the social power attached to extreme religiosity; something that Mme Guyart had experimented with after her husband's death and as she worked in her brother-in-law's house as a glorified servant. And she is recorded as having won at least one battle in the convent in Quebec by employing extreme humility (Sp 2 : 304). But the link between Marie de l'Incarnation and Teresa of Avila is more immediate and direct than this would suggest. Teresa of Avila developed with her confessor, Garcia de Toledo, the concept of *the vow of the most perfect*. It was this vow that Jérôme Lalemant suggested Marie de l'Incarnation might adopt in her relations with him ([898]). This vow, as described by Marie de l'Incarnation to her son, essentially set up a formula of at once enhanced power for the confessor and limited obedience for the penitent. The vow stated that she could absolutely follow the advice of her confessor (over all other advice or regulation) provided that three conditions were met:

1. The confessor was aware of this vow and that the person had made it.
2. That the penitent be the one to propose to the confessor those things which seemed to her to be of the greatest perfection. That she sought his view on them, which view would serve as an order to her.
3. That the thing specified to her was indeed of greater perfection, as far as she was concerned ([899]).

This formula gave the confessor power verging on that advocated by Molinos while keeping the penitent very clearly in control in the relationship. It was the penitent who proposed. The confessor could agree or dissent. But it was the judgment of the penitent about her own path to perfection that determined whether his advice would be followed or not.

The way this played out is very clear in an incident of great importance to Marie de l'Incarnation and central to the whole Catholic dilemma with the problem of a personal and direct relationship with God. Sometime after taking over her spiritual direction, Jérôme Lalemant objected to the very familiar way in which Marie de l'Incarnation addressed God in her prayers (Sp 2: 300). There was an intimacy in some of her conceptualizations of God that is startling. In her second meditation from her spiritual writings in Tours she referred to her experience as the wife of Christ in the following terms:

> All of his favours and particularly that of spouse were so vividly represented to my spirit that I could do nothing else but abandon myself in the arms of he whom I knew as my God, my Father, and my Spouse (Sp 1: 460-461).

And from her 1654 Relation:

> There is nothing in the realm of the senses that approaches this divine process but I must express myself in terrestrial terms, as we are composed of that matter. It was by divine touches and penetrations of him in me and in an admirable fashion reciprocally of me in him, such that, being no longer myself, I became him through intimacy of love and of union so that being lost to myself, I could no longer see myself, having become him by participation (Sp 2: 138).

Or in the form of a more direct address (in part drawn from the *Song of Songs* 8:1):

> But come, oh my Love, so that I may pour into you through a reciprocal love, in so far as my lowness allows, and that you, Love, can suffer it. *This is why I wanted to see you, my little brother, sucking the breasts of my* mother, o adorable Verbe Incarné, *in order to embrace you at my ease, and let no one be scandalized by it.* Because you were rendered thus for this, and that is why I want you (Sp 2: 126).

Elsewhere she described her communication with God as 'mouth to mouth' (Sp 2: 141) and as the 'unspeakable' communication of 'spirit to spirit' (Sp 2: 155). In response to Jérôme Lalemant's objection, Marie de l'Incarnation did just exactly what the third condition of the vow of the most perfect allowed: she turned to prayer and God and, finding that she could address God in no other way, declined to obey her confessor. Or, as Marie de l'Incarnation put it:

> In this state of union with God, it is impossible to subsist with any design that might create an opposition that is contrary to its operation (Sp 2: 301).

Her confessor could only concede in the face of the will of God.

Mysticism, at the heart of which lay a rejection of the earthly life, symbolized by the body, bestowed on Marie de l'Incarnation a freedom and an authority to which even the powerful Jesuits bowed. As a religious person she was, as she pointed out, 'an entirely other creature' (Sp 2: 318). Mysticism even leant her sufficient authority, within her conversations with God, to make declarations about the nature of God: 'No my Love, you are not fire, you are not water, you are not what we say you are. You are what you are in your glorious eternity' (Sp 1: 388). Perhaps the Church was right in rejecting mysticism toward the end of the seventeenth century (Sp 1: 534). Potentially its threat to Church hierarchy and authority was as great as was that of Protestantism.

Conclusion

This paper is a first tentative step toward writing a place-rich historical geography of the religious orders of early New France. It makes particular use of Malpas' conceptualization of the meaning of place. Although Malpas is very aware of scholarship in geography concerning place and space, he is not particularly concerned with the project of historical geography. He is, however, fascinated with Proust and *Remembrance of Things Past*. What draws Malpas to Proust and, particularly, to that famous section in which the taste of a crumb of a madeleine dipped in tea transports Marcel back to the Combray of his childhood, is the way Proust unfolds 'the multiplicity and unity of experience, and so of the world,' through place (Malpas 1999, 163). In Proust's reconstruction place is an inherently complex concept. It is a structure comprising spatiality and temporality (past, present and anticipated future), subjective experience and physical existence, the self and the social. The elements of the structure are each an essential element of place. Together, in their interaction, they create place. Proust is interested in the role of place in the construction (reconstruction) of the self whereas I am interested in reconstructing place through an analysis of the self or identity. My analysis begins in this paper with bodily experience of the world, and then explores the networks of contact, exchange and knowledge that do much to form the nested places of one individual in particular. In subsequent papers I plan to explore the narratives that individuals used to make sense of their place in the world, the shared narratives (churches, schools, properties, roads, etc., and their uses) that structured their everyday lives, and to delve further into the spiritual aspirations that, for some, gave ultimate meaning to identity. It is the unity of experience, identity and place that makes place so powerfully evocative for us. Glimpses of this unity are captured in film and novels but are so rarely captured in histories. I believe that this unity and complexity can serve as the very challenging foundation of a rigorous (historical) geography.

It may seem strange to base this first article on a seventeenth-century nun, but Marie de l'Incarnation is of interest on a number of levels. Her rich correspondence and spiritual writings provide a solid beginning for the study of place and its meaning in an environment of significant social and especially gender control. Her contacts with the Jesuits and the indigenous people of northeast North America also open the possibility of studying contrasting spiritual narratives – although that is not an issue that has been explored in this paper. She is of interest as a strongly independent and determined woman living with and through a powerful hierarchical institution and as a woman who in the face of considerable social obstacles and constraining social power, developed a strategy for social and spiritual freedom.

Her strategy was well tailored to the age in which she lived and to the narratives that structured the lives of the people with whom she interacted. It was through the rejection of the body, through extreme asceticism, that she was able to turn the prison of the body into a means of freedom. It was not only the body, but in theory, anyway, the reasoning mind, the self and her will that had to be rejected. But without will, there is no freedom, no power and no choice. Marie de l'Incarnation subordinated her will to the will of God and thereby won herself considerable power amongst her contemporaries.

Thus, although Marie de l'Incarnation was enclosed in a convent for many years, through her role as foundress and her various roles within the Quebec convent – superior, supply manager, instructor and spiritual guide to the novices – she acquired significant social power and an extensive network of interactions that made her place in the world far larger and far more significant than one might suppose. Through her daily activities she became widely known within the colony: to indigenous visitors, pupils and parents; to the French settlers; to the missionary fathers; and through her own letters and her son's posthumous publication of her spiritual writings, to a far larger readership in Europe. Through her letter writing, she maintained her contacts in France, kept up to date on key political events within Europe and especially within the Church, collected money and recruits for the Quebec convent, and probably for the first time developed a close and satisfying relationship with her son. There is ample evidence that she was aware of the liberating power of correspondence and also of the advantage it gave the Europeans vis-à-vis the indigenous peoples of Canada. Through her writing and reading but more through her personal interaction with her confessors, Marie de l'Incarnation maintained close contact with some of the most powerful ecclesiastics of the seventeenth century, especially the Jesuit fathers with whom she shared a mission, a narrative, and the places and spaces of God.

Note

1   Marie de l'Incarnation's spiritual writings have been published by Albert Jamet, ed., [Marie de l'Incarnation (1599-1672)]. *Écrits spirituels et historiques,* two volumes (Québec-Paris, 1929-39). They will be referred to as: Sp volume number: page number. References to her correspondence which has also been published by Guy-Marie Oury, ed., [Marie de l'Incarnation (1599-1672)]. *Correspondance* (Solesmes: Abbay Saint-Pierre, 1971) will be referred to as: [page number].

References

Ariès, P. (1960) *L'Enfant et la vie familiale sous l'ancien régime.* Plon, Paris.
Armstrong, K. (1986) *The Gospel according to woman. Christianity's creation of sex war in the West.* Elm Tree Books, London.
Buttimer, A. (1983) *The practice of geography.* Longman, London and New York.
Buttimer, A. and Seamon, D. (eds) (1980) *The human experience of space and place.* Croom Helm, London.
Davis, N.Z. (1995) 'Marie de l'Incarnation. New worlds' in *Women on the margins. Three seventeenth-century lives.* Harvard University Press, Cambridge and London.

Deleuze, G. (1988) *Foucault.* The Athlone Press, London.

Entrikin, J.N. (1991) *The betweenness of place.* Johns Hopkins University Press, Baltimore.

Foucault, M. (1978) *The history of sexuality.* Pantheon Books, New York.

Foucault, M. (1979) *Discipline and punish: the birth of the prison.* Vintage Books, New York.

Gutman, H. (1988) 'Rousseau's *Confessions.* A Technology of the self' in Martin, L.M., Gutman, H. and Hutton, P.H. *Technologies of the self. A seminar with Michel Foucault.* University of Massachusetts Press, Amherst, pp. 99-120.

Hutton, P.H. (1988) 'Foucault, Freud and the technologies of the self' in Martin, L.M., Gutman, H. and Hutton, P.H. *Technologies of the self. A seminar with Michel Foucault.* University of Massachusetts Press, Amherst, pp. 121-144.

Lapointe, G (1974) *Constitutions et reglements des premières Ursulines de Québec par Père Jérôme Lalemant S.J. 1647.* Monastère des Ursulines de Québec, Quebec.

Malpas, J. (1999) *Place and experience: a philosophical topography.* Cambridge University Press, Cambridge.

Sassen, S. (2002) At the *Network Worlds Symposium* held at Queen's University, Kingston, May-June 2002.

Thwaites, R.G. (ed.) (1846-47) *The* Jesuit Relations *and allied documents. Travels and explorations of the Jesuit missionaries in New France, 1610-1791. The original French, Latin, and Italian texts, with English translations and notes; illustrated by portraits, maps, and facsimiles* (volumes 1-7). The Burrows Brothers Company, Cleveland.

Trudel, M. (1999) *Les écolières des Ursulines de Québec, 1639-1686: Amérindiennes et Canadiennes.* Les Cahiers du Québec, Quebec.

Chapter 11

Placing the Holy

Gunnar Olsson

A philosophical problem has the form: 'I don't know my way about.'
Ludwig Wittgenstein

What does it mean to be human? A godly question as impossible to answer as not to pose, a monstrous creature too evasive to capture, too forbidden to leave alone. In the meantime the river runs, and by a commodius vicus of recirculation we are once more brought back to the swerve of shore and bend of bay, to that magic theater on whose stage the human actors H.C. Earwicker (alias Here Comes Everybody) and Anna Livia Plurabelle are found again, a chaosmos in which categories ride the surf, traces are erased, drops turn to clouds, clouds to drops.

The cosmological constellations flash back and forth, not the least to the *Enuma Elish*, the oldest creation epic yet to be discovered, a drama originally projected in cuneiform writing onto seven clay tablets now preserved in several editions and various conditions of readability, the first mapping of what it means to be human. In the hazy beginnings of those Babylonian beginnings from perhaps four millennia ago,[1] nothing exists except two positions, one above, the other below. So primeval is this situation that it takes place at a time 'when there was no heaven, no earth, no height, no depth, no name' (Sandars 1971, 73). No nothing, not even, as in the younger *Genesis*, an earth without form and void; in the conception of the Babylonian poet a void is obviously a void-in-itself, not merely a void of form.

Even though time has treated the tablets harshly, there is enough to suggest that the originator(s) understood that without names there may well be shapeless matter (more precisely fluid water rather than solid earth) but never meaning-filled meaning. Thus, in this beginning before the beginning

When on high the heaven had not (yet) been named,
(And) below the earth had not (yet) been called by a name;
(When) Apsu, primeval, their begetter,
Mummu, (and) Tiamat, she who gave birth to them all,
(Still) mingled their waters together,
And no pasture land had been formed (and) not (even) a reed marsh was to
be seen;
When none of the (other) gods had been brought into being,
(When) they had not (yet) been called by (their) name(s, and their) destinies had not (yet)
been fixed.
(*Enuma Elish* trans. Heidel 1951, Tablet I, 1-8)

Out of this watery cocktail of Apsu and Tiamat – neither stirred, nor shaken – eventually rose the undefined mist of their son Mummu, later described as 'the creator of heaven (and) earth, who directs the c[louds]; ... To whom no one among the gods is equal in power' (Heidel 1951, Tablet VII, 86, 88).

To reread the *Enuma Elish* is to encounter an early example of cartographical reason, to be reminded of the fact that we find our way through the unknown by constructing invisible maps of the invisible and by following the direction of culturally specific compasses whose needles point not towards the magnetic north pole but towards the socially taken-for-granted. For instance, it has already been shown that the very first words of the first line of *Enuma Elish* – the two words that have given the epic its name – are usually translated as 'When on high' or 'When above,' while the first words of the second line are rendered as '(And) below.' Indeed it is these fundamental concepts of 'above,' 'below,' 'when' and 'and' that form the meshes of the coordinate net in which the world is both captured and created. At least in this sense Edward Casey is correct: 'There is no creation without place' (Casey 1997, 16; also Casey 1993 and Smith 1987).

Here, as in any work of reconstruction, memory serves as the screen of projection par excellence, for memory is longer than we think. While to James Joyce history was a nightmare from which he was trying to awake, to Immanuel Kant the ideas of space and time were not merely 'intuitions,' but '*a priori* intuitions,' the seedbed of what for him formed the synthetic *a priori*. To these giants, as to everyone else, memory is meaning accumulated, culture transferred,[2] for the world we encounter is always a mediated world, never a collection of things-in-themselves; just as Bertrand Russell used to argue that our ordinary language conveys the metaphysics of the Stone Age, so twentieth century geneticists notice that our bodies are programmed for a life distinctly different from our own. Put differently, memory is 'the intersection of mind and matter' (Bergson 1991, 13), a saying which says that the thought (not necessarily the reality) of a *tabula rasa* is literally unimaginable;[3] were it not for the resistance of a reflecting wall there would be no images to capture, hence nothing to understand either. According to Søren Kierkegaard's aphorism, we live forwards and understand backwards, negotiating the abyss between the two worlds of being and understanding in the process.

The classical questions of Thought and Extension – the relations between invisible ideas and concrete matter – keep recurring, albeit always in another guise.

*

To understand is to be immersed in language, to be lodged in the connective between one expression and another. At least in that context, it is literally true that in the beginning is the word; as every Christian knows, the *Enuma Elish* is not alone in preaching that without names there may well be ups and downs, but neither gods nor rocks, neither knowledge nor understanding. Hence no humans either, for to be human is to be a very peculiar animal, member of a species whose individuals are kept together and apart by their use of signs. It is through them that we define and redefine who and what we are, it is through them that individual and society are connected. And yet. Speaking and understanding is never enough. I want to be believed as well, to share my understanding with others. As Immanuel Kant put it in one of the most profound statements ever to be made: 'in the synthetic original unity of apperception, I am conscious of myself, neither as I appear to myself, nor as I am in myself, but only that I am. This *representation* is an act of *thought*, not an *intuition*. [It follows] that I have no knowledge of myself as I am, but only as I appear to myself ... As for the knowledge of an object different from myself, I require ... an intuition also of the manifold in me' (Kant 1966, B 157). The upshot is that I cannot deny that I partake of multitude, that the many in me are one.

The only problem with that attitude is that there seems to be an explanatory gap – a sort of excluded middle – between what might be called a 'mind state' and a 'brain state,' the former an issue of understanding human understanding, the latter of understanding insensate molecules (itself, of course, a form of understanding, see Papineau 2002). Thus, while it is the business of imagination to connect, it is the purpose of understanding to make the principles of imagination explicit by relating them to the thesis that 'the consciousness of my own existence is, at the same time, an immediate consciousness of the existence of other things outside me' (Kant 1966, B 276). To explicate this 'thesis of the necessary unity of consciousness' is to specify the translation function through which I mirror myself in you and you in me. Kant's concern – like mine – is not with the specification of objective rules, but with judgment, i.e. with the 'subjective principles which are derived, not from the quality of an object, but from the interest which reason takes in a certain possible perfection of our knowledge of an object' (Kant 1966, B 694). Just as the map (of which the sign is a special case) is our privileged means for finding the way, so the travel story is the most effective device for transporting our imaginations from here to there; if we want to achieve objectivity in our observations, we have no choice but to move, not only our minds but our bodies as well. It cannot be said more clearly: by sharing our common understandings we show who we are.

Such is the ideal. As always, reality is different. For as soon as I set out on my epistemological journey, I quickly realize that what I *want* and what I *can* are never one and the same. Pushed to its limits by the joint forces of modal verbs, language itself reveals that the idea of perfect translation is as impossible as the construction of a perfect map. To be human is consequently not to be a semiotic animal pure and simple, but to be an epistemological and existential mongrel, a bastard sprung from two cousins of unclear ancestry, semiotic on the spear side, rhetorical on the spindle side.

Different as these modes of communication may be, they are nevertheless alike in the sense that both are rooted in desire – a self-referential desire so desirous that it can never be satisfied, a desire desiring not this or that but a desire desiring itself, a desire living off itself, a desire so strong that it devours its own children. Driven by its capacity of imagining what lies beyond its body, the semiotic animal tries to be what it knows it cannot be: a perfect sign perfectly communicated. Not like other apes steered solely by evolutionary chance and survival of the fittest, but a species uniquely molded in the processes of cultural transmission of purposeful action (Tomasello 1999; Boehm 1999). And so it is that to *homo sapiens* language is not merely a medium of reflection but the very foundation of self-knowledge, not merely a mirror but the taint of the mirror (see Gasché 1986).

The fact that chimpanzees know how to lie, does not mean that they also know how to write their autobiographies or how to paint their self-portraits. In parallel herewith, a dog may well expect his master to come home, but not that he will come home next week. But I, in my capacity as a human being, I know not only how to write my autobiography and paint my self-portrait, but I also know that at ten o'clock tomorrow morning I must deliver a lecture in the aula of the University building. The difference is that among all animals the humans are the only ones capable of realizing (and of saying) not only that we *have* a body but also that we *are* a body. Look, I am invisible! Invisible like the fish down there in the water, the prey that I have come down here to the bridge in order to tempt with my bait and catch on my hook, the thin fishing line and my power of imagination the only connections between us. The red-and-white float bobbing up and down, an abstract dance of life and death.[4] What a fantastic performance, what an incredible illustration of the fact that the human being is the only creature capable of killing for the sake of an idea! Power is likewise, for power is by definition a magic game of ontological transformation. Dreams are such stuff that we are made of.

Once again Søren Kierkegaard knew how to lay bare these theatrical roots of irony, lessons that the later existentialists were eager to adopt as their own: essence and phenomena misrelated; the world always beyond reach; suicide the only way to get away. As so often, etymology is instructive, for the term 'irony' has its roots in the Greek comic character *eirōn*, a clever underdog who by his wits triumphs over the boastful *alaōn*, thereby proving that even though comedy is often more revealing than tragedy, the two genres are related like the two sides of the same coin. Thus, while Aristotle held that tragedy imitates men who are better than the average, comedy those who are worse, Quintilian claimed that ethos is akin to the comic, pathos to the tragic, the former an everyday emotion for ordinary men, the latter a temporary state which strikes extraordinary men under extraordinary circumstances. In both cases an incongruity between ideal concepts and objective reality.[5] Kierkegaard, on his part, saw everything as contradiction, albeit with the difference that whereas tragedy is about suffering contradiction that culminates in irreconcilable ethics, comedy is about painless contradiction that culminates in an overflowing aesthetics. No wonder that we read tragedy alone and comedy together, for the former is concerned with individuals, the latter with classes. No wonder that we release our tensions by crying in private and laughing in public, no wonder that in the mosque the believer first prays alone, then in the company of his brethren.

Two types of tears. Two modes of socialization and two forms of cathartic cleansing. Two provinces in the Realm of Power.

*

In the map of power naming is the dematerialized point into which all understanding is condensed.

And for that reason of cartography we must now return to the *Enuma Elish*, the Babylonian creation epic which some have claimed is not a creation epic at all,[6] merely a well rehearsed story of how the god Marduk gains and retains his elevated position as LORD of lords. Regardless of how it is categorized, this ancient text is nevertheless a crucial document, for what it does is to lay bare the rhetorical techniques through which undifferentiated chaos is turned into ordered and differentiated cosmos – then as well as now. Since it was intended for recitation, it was cast into a type of poetry in which each line forms a distich such that the two halves typically stand in contrast to each other;[7] the Derridean sense of difference deferred is generated through an ingenious use of meter rather than rhyme, the latter form of rhythm unknown not only to Babylonian but to Hebrew writers as well.

On this rendering the *Enuma Elish* emerges as nothing less than a search for the *topos* of *topoi*, a text more concerned with the creation of people than with the creation of things, a foundational treaties on politics rather than religion. Its premise is that in the beginning of the beginning nothing has yet been formed, because in the beginning of the beginning no things have yet been named. All that exists are the two coordinates of above and below, cardinal positions waiting to be inundated by the fluids of Apsu and Tiamat, the former fresh, the latter bitter. And, as if to underline the spatiality of its own structure, the term 'Apsu' literally means 'abyss' and 'uttermost limit,' by linguistic coincidence connectable also to 'the great deep,' 'the primal chaos,' 'the bowels of earth,' 'the infernal pit' (Sandars 1971, 25 f); a perfect example of proper name and definite description merged into one, *the* place of the holy.

From the mingling of Apsu and Tiamat there eventually emerged three pairs of gods, each generation wiser and more intelligent than its predecessor, the figure of Nudimmud-Ea the most outstanding of them all. But noise annoys and as typical teenagers these youngsters eventually became such a nuisance that

> Apsu made his voice heard
> And spoke to Tiamat in a loud voice,
> 'Their ways have become very grievous to me,
> By day I cannot rest, by night I cannot sleep.
> I shall abolish their ways and disperse them!
> Let peace prevail, so that we can sleep.'

When Tiamat heard this she got very upset, furiously shouting

> How could we allow what we ourselves created to perish?
> (*Enuma Elish* trans. Dalley 1998, 234)

But Apsu receives contrary advice from his vizier Mummu and ignoring Tiamat's objections he decides to go ahead. Barely in time to avert the pending catastrophe the god Ea, superior in understanding, learns about the plot. Through a magic spell he puts Apsu to sleep, removes his crown and places it on himself. Thus having established his position, Ea kills Apsu.

On top of the corpse, i.e. across the abyss, Ea then builds a splendid house for himself and his wife Damkina. There, in the heart of holy Apsu, in the Chamber of Destinies, their son Marduk is conceived, the most awesome being ever to be.[8] When his father saw him he beamed with pride and his heart was filled with joy. Indeed

> He made him so perfect that his godhead was doubled.
> Elevated far above them, he was superior in every way.
> His limbs were ingeniously made beyond comprehension,
> Impossible to understand, too difficult to perceive.
> Four were his eyes, four were his ears;

When his lips moved, fire blazed forth.
The four ears were enormous
And likewise his eyes; they perceived everything.
Highest among the gods, his form was outstanding.
(*Enuma Elish* trans. Dalley 1998, 235 f.)

Already from birth Marduk is held to be everything, including a premonition of
Janus, the Roman gate-keeper who from his watch-tower at the middle of the bridge
controls both sides of the abyss at the same time. To impose his will on the
troublemakers, Marduk now blows a devastating flood-wave. But instead of calming
the rioters this act upsets them so much that they convene an assembly in which
Tiamat is authorized to declare war on whoever was responsible for the killing of
sweet Apsu. To that end she puts together a terrifying army of snakes with venom
instead of blood in their veins, a phalanx of dragons, a sphinx, a horned serpent, a
rabid dog, a scorpion man, a fish-man, a bull-man, and eleven more of the same kind.
As commander in chief she appoints Kingu, an upstart in whom she invested absolute
power and forced to share her bed lest he run out of control.

At this stage Ea learns about Tiamat's preparations. Charged with the task of
talking Tiamat out of her plans, the wise Anshar is dispatched as an emissary to her
court. Terrorized by what he sees, Anshar returns with the report that not only has he
himself failed to complete his mission but also that 'no one else can face her and
come back' (Sandars 1971, 81) Things are obviously getting out of hand and everyone
agrees that diplomacy will no longer do. Ea consequently summons Marduk to his
private quarters asking him to step forth, to show who he is by being equal to himself.
Marduk, who for the first time is now addressed as 'The LORD,'[9] first rejoices then
immediately proceeds to list the conditions under which he is going to accept the
challenge. These are his words, for emphasis three times repeated:

> Lord of the gods, destiny of the great gods,
> If I am indeed to be your avenger,
> To vanquish Tiamat and to keep you alive,
> Convene the assembly and proclaim my lot supreme.
> When ye are joyfully seated together in the Court of Assembly,
> May I through the utterance of my mouth determine the destinies,
>   instead of you.
> Whatever I create shall remain unaltered,
> The command of my lips shall not return (void), it shall not be
>   changed.
> (*Enuma Elish* trans. Heidel 1951, 29-30, Tablet II:122-129)

*

Through the utterance of my mouth shall I determine your destinies! In their
acceptance of this decree, the friendly gods demonstrate how they differ from Tiamat;
in her case words have lost their power and politics must be conducted by other
means. Since every audience demands its own propaganda, Marduk chooses not to
threaten the electorate with new taxes but invites them to a feast instead. For this,

They gathered together and departed,
All the great gods who determine [the destinies].
They entered the presence of Anshar and filled [the court of
     Assembly];
They kissed one another [as they came together] in the assembly,
They conversed (and) [sat down] to a banquet.
They ate bread (and) prepared w[ine].
The sweet wine dispelled their fears;
[Their] bod[ies] swelled as they drank the strong drink.
Exceedingly carefree were they, their spirit was exalted;
For Marduk, their avenger, they decreed the destiny.
(*Enuma Elish* trans. Heidel 1951, 35-36, Tablet III:129-138)

And out of the ashes from that revel, Marduk emerges as a splendid Phoenix, LORD of lords, God of gods. They erect for him a throne, they give him sovereignty over the whole universe and, as a premonition of what was later to reappear in the Christian confession, they observe that to trespass is to sin; more precisely

None of the gods shall transgress your limits,
(*Enuma Elish* trans. Dalley 1998, 250)

in the original rhetorically bejewelled, embellished and enhanced through the poetics of alliteration: *itukka la ittiq.*

Uncertainty nevertheless breeds suspicion and Marduk is now asked to prove himself. Accordingly

They sat up in their midst one constellation,[10]
And then they addressed Marduk their son,
'May your decree, O Lord, impress the gods!
Command to destroy and to recreate, and let it be so!
Speak and let the constellation vanish!
Speak to it again and let the constellation reappear.'
He spoke, and at his word the constellation vanished.
He spoke to it again and the constellation was recreated
(*Enuma Elish* trans. Dalley 1998, 250 and footnote 15, p. 275)

A magnificent show of ontological transformations. The magician of power in outstanding performance. Let there be! And there is. Ovations rising to the sky, on this particular occasion in Babylon, later in Jerusalem. When the gods (literally his fathers) see how effective his utterance is, they all rejoice, bless their LORD and chant MARDUK IS KING, OUR KING IS MARDUK!

Bestowing upon him the scepter, the throne and the royal robe, the assembly then send him off to 'pour out the life' of Tiamat. His weapons are numerous, but most decisive is the net in which he intends to catch her and the four winds by which he plans to blow her up. Kingu runs away in panic. Tiamat looses her mind, in anger her wits are scattered. And then,

Face to face they came, Tiamat and Marduk, sage of the gods.
They engaged in combat, they closed for battle.

The LORD spread his net and made it encircle her,
To her face he dispatched the [Evil] wind, which had been behind:
Tiamat opened her mouth to swallow it,
And he forced in the [Evil] wind so that she could not close her lips.
Fierce winds distended her belly;
Her insides were constipated and she stretched her mouth wide.
He shot an arrow which pierced her belly,
Split her down the middle and slit her heart.
Vanquished her and extinguished her life.
He threw down her corpse and stood on top of her.
(*Enuma Elish* trans. Dalley 1998, 253)

[For the rest of the troops]
He imprisoned them and broke their weapons.
In the net they lay and in the snare they were
(*Enuma Elish* trans. Heidel 1951, 41)

What a timeless text, the LORD spreading his net! If the former dictator of Iraq ever wanted an instructor in biological warfare, he need not go far to find him. If the generals of the Pentagon ever needed to know more about the relations between cartographical coordinates and the targeting of smart bombs, they need not experiment on the Chinese in Belgrade but may consult the *Enuma Elish* instead. Likewise with the priniciples of Derridean deconstruction. For when it comes to the handling of Tiamat's dead body, Marduk finds himself lodged in the abysmal interface between *de*-struction and *con*-struction. Like a professional butcher

He turned back to where Tiamat lay bound, he straddled the legs and smashed her skull (for the mace was merciless), he severed the arteries and the blood streamed down the north wind to the unknown ends of the world. ...
The LORD rested; he gazed at the huge body, pondering how to use it, what to create from the dead carcass. He split it apart like a cockle-shell; with the upper half he constructed the arc of sky, he pulled down the bar and set a watch on the waters, so they should never escape.
(*Enuma Elish* trans. Sandars 1971, 91-92)

Then the most decisive step of all. For when the epic now moves from the power struggles of the gods to the creation of the world, Marduk enters the stage disguised in new clothes, no longer dressed up in the warrior's coat of mail but in the uniform of the land surveyor. The implements changed as well, the magic net no longer a tool for capturing monstrous rivals but a device for ordering thing-like entities as stars, towns and people. The abyss nevertheless remains the ruler's privileged *topos*, for as

He crossed the sky to survey the infinite distance; he stationed himself above apsu, that apsu built by [Ea] over the old abyss which now he surveyed, measuring out and marking in.
(*Enuma Elish* trans. Sandars 1971, 92)

*

Measuring out and marking in, Marduk now proceeded to the construction of what eventually was to be a celestial globe. Onto the sky he projected positions for the great gods and through a kind of triangulation he set up three constellations for each of the twelve months. As the pole of the universe he chose the planet Jupiter (possibly Mercury), Tiamat's liver he made the mark of the zenith.

Once these heavenly bodies were in place, Marduk returned to Tiamat's corpse cutting it up into pieces to which he allotted new functions: out of her eyes welled Euphrates and Tigris, her paps became mountains, her crotch the fulcrum of the sky. The construction of the physical universe completed, he went on to the building of yet another temple and to the arrangement of yet another banquet. All in honor of himself. As usual under the influence of sweet wine the assembly shouted in unison, GREAT LORD OF THE UNIVERSE!

In response the Absolute delivered a speech from the throne, his four eyes and big ears simultaneously scanning the above and the below, the past and the future, no member untouched.[11] This is what he said to the gods his fathers:

> In the former time you inhabited the void above the abyss, but I have made Earth as the mirror of Heaven, I have consolidated the soil for the foundations, and there I will build my city, my beloved home.... It shall be BABYLON. [And there the fallen gods, our defeated enemies, the supporters of Tiamat, will be serving us day after day].
> (*Enuma Elish* trans. Sandars 1971, 95-96)

Yet another case of ontological transformation, yet another instance of invisible ideology turned into touchable stone, the Earth made a mirror of Heaven, the Heaven itself a material projection of social relations. Babylon the most marvelous of cities, the center of the world, its own street pattern drawn in the shape of a net.

And yet. For absolute rulers enough is never enough. Moved by the desire of creating a work of consummate art, Marduk therefore conceives a plan designed to solidify his position and perpetuate his rule. To placate the supporters he promises that they will never need to work again, that after the next election leisure will be their lot. Marduk is wise enough to honor the pledge and to that effect he swiftly creates a primeval man, the prototype of you and me, by definition slaves of the gods. Not an invention formed in the image of the Almighty but a savage concoction stirred together from the blood of the slaughtered Kingu, mankind a dish of *Boudins à la Mésopotamie*. Nothing like a perfect copy of the perfect original, merely a black sausage. A deed appropriately described as a deed impossible to describe.

Then Marduk – at this final stage called 'the King' – divided the gods into two groups, each vassal assigned his proper place in the feudal system. Three hundred he stationed as watchers of Heaven, an equal number as guardians of the Earth. In return for their new-won freedom the once defeated erected yet another temple above the abyss, this time in honor of their conqueror turned benefactor. As a token of appreciation the latter in return invited them to yet another celebration, for the first time featuring nothing but black-headed waiters. And the next morning the entire congregation was taken on a tour through the Hall of Armory, where the intricate construction of the net was explained and the workmanship admired.

To top it off, the banquet closed with the performance of a liturgy in which the occult was explained in a way that no one could misunderstand. At the end, which in effect is not the end but another beginning, Marduk is the King, Marduk is the Absolute, Marduk is the LORD of lords. Hail to the Chief! Fifty were his names, so numerous that if ever attacked he could always hide behind another alias. Never catchable as the specific this *or* that, always on the move as an ambiguous this *and* that.[12] Never a logical either-or, always a dialectical both/and. Many in one, one in many. Ungraspable multiplicity, like the Hebrew JHWH a tautology defining himself as a being who is who he is. Jacques Lacan was obviously neither the first nor the last to understand that the real is given its structure by the power of the name; as Gilles Deleuze and Félix Guattari aptly put it, 'the proper name does not indicate a subject [but] designates something that is of the order of the event, of becoming or of the haecceity. It is the military men and meteorologists who hold the secret of proper names, when they give them to a strategic operation or a hurricane. *The proper name ... marks a longitude and a latitude*' (Deleuze and Guattari 1987, 264, emphasis added). Heaven above and Hell below, the sinister to the left the righteous to the right. Cartographical reason *in nuce*.

*

In the mist-enveloped regions of religion naming is the name of the game, an exercise in ontological transformations where earthly people appear as projections of heavenly gods, social relations as signs in the sky. And to me this is the central message of the *Enuma Elish*: the world created is the mirror image of the Court of Power, literally a representation of a representation, a map of a map, a signified meaning searching for its own coordinates; nascent sociology turned to mythical astrology, mythical astrology to pre-scientific astronomy, time-bound star-gazing to space-bound exploration. Different ways of finding the way.

And once every twelfth month (on the fourth day of the New Year's celebrations held at the time of the spring equinox) the epic was recited, the ruling King first divested of his insignia, then – following the negative confession that the sinner was not he but his enemies – he was dressed up again in his robe of stars and stripes, an embodied constellation of points and lines projected onto the sky itself. Yet the real *topos* of power is always lodged in the abyss, for just as the primeval Apsu occupied the empty space between categories so do his contemporary descendants.

There it is, the altar of alterity, the holiest of the holy. Heavenly Chamber of Destinies off limits to everyone except the King and the High Priest. And out of the blue emerges a Duchampian nude descending a staircase, *ma Sœur* turning *Madame* in the process. Yet another re-Joycean miracle, yet another instance of how a commodius vicus of recirculation brings us back to Howth Castle and Environs. All done as a tribute to Anna Livia Plurabelle.

Notes

1   The exact dating is highly disputed. While the oldest manuscripts thus far discovered stem from the ninth century BC, there is general agreement that the original is considerably older, possibly even from the time of Hammurabi, who ruled from 1792 to 1750 BC. As will eventually become clear, the translations are many and sometimes quite diverging, not

only in form but in content as well. See translations of *Enuma Elish* in Dalley (1998), Heidel (1951), Sandars (1971), Westenholz and Westenholz (1997).

2   The Greek roots reach deeper into the western unconscious than we often care to admit, a point well made by Jaeger (1934, 1944, 1947). A more recent excavation of similar themes is Zeruneth (2002). At the same time I am fully aware that herein lies not only insight but blindness as well. Thus, for comparisons between Greek and Chinese modes of understanding see Jullien (1998).

3   For a conflicting view witness the following quotation from Norman Malcolm (1993, 65 f.): 'A conspicuous feature of philosophical thinking about memory has been the insistence that in remembering there must be an image, or copy, or picture, or mental representation, of what is remembered. Russell maintained that memory demands an image. In fact [though], images are sometimes present and sometimes not, when we remember or recall situations, events or objects that we have perceived, witnessed or experienced in the past. Why is there this demand that there must be an image, or 'something like an image,' in remembering?'

4   A wonderful variation on this theme is in the autobiographical tale of Reuterswärd (1989).

5   On this incongruity see Schopenhauer (1966, esp. Vol. II, Ch 37), ostensibly about poetry, in fact about tragedy. On related themes in comedy see Bergson (1917). Also see Malantschuk (1971), and, especially, Hegel (1967, 725-749).

6   For an extreme position see the pamphlet by Rabbi Dr. I. Rapaport (OBE) (1979).

7   For illustrative examples see the introduction in Heidel (1951, 15-17). Since this type of poetry is literally non-translatable – and since my own interpretations build entirely on various renderings into English, Danish and German – much will clearly be lost. Trying to find my way in the wilderness I have picked my quotations not from one but from several of the classical translations. About Babylonia I am fully aware that I am saying nothing and my only hope is that the experts will not be too offended by the inevitable mistakes. A useful introduction to the historical setting of the various poems is in the introductory chapter of Lambert (1960).

8   In this context it is worth recalling the Gnostic world of Valentinus, in which the sacred place of radical transcendence is located in the Bythos, by definition a bottomless abyss containing nothing but the nothingness of radical silence; indeed ignorance is raised to an ontological position of the first order. According to this heretical theory formulated around 150 AD, the beginning of the beginning was in the primal Father named 'Bythos' (literally 'abyss'), uncontainable and invisible. Sometimes this Pre-Beginning acts alone, sometimes in concert with his own thought variously called 'Grace' and 'Silence' but always given in the feminine gender. Together these forces eventually produce an extended family of offsprings among them the monster Achamonth, herself the child of the Aeon Sophia and a power called 'Limit,' the latter sometimes confirming and strengthening, sometimes dividing and delimiting. It was this misfit who in her wickedness brought forth the Demiurge, purportedly made in the image of the primal Father, but in effect a false and ignorant copy of the original Bythos. In the mistaken belief that he himself was the abyss, the Demiurge (by Valentinians understood not as the primal Father of Bythos but as a strange mixture of Plato's 'creator of the world,' as described in the *Timæus*, and the Jewish God, as presented in the Old Testament) then crafted heaven and earth and everything else there is. Perhaps an echo from pre-Christian Greece, perhaps an attempt to tie the evil in the world to a premundane fault, perhaps an early case of iconoclasm. In my own imagination a fantastic troupe of categorial jugglers performing their ontological tricks on the floor of the Saussurean Bar–in–Between. For background discussions see Grant (1961, esp. Ch. 5); Jonas 1963; Olsson 2000, esp. pp. 44-54; Besançon, 2000, esp. p. 86 ff.

9   On this important point all translations seem to agree. As proof see Heidel (1951, 29, Tablet II:102 and Tablet II:120); Pritchard (1955, 64, Tablet II:102 and Tablet II:120);

Sandars (1971, 81-82). In Danish the equivalent of the LORD is HERREN, which consequently is the term used in Westenholz and Westenholz (1997, 206)

10   The term here translated as 'constellation' has often been rendered as 'garment,' sometimes even as 'a kind of apparition,' perhaps a warning against the temptations of nudity; neither gods nor emperors may be seen naked. I nevertheless prefer the word 'constellation,' because that term reflects the importance of astronomy and astrology in Babylonian culture. The interpretational problem has its roots in the cuneiform signs *ba* and *ma*, which are exceedingly difficult to distinguish from each other.

11   Here it should be noted that Marduk's enormous ears were considered true signs of wisdom. In this respect there are obvious connections between the *Enuma Elish* and the *Hebrew Bible*, for in both cases hearing was taken to be the outstanding mode of understanding. With the Greeks and the advent of Jesus Christ this preference was changed and sight became the privileged sense, the knowledgeable seers literally leading the ignorant blind.

12   For detailed and fascinating discussions of Marduk's 50 names – not the least as they reflect the invention of counting in tens – see Böhl (1936, 191-218) and Bottéro (1977).

References

Bergson, H. (1917) *Le rire: essai sur la signification du comique*. Félix Alcan, Paris.

Bergson, H. (1991) *Matter and memory*. Zone Books, New York.

Besançon, A. (2000) *The forbidden image: an intellectual history of iconoclasm*. University of Chicago Press, Chicago.

Boehm, C. (1999) *Hierarchy in the forest: the evolution of egalitarian behavior*. Harvard University Press, Cambridge, MA.

Böhl, F.M.T. (1936) Die fünfzig Namen des Marduk, *Archiv für Orientforschung* 11.

Bottéro, J. (1977) 'Les noms de Marduk, l'écriture et "la logique" en Mésopotamie ancienne' in Jong, E.M. de (ed.) *Essays on the ancient Near East in memory of Jacob Joel Finkelstein*. Archon Hamdon, Connecticut.

Casey, E.S. (1993) *Getting back into place: toward a renewed understanding of the place-world*. Indiana University Press, Bloomington.

Casey, E.S. (1997) *The fate of place: a philosophical history*. University of California Press, Berkeley.

Dalley, S. (1998) *Myths from Mesopotamia: creation, the flood, Gilgamesh, and others*. Oxford University Press, Oxford.

Deleuze, G. and Guattari, F. (1987) *A thousand plateaus: capitalism and schizophrenia*. University of Minnesota Press, Minneapolis.

Gasché, R. (1986) *The taint of the mirror: Derrida and the philosophy of reflection*. Harvard University Press, Cambridge, MA.

Grant, R.M. (ed.) (1961) *Gnosticism: a source book of heretical writings from the early Christian period*. Collins, London.

Hegel, G.W.F. (1967) *The phenomenology of mind*. Harper and Row, New York.

Heidel, A. (1951) *The Baylonian Genesis: the story of creation*. Second edition, University of Chicago Press, Chicago.

Jaeger, W. (1934) *Paideia: Die Formung des Griechischen Menschen* (Erster Band). Walter de Gruyter, Berlin.

Jaeger, W. (1944) *Paideia: Die Formung des Griechischen Menschen* (Zweiter Band). Walter de Gruyter, Berlin.

Jaeger, W. (1947) *Paideia: Die Formung des Griechischen Menschen* (Dritter Band). Walter de Gruyter, Berlin.

Jonas, H. (1963) *The gnostic religion: the message of the alien God and the beginnings of Christianity.* Beacon, Boston.

Jullien, F. (1998) *Un sage est sans idée: Ou l'autre de la philosophie.* Seuil, Paris.

Kant, I. (1966) *Critique of pure reason.* Anchor Books, Garden City, NY.

Lambert, W.G. (1960) *Babylonian wisdom literature.* Clarendon Press, Oxford.

Malantschuk, G. (1971) *Kierkegaard's thought.* Princeton University Press, Princeton.

Malcolm, N. (1993) *Wittgenstein: a religious point of view.* Routledge, London.

Olsson, A. (2000) *Läsningar av INTET.* Albert Bonniers, Stockholm.

Papineau, D. (2002) *Thinking about consciousness.* Oxford University Press, Oxford.

Pritchard, J.B. (ed.) (1955) *Ancient Near Eastern texts relating to the Old Testament.* Princeton University Press, Princeton.

Rapaport, I. (OBE) (1979) *The Babylonian poem Enuma Elish and Genesis chapter one: a new theory on the relationship between the ancient cuneiform composition and the Hebrew Scriptures.* Hawthorn Press, Melbourne.

Reuterswärd, C.F. (1989) *Titta, jag är osynlig!* Gedins Förlag, Stockholm.

Sandars, N.K. (1971) *Poems of Heaven and Hell from ancient Mesopotamia.* Penguin, Harmondsworth.

Schopenhauer, A. (1966) *The world as will and representation.* Dover, New York.

Smith, J.Z. (1987) *To take place: toward theory in ritual.* University of Chicago Press, Chicago.

Tomasello, M. (1999) *The cultural origins of human cognition.* Harvard University Press, Cambridge, MA.

Westenholz, U. and Westenholz, A. (1997) *Gilgamesh og Enuma Elish: guder of mennesker i oldtidens Babylon.* Spektrum, København.

Wittgenstein, L. (1997) *Philosophical investigations.* Blackwell, Oxford.

Zeruneth, K. (2002) *Træhesten: fra Odysseus til Sokrates.* Gyldendal, København.

Chapter 12

Masquing and Dancing the Theater of State: The 'Invention' of Britain as a Landscaped Body Politic

Kenneth R. Olwig

Introduction

Britain has long been recognized as being of world importance in the creation of the scenic landscape garden. It is less known that Britain, in a certain sense, was itself a creation of the 'invention' of landscape as scenery. In 1603, when King James VI of Scotland (1566-1625) became James I of England, the idea of uniting the two kingdoms under one royal head as 'Britain' was foreign to most people in both countries. James wished to have himself crowned king of 'Britain,' but the English parliament, whose power was legitimated by the specific force of English custom and common law, was opposed, just as the idea also was questioned Scotland. The Stuart Court replied, in part, by creating landscape scenic 'inventions' (as they were called at the time) for the state's theater, to foster the idea that the geographical body of Britain was a natural body politic, with James as its masculine head and a populace that was encompassed by his body and danced to his drum. This mode of representation provided a means of visualizing the incorporation of England and Scotland within one state as a form of marriage between James, the *head* of state, with his *body* politic. James' wife, Queen Anne of Denmark (1574-1619), played a central role in the orchestration of this vision of landscape through her production of fabulously expensive theatrical productions, called *masques*, which combined text, music, dance, rhythm and spectacular scenery. Anne was not the dumb blond often depicted in British historiography. She had a key role in developing the cultural politics of the Stuart court, drawing upon experience (and talent) garnered from the courts of her Northern European home region. In the first such production, *The Masque of Blackness* from 1605, Anne played the lead female role as a black African princess made white (*fair*/beautiful) by the mild male Stuart sun of Britain – thus linking ideas of race, gender and insularity to that of the then new concept of landscape as scenery. These inter-linked ideas continue to be relevant to the discourse of landscape, and continue to underline the political role of landscape as the expression of both a geographical body and a body politic. This fusion of the notion of landscape with that of the body is a singularly powerful example of the geographical use of organic metaphor, as described by Anne Buttimer in her influential analyses of the use of root metaphors in geography (Buttimer 1982; 1993).

Making Believers with Landscape Make-Believe

The rise of landscape as 'make-believe' is succinctly chronicled in this statement by Yi-Fu Tuan, one of the early pioneers, together with Anne Buttimer, of humanistic geography:

> Scenery and landscape are now nearly synonymous. The slight differences in meaning they retain reflect their dissimilar origin. Scenery has traditionally been associated with the world of illusion which is the theater. The expression 'behind the scenes' reveals the unreality of scenes. We are not bidden to look 'behind the landscape,' although a landscaped garden can be as contrived as a stage scene, and as little enmeshed with the life of the owner as the stage paraphernalia with the life of the actor. The difference is that landscape, in its original sense, referred to the real world, not to the world of art and make-believe. In its native Dutch, 'landschap' designated such commonplaces as 'a collection of farms or fenced fields, sometimes a small domain or administrative unit.' Only when it was transplanted to England toward the end of the sixteenth century did the word shed its earthbound roots and acquire the precious meaning of art. Landscape came to mean a prospect seen from a specific standpoint. Then it was the artistic representation of that prospect. Landscape was also the background of an official portrait; the 'scene' of a 'pose.' As such it became fully integrated with the world of make-believe (Tuan 1974, 133).

This statement raises the question, what is it that the 'English' were 'making believe' at the time landscape was transformed into scenery? An answer to this question might well be that they were making believe that England, Wales and Scotland formed a polity called 'Britain.' The crowning of James VI of Scotland as James I of England brought the possibility of creating a united British kingdom near to fruition, but it did not become a reality – Parliament balked.

Time and time again, despite James' almost pathetic pleading to the Parliament, his will was frustrated by the recalcitrant MPs (Camden 1806 [1607], vol. iv, 1-5; James 1918a [1616], 286; James 1918c [1616], 293-305; James 1918d [1616], 271-273). James considered it to be 'a foolish Querke of some Iudges, who held that the Parliament of *England,* could not vnite *Scotland* and *England* by the name of *Great Britaine,* but that it would make an alteration of the Lawes' (James 1918b [1616], 329). The judges and Parliament, however, stood by their quirks, and they had 'good reasons' for doing so, because 'such a fusion of the laws and Governments of the two nations,' to use the historian G.M. Trevelyan's words, would have 'rendered the King independent alike of Scottish nobles and of English Parliaments' (Trevelyan 1960 [1925], 100-101). The idea of Britain, though popular with the court, apparently was not just unpopular with English (and Scottish) parliament; it also was broadly unpopular with the masses. This is suggested by an anonymous popular writer who wrote, in a message directed to James, that the people:

> make a mock of your word 'Great Britain,' and ofter to prove that it is a great deal less than little England was wont to be, less in reputation, less in strength, less in riches, less in all manner of virtue (quoted in Trevelyan 1960 [1925], 123).

Landscape: From Place and Polity to Scenery

The transformation of landscape into a realm of scenic illusion can be seen to occur in the introduction to *The Masque of Blackness*. This passage was written by Ben Jonson (1573-1637) and staged by Inigo Jones (1573-1652). This passage not only contains one of the earliest uses of landscape in the modern English language, it also describes what may have been the earliest use of perspective scenery in Britain. The passage reads:

> First, for the Scene was drawn a Landtschap ... which falling, an artificial sea was seen to shoot forth, as if it flowed to the land raised with waves which seemed to move, and in some places the billow to break, as imitating that orderly disorder which is common in nature ... the scene behind seemed a vast sea, and united with this that flowed forth, from the termination or horizon of which (being on the level of the state, which was placed in the upper end of the hall) was drawn, by the lines of perspective, the whole work shooting downwards from the eye ... So much for the bodily part, which was of Master Inigo Jones his design and act. (Ben Jonson, *The Masque of Blackness*, 1605)

Jonson's description of Jones' scenery for *The Masque of Blackness* contains important clues to how theater scenery could be made to appear to embody the body politic of Britain and thus provide the basis by which, as Régis Debray has remarked, the nascent state could make itself visible and let itself be heard. 'It is,' as he put it, 'the theater of the state which creates the state, just as the monument creates memory' (1994). The theater was so constructed that (to use Ben Jonson's words) the 'termination or horizon' of the scene was at 'the level of the state which was placed in the upper end of the hall.' This horizon is so 'drawn by the lines of perspective' that 'the whole work' appears to be 'shooting downwards from the sky' (Jonson 1969, 48-50). The focal point for the lines of perspective is the elevated seat of 'state' or royal throne.[1] This throne of state was elevated above the public upon a palisade so that the King's eye is at the focal point of the lines of perspective that gives the scenery the illusion of three-dimensional depth. At the one end of the hall we have the royal spectator, the *head* of state, commanding the scene from his seat of power. At the other end we have the spectacle of the lightly clothed bodies of the Queen, and her consort, forming a living, moving and dancing dimension of the stage scenery – the '*bodily part,*' as Jonson called it, using a metaphor that was no mere figure of speech, but a 'root metaphor' (Buttimer 1982; 1993), which played a major role in the legitimization of James' idea of a British state.

During the era of James' regal predecessor, Queen Elizabeth, the monarch had made regular cyclical journeys, called progresses, from place to place in her lands in order to be bodily present among the bodies of her subjects, as a vital method of sealing a *corp*orate pact between them. The theater, however, changed this relationship. Now it was the king, upon a lofty palisaded throne of state, who commanded the scenery of the 'bodily part' with his abstracted gaze. The progressive movement now took place as stage after stage of scenery was changed as an expression of the linear vision of the monarch (Olwig 2002b). The masques represented a radical distancing of the spectator's eye from the body of the players, which would have been foreign to earlier forms of theater. This distance, however, was broken in the course of the play when, at a signal, the players were wedded to the spectators in a measured, geometric dance.

King James followed, as head of the state church, the tradition of the Christian church in defining Christ's church in terms of a mystic body, incorporating its member parts. In the words of Francis Bacon the monarch was as 'a body corporate in a body natural, and a body natural in a body corporate' (*corpus corporatum in corpore naturali, et corpus naturale in corpore corporato*) (quoted in Kantorowicz 1955, 80). The theater made it possible for James to make believe that the polity of the *Landtschap* is embodied, or incorporated, within the illusory bodily space of his state as body politic. Or, as jurists of the time were wont to put it, the ruler has:

> two Bodies, the one whereof is a Body natural . . . and in this he is subject to Passions and Death as other Men are; and the other is a Body politic and the Members thereof are the subjects, and he and they together compose the corporation, and he is incorporated with them and they with him, and he is the Head, and they are the Members (quoted in Kantorowicz 1955, 91).

King James saw himself as the monarch of a unified Britain. The space of the landscape scenery made for him by Inigo Jones, encompassed a similarly expansive vision. James loved to make believe that he was the head of a body politic defined by the geographical body of Britain, but he never actually succeeded in forming a single state (this took place a century later). James saw the amalgamation of once independent countries or lands into sub-units of a progressively larger state as an ongoing process that would soon lead to the creation of the unified state of Britain. When James became king of England in 1603, he not only unified the crowns of the greatest British Isle but also was put in the position of being able to unite the isles of Britain through the 1603 defeat of the Irish – which had a somewhat anomalous position because it belonged to a separate island, and because James' only right to rule the Irish was the right of conquest. The new Stuart monarch could thus complete the work of union begun by Henry VIII, who had taken the title king of Ireland in 1541 and who had effectively brought Wales within the English realm by 1543. The subjection of Scotland to England, according to a 1607 speech by James to Parliament, would thus mean that Scotland would 'with time become but as Cumberland and Northumberland, and those other remote and Northerne Shires.' Scotland, like the 'other Northerne Countreys' would be 'seldome seene and saluted by their King, and that as it were but in a posting or hunting ioruney' (James 1918c [1616], 294).

It was not before 1707 that the English and Scottish states actually coincided, through the union of parliaments, with the geographical body of Britain. It took a long time for this idea to mature, but the scenic illusion of landscape facilitated the idea that differing, historically constituted, polities and places can be unified within the space of a body politic as embodied by geographical body. In this way the separate *lands* or countries of *Scotland* and *England* could be unified within a body of dry *land*, bound by a ring of water – an element highly symbolic of wholeness and totality (Buttimer 1985). The *land* in *land*scape is hereby transformed from a country and a commonwealth, to the scenic *land* surface upon which the drama of the commonwealth is not only acted out, but also danced, by the bodies of the body politic. The point of the landscape scene at the opening of *The Masque of Blackness*, in fact, is to set the stage for just such drama and ballet.

The Wedding of Body and Landscape

When incorporated into the bounds of the geographic body of Britain, the polity transmogrifies into a natural national body, bound by bonds of soil and blood, analogous to the mystical body and blood of Christ in the Christian communion. Through such imagined bonds of blood a community of diverse peoples, embodied by the blessed isles of Britain, was envisioned by the architects of the Stuart court as a single new British nation state. The theater in which James sat upon his elevated throne of state, from which his eyes commanded a unified view of the body politic, thus represented a unity – the theater of state – that, for him, was truly corporeal. For James, the union of England and Scotland under his headship was not simply a political matter, it was corporeal, as in a marital union which, as he put it in a speech to the English Parliament, was 'made in my blood, being a matter that most properly belongeth to me to speake of, as the head wherein that great Body is vnited' (James 1918d [1616], 271). James' 1603 speech to the English Parliament perfectly captures this argument:

> Do we not yet remember, that this Kingdome was diuided into seuen little Kingdomes, besides Wales? And is it not now the stronger by their vnion? ... But what should we sticke vpon any naturall appearance, when it is manifest that God by his Almightie prouidence hath preordained it so to be? Hath not God first vnited these two Kingdomes both in Language, Religion, and similitude of maners? Yea, hath hee not made vs all in one Island, compassed with one Sea, and of it selfe by nature so indiuisible ... These two Countries being separated neither by Sea, nor great Riuer, Mountaine, nor other strength of nature ... And now in the end and fulness of time vnited, the right and title of both in my Person, alike lineally descended of both the Crownes, whereby it is now become like a little World within it selfe, being intrenched and fortified round about with a naturall, and yet admirable strong pond or ditch ... What God hath conjoined then, let no man separate. I am the Husband, and all the whole Isle is my lawfull Wife; I am the Head, and it is my Body[2] ... (James 1918d [1616], 271-273).[3]

The Queen's Masque

As the surveillant head upon the elevated throne of state, James was given a privileged commanding view of the stage, or 'bodily part.' The queen, however, also played an important role as the female object of the king's male gaze, playing upon centuries-old gendered discourses counterpoising a male celestial sphere, to that of a female earth under the watchful eye of an Apollonian godhead (Olwig 1993; Cosgrove 2001). This was a role that this independent minded queen, who kept her own court, gave herself, thereby making herself, like Elizabeth on a progress, the object of the spectator's gaze.[4] *The Masque of Blackness* was also called 'The Queen's Masque' because it was Anne who conceived the idea for it, arranged to have it produced with a text by Ben Jonson and scenery by Inigo Jones, and played the leading role. Inigo Jones, though

English, had worked for Anne's illustrious brother, King Christian IV of Denmark, and had helped bring Renaissance Italian ideas of design to the Danish court. Anne now headhunted him for the Stuart court to create the scenography and costumes for a long series of masques, many with texts by Jonson – she also mentored Jones as the architect of her palace at Greenwich, helping him to become Britain's pre-eminent pioneering Renaissance architect.

The Masque of Blackness, together with its sequel, *The Masque of Beauty*, is about a bevy of black African princesses (played by Queen Anne and her entourage made up in blackface and body) who have set sail for a place called Britannia, where they believe the mild sun (symbolic of its Apollonian Stuart monarch) will make them fair, and hence beautiful. The dramatic meeting of water and land in this opening scene is intended to set the stage for these princesses' eventual landing on 'British' soil. The fact that this is 'British,' and not 'English,' soil was laden with political meaning at a time when the ancient Roman era name of 'Britain' was undergoing a revival at the hands of supporters of the unification of Scotland and England/Wales as one state.

The Masque of Blackness used great sums of money to create central point perspective techniques that would trick the eye into seeing the scenery as a three-dimensional, bodily whole. The masque was directed to the elite of English society, and gave an aura of reality to the idea of Britain as a unified territory that could be encompassed by the superior vision of a single state. It thus provided a unified visual means of representing a British body politic, which would counter the fragmented discursive representation of the English (and Scottish) Parliament. The scenic illusion was created through the use of techniques developed in cartography, by which a panoramic view could be created. The scenery described by Jonson at the outset represents the line between Britain and the sea that defined Britain as a geographical body and a *land*mass. The 'land' depicted in this *land*scape is made up of rock and soil, unlike the *land* in the words Eng*land* and Scot*land* which are the places, or countries, of a people (Olwig 1993; 2001). The princesses in this masque do not become English by entering England and adopting its culture and language. They become British by crossing the line between water and dry land that enables them to become physically white, the mark of a person of the British soil.

The British Name

Ben Jonson makes a great issue of the British name in the masque's text. The revival of the British name had been given great prominence by the supporters of the union of the kingdoms in anticipation of James' crowning and they continued to do so, of course, after the crowning. There can be little doubt that Jonson derived his conception of Britain from his mentor and teacher, William Camden (1551-1623). Jonson felt that he owed to Camden: 'All that I am in arts, all that I know' (Jonson 1985, 226).[5] Camden was a founding figure in British regional geography (chorography), antiquarian studies and 'civil' history more generally (Taylor 1934, 9-13; Trevor-Roper 1971; Cormack 1997). His scholarly fame was based upon his 'chorographical' study *Britannia*, which was first published in Latin in 1586. Later editions came with a frontispiece in which a map of the blessed British isle is flanked by Neptune and an Albion-like figure (Camden 1806 [1607]). The idea that terra firma was made up of organically shaped bodily parts was built into the 'choreographic' approach to geography. According to Camden's preface to *Britannia* he 'entered the theatre of this learned age' in order to illustrate 'the antient state of my native *country* of Britain' (Camden 1806 [1607], xxxv, emphasis mine). His goal was 'to restore antiquity to Britain, and Britain to her own antiquity' (quoted from the 1586 edition of *Britannia* and translated by Boon 1987, 3). Jonson felt that it was principally to Camden that 'my country owes the great renown and name wherewith she goes.'[6] In the 1607 edition of *Britannia*, however, Camden modestly (and wisely) gave the credit for the revival of Britain to James I. It is dedicated to 'James, King of Great Britain, France, and Ireland; Defender of the Faith; born for the eternity of the British name and Empire' (Camden 1806 [1607]).

Masquing Britain

Britain plays a central role in *The Masque of Blackness*. The plot revolves around the fact that a group of black princesses are told by the reflection of the moon in the Nile river that they can be made white, and hence *fair* (beautiful), if they can sail to a place that has the suffix *–tania*. They sail to a variety of *tanias*, such as Mauritania and Lusitania, before they finally arrive at their destination:

> This blessèd isle doth with that *-tania* end,
> Which there they [the princesses] saw inscribed, . . .
> Britannia, which the triple world [heaven, earth and underworld] admires, . . .
> With that great name Britannia, this blessed isle
> Hath won her ancient dignity and style [due to the crowning of James as king of
> both Scotland and England/Wales]
> (Jonson 1969, 55, 204-217).

The landscape of Britain, under the rulership of its mild sunlike Apollonian Stuart king, could not only turn a black princess' body white, but even transform the body politic of society itself. This is exemplified by a scene in a masque where the entrance of the monarch is the signal for a change of scenery in which:

> The scene is varied into a landscipt in which was a prospect of the King's palace of Whitehall and part of the city of London seen afar off, and presently the whole heaven opened, and in a bright cloud were seen sitting persons representing Innocency, Justice, Religion, Affection to the Country, and Concord, being all companions of Peace (reprinted in Orgel and Strong 1973, 457, lines 330-345).

The landscape envisioned here is that of a country, united within the space of the landscape scene, making progress under an enlightened monarch (Olwig 2002b).

The Political Landscape

Landscape was a word laden with political meaning, and contention, in the Europe of the sixteenth century, when the word entered the modern English language. At the heart of the contention lay differing understandings of the meaning of the *land* in *land*scape. Was the *land* a physical body of soil, represented as landscape scenery, upon whose stage-like surface the drama of state is played and danced under the rule of the lord? Or was *land* the place and polity of a people, an arena or place of activity, represented, according to estate, in parliamentary institutions? In the first case it is the statutory law of the lord that shapes the land, in the second it is the common law of the people, based on custom, as represented by a parliamentary body, that gives the land its shape. In the first case the land becomes one with the body of the lord, in the second the land is represented by a parliamentary body made up of numerous distinct bodies. When *land* means place and polity, the meaning of the *land*- in *land*scape[7] is very close to that of *country* (and 'county') in English, and equivalent words in the Romance languages such as the *pays*- in the French *paysage*.[8] Otto Brunner has studied the historical role of landscape in continental European politics. Though his evidence is from the Alps, it would also have applied to the landscape polities under the domain of Queen Anne's father, King Frederik II of Denmark, and her brother, King Christian IV of Denmark (not to mention other noble relatives) (Olwig 1996). Brunner describes the situation in the following terms:

> The *Land* comprised its lord and people, working together in the military and judicial spheres. But in other matters we see the two as opposing parties and negotiating with each other. Here the Estates appear as the 'Land' in a new sense, counterposed to the prince, and through this opposition they eventually formed the corporate community of the territorial Estates, the *Landschaft* [*landscape* in English]. At this point the old unity of the *Land* threatened to break down into a duality, posing the key question that became crucial beginning in the sixteenth century: who represented the *Land,* the prince or the Estates? If the prince, then the *Landschaft* would become a privileged corporation; if the *Landschaft,* then it would become lord of the *Land* (Brunner 1992 [1965], 341).

The ascendancy of James VI of Scotland to the throne of England as James I in 1603 brought latent conflicts of a similar sort to a head in England. This was particularly due to his belief that it was his natural born destiny, as the inheritor of both crowns, to amalgamate his two separate kingdoms of Scotland and England into one British State. This amalgamation would transcend the country of England as customarily represented, according to estate, by Parliament. A somewhat similar situation could be found in Scotland, where James had established his rule, but it exhibited much greater tensions in England, where James was new to the throne, and where he was seen to be a foreigner. The parallels between the struggle between *court* and *country* in Britain (as it eventually came to be known), and the struggle between *lord* and *Landschaft* on the continent, would not have been lost on a cosmopolitan Renaissance English court that included Anne as one of its most important players on the cultural scene.[9]

The Matter of Words

The Stuart court might be seen as having successfully appropriated the idea of landscape by establishing its English meaning as a form of scenery. Their use of landscape, as in *The Masque of Blackness*, helped create, in fact, the insular British mentality that would eventually help close the British mind to other languages. The scenic understanding of land and landscape certainly dominates much British discourse on landscape to this day, but a glance at the *Merriam-Webster* dictionary suggests that the picture is somewhat more complex. Notions of 'political landscape,' closer to that described by Brunner, in which landscape is an area of activity or a 'scene,' in the sense of a 'place of an occurrence or action,' are very much alive (Merriam-Webster 1995: scene). In fact, it is the idea of landscape as 'vista' and 'prospect' which is becoming obsolete:

> *land scape* Etymology: Dutch landschap, from land + -schap –ship, Date: 1598. **1 a** : a picture representing a view of natural inland scenery **b** : the art of depicting such scenery. **2 a** : the landforms of a region in the aggregate **b** : a portion of territory that can be viewed at one time from one place **c** : a particular area of activity : SCENE (political landscape). **3** obsolete : VISTA, PROSPECT (Merriam-Webster 1995: landscape).

Representing the Political Landscape

The Stuart court's representation of Britain in terms of the landscape scenery of a geographic body was one way to represent a particular political vision. The representation of the English and Scottish people by Parliament was another. The first mode of representation made use of the geometric laws of single point perspective to show how the British state under its head could organize a harmonious and fertile land. To this day landscape architects and designers make use of such techniques to create esthetically beautiful landscape scenes. Parliamentary representation, on the other hand, was rooted in English common law that, in turn was rooted in English custom. According to James' great foe in support of the rights of Parliament, Chief Justice Edward Coke (1552-1634) put it, 'Custom lies upon the land.'[10] The customary law of the land binds and shapes the land (Gurevich 1985, 157-159; quotation from Thompson 1993, 97-98). To this day, when studying landscape, we must take account not just of the designs made by professionals upon the land. The shaping of the land as place also occurs through a people's 'habitus' and place 'practice' (Bourdieu 1977; Certeau 1984) as this is expressed through the rhythms of dwelling in a land, including the rites, such as processions, parades and the beating the bounds, by which custom is inscribed upon the land through the cyclical marching and dancing of people through the political landscape (Harrison 1988; Olwig 2002a). The *land* in *land*scape is not just the surface of a geographical body, it is also the practiced place of a body politic.

Notes

1 The word for the 'state' was synonymous for the throne, the monarch's seat. A primary meaning of *seat* is 'a special chair (as a throne) of one in eminence; also: the status of which such chair is an emblem' (Merriam-Webster 1968: seat, 1a). The standing attributed to this elevated seat was emphasized by the fact that most guests at a court masque had to stand up, and the best standing position was closest to the monarch. The words *estate/state* likewise suggested position, status or station in a hierarchy which involves a form of graduated relation between a number of parts (Dowdall 1923, 102). The word stage has

similar etymologically origins to words like status, station and state and still carried such meanings during the Stuart era, whereas the application of the word stage to the platform upon which players perform was of relatively recent date (1551) (OED 1971: stage, 2, 5). The theater stage provided a framework within which the state could be envisioned. On the construction of the masque theater, and its political implications, see: Orgel and Strong 1973, 1-14; Orgel 1975.

2   This was an idea that was well established in the tradition that conceived of the state as the secular parallel to the church as mystical body of Christ. The mid-fourteenth century Neapolitan jurist, Lucas de Penna thus used the marriage metaphor to compare the relationship of Christ to his church with that of the prince to his state: 'And just as men are joined together spiritually in the spiritual body, the head of which is Christ . . . , so are men joined together morally and politically in the *respublica*, which is a body the head of which is the Prince' (quoted in Kantorowicz 1957, 216). The idea of the state as a body politic with the King as head was established by the era of Henry VIII, who used it to refer to 'This Realm of England' (OED 1971: body).

3   These ideas were clearly 'in the air' at this time. James may well have derived some inspiration from 'Gaunt's' famous lines about England from Shakespeare's *Richard II* (Act II, 40-50), which were much admired at the time and quoted in *England's Parnassus* (1600).

4   There is much to suggest that Anne, unlike the virginal Elizabeth, rather deliberately flaunted the male gaze. The scanty, often bare breasted, clothing worn by the masquers scandalized not only the puritans, but also some of the privileged elite that attended the masques. Anne's appearance in the first court production of the new reign as a bare armed black skinned Negro caused considerable comment. This was literally Anne's mask of blackness. Though the masque combines elements of what later was to be known as blood and soil nationalism (and James may have been initially attracted to Anne's pale blondness by nascent racial motives), Anne herself does not appear to have been a racist. She later, for example, befriended the American Indian princess, Pocahontas, and treated her like an equal. Anne's mother was a powerful figure at the Danish court and Anne showed considerable independence throughout her marriage to James, who was something of a misogynist.

5   In his conversations with William Drummond, Jonson reported that: 'He himself was posthumous born a month after his father's decease; brought up poorly, put to school by a friend (his master Camden)' (Drummond 1985 [1619], 600, lines 194-196). Camden was usher and then headmaster of Westminster School. He was the son of a painter and possibly lost his father at an early age, which allowed him to be admitted to school as an orphan. This might have disposed him to take Jonson under his wing since Jonson was the fatherless adopted son of a bricklayer (Gough 1806, ix-x).

6   In the poem *To William Camden* Jonson wrote (Jonson 1985, 226): 'Camden, most reverend head, to whom I owe/All that I am in arts, all that I know,/How nothing's that? to whom my country owes/The great renown and name wherewith she goes.'

7   The word which we now spell *landscape* originated from the Germanic family of languages, to which the English language belongs. The Dutch have spelled the word *landschap*, the Danes *landskab*, the Swedes *landskap*, the Germans *Landschaft* and the Anglo-Saxons *landscipe* (Onions 1966: landscape; Bosworth and Toller 1966-72 [1898-1921]: landsceap, landscipe; Klein 1967: landscape; OED 1971: landscape).

8   *Paysage* appends the suffix *age* to pays in much the way as *schaft* is appended to *Land* in the Germanic languages, or *ship* to *town* in English. *Pays* carried essentially the same connotations of areal community and people as country and land. The equivalent Italian terms, *paése* and *paesàggio* carry the same meaning (Gamillscheg 1969: pays, paysage; Battisti and Alessio 1975: paése, paesàggio; Robert 1980: pays, paysage).

9   The conflicts emerging from the Stuart desire for the absolute power of continental European monarchs is linked to the contradictions which were to emerge later in the

century, and which are still the object of controversy amongst historians, known as the conflict between *court* and *country* (Zagorin 1969).

10 An expression which is very much in agreement with the derivation of the concept of law from the Old English *lagu* which was akin to *licgan* meaning 'to lie.'

References

Battisti, C. and Alessio, G. (1975) *Dizionario Etimologico Italiano*. Instituto Di Glottologia, G. Barbèra, Firenze.

Boon, G.C. (1987) Camden and the Britannia, *Archaeologia Cambrensis* 136, pp. 1-19.

Bosworth, J. and Northcote Toller, T. (1966-72 [1898-1921]) *An Anglo-Saxon dictionary*. London, Oxford University Press.

Bourdieu, P. (1977) *Outline of a theory of practice*. Cambridge University Press, Cambridge.

Bourdieu, P. (1991) *Language and symbolic power*. Cambridge University Press, Cambridge.

Brunner, O. (1992 [1965]) *Land and lordship: structures of governance in Medieval Austria*. University of Pennsylvania Press, Philadelphia.

Buttimer, A. (1982) Musing on Helicon: root metaphors and geography, *Geografiska Annaler* 64 B, pp. 89-96.

Buttimer, A. (1985) 'Nature, water symbols, and the human quest for wholeness' in Seamon, D. and Mugerauer, R. (eds.) *Dwelling, place and environment*. Martinus Nijhoff, Dordrecht, pp. 259-280.

Buttimer, A. (1993) *Geography and the human spirit*. The Johns Hopkins Press, Baltimore.

Camden, W. (1806 [1607]) *Britannia*. John Stockdale, London.

Certeau, M. de (1984) *The practice of everyday life*. University of California Press, Berkeley.

Cormack, L.B. (1997) *Charting an empire: geography at the English universities, 1580-1620*. University of Chicago Press, Chicago.

Cosgrove, D.E. (2001) *Apollo's eye: a cartographic genealogy of the earth in the western imagination*. The Johns Hopkins University Press, Baltimore.

Debray, R. (1994) *L'Etat séducteur: Les révolutions médiologiques du pouvoir*. Gallimard, Paris.

Dowdall, K.C. (1923) The word 'state', *The Law Quarterly Review* 39, pp 98-125.

Drummond, W. (1985 [1619]) 'Conversations with William Drummond of Hawthornden' in Donaldson, I. (ed.) *Ben Jonson*, Oxford University Press, Oxford, pp. 595-611.

Gamillscheg, E. (1969) *Etymologisches Wörterbuch Der Französischen Sprache* (volume two). Winter, Heidelberg.

Gough, R. (1806) 'The Life of Mr. Camden' in Gough, R. (ed.) *Britannia by William Camden*. John Stockdale, London xi-xxx.

Gurevich, A.J. (1985) *Categories of medieval culture*. Routledge and Kegan Paul, London.

Harrison, M. (1988) 'Symbolism, 'ritualism' and the location of crowds in early nineteenth-century English towns' in Daniels, S. and Cosgrove, D. (eds.) *The iconography of landscape: essays on the symbolic representation, design and use of past environments*. Cambridge University Press, Cambridge.

James, I. (1918a [1616]) 'A Speach in the Parliament Hovse, As Neere The Very Words as Covld Be Gathered At The Instant (1605)' in McIlwain, C.H. (ed.) *The political works of James I*. Harvard University Press, Cambridge, Mass., pp. 281-289.

James, I. (1918b [1616]) 'A Speach In The Starre-Chamber, The XX. of Jvne. Anno 1616' in McIlwain, C.H. (ed.) *The political works of James I*. Harvard University Press, Cambridge, Mass., pp. 326-345.

James, I. (1918c [1616]) 'A Speach To Both Hovses of Parliament, Delivered In The Great Chamber At White-Hall, The Last Day of March 1607' in McIlwain, C.H. (ed.) *The political works of James I*. Harvard University Press, Cambridge, Mass., pp. 290-305.

James, I. (1918d [1616]) 'A Speach, As It Was Delivered In The Vpper Hovse Of The Parliament To The Lords Spiritvall And Temporall, And To The Knights, Citizens And Burgersses There Assembled, On Mvnday The XIX. Day of March 1603. Being The First Day of The First Parliament' in McIlwain, C. H. (ed.) *The political works of James I.* Harvard University Press, Cambridge, Mass., pp. 269-280.

Jonson, B. (1969) *The complete masques.* Yale University Press, New Haven.

Jonson, B. (1985) *The Oxford authors: Ben Jonson.* Oxford University Press, Oxford.

Kantorowicz, E.H. (1955) Mysteries of State: an absolutist concept and its late mediaeval origins, *The Harvard Theological Review* 48, pp. 65-91.

Kantorowicz, E.H. (1957) *The King's two bodies: a study in mediaeval political theology.* Princeton University Press, Princeton.

Klein, E. (1967) *A comprehensive etymological dictionary of the English language.* Elsevier, Amsterdam.

Kunst, C. (1994) *Römische Tradition und englische Politik: Studien sur Geschichte der Britannienrezeption zwischen William Camden und John Speed.* Georg Olms, Hildesheim.

Merriam-Webster (1968) *Webster's Third New International Dictionary of the English Language, Unabridged.* G. and C. Merriam, Springfield, Mass.

Merriam-Webster (1995) *Collegiate Dictionary* (tenth edition) Merriam-Webster, Springfield, MA.

OED (1971) *Oxford English Dictionary.* Oxford University Press, Oxford.

Olwig, K.R. (1993) 'Sexual cosmology: nation and landscape at the conceptual Interstices of nature and culture, or: what does landscape really mean?' In Bender, B. (ed.) *Landscape: politics and perspectives.* Berg, Oxford, pp. 307-343.

Olwig, K.R. (1996) Recovering the substantive nature of *Landscape, Annals of the Association of American Geographers* 86, pp. 630-653.

Olwig, K.R. (2001) 'Landscape as a contested topos of place, community and self' in Hoelscher, S., Adams, P.C. and Till, K.E. (eds.) *Textures of place: exploring humanist geographies.* The University of Minnesota Press, Minneapolis, pp. 95-117.

Olwig, K.R. (2002a) *Landscape, nature and the body politic: from Britain's Renaissance to America's New World.* University of Wisconsin Press, Madison.

Olwig, K.R. (2002b) 'Landscape, Place and the State of Progress' in Sack, R.D. (ed.) *Progress: geographical essays.* Johns Hopkins University Press, Baltimore, pp. 22-60.

Onions, C.T. (1966) *The Oxford dictionary of English etymology.* Oxford University Press, Oxford.

Orgel, S. (1975) *The illusion of power: political theater in the English Renaissance.* University of California Press, Berkeley.

Orgel, S. and Strong, R. (1973) *Inigo Jones: the theatre of the Stuart court.* University of California, Berkeley.

Ortelius, A. (1595 [1570]) *Theatrum Orbis Terrarum.* Antwerp.

Robert, P. (1980) *Dictionnaire alphabétique et analogique de la langue française.* Le Robert, Paris.

Taylor, E.G.R. (1934) *Late Tudor and Early Stuart geography: 1538-1650.* Methuen, London.

Thompson, E.P. (1993) *Customs in common.* Penguin, London.

Trevelyan, G M (1960 [1925]) *England under the Stuarts* (volume two). Penguin, Harmondsworth.

Trevor-Roper, H (1971) *Queen Elizabeth's first historian: William Camden and the beginnings of English 'civil history'.* Jonathan Cape, London.

Tuan, Y-F. (1974) *Topophilia: a study of environmental perception, attitudes, and values.* Prentice-Hall, Englewood Cliffs, New Jersey.

Yates, F.A. (1969) *Theater of the world.* RKP, London.

Zagorin, P. (1969) *The Court and the Country: the beginning of the English Revolution.* Routledge and Kegan Paul, London.

Chapter 13

Heritage Landscapes, Geographical Imaginations and Material Cultures: Tracing Ulster's Past

Nuala C. Johnson

A display in a museum may simply be telling a story,
but the existence of a museum has a story to tell.
Robert Hewison

The proliferation of heritage sites, theme parks, museums and interpretative centers over the past two decades has stimulated a considerable debate about the role of heritage in the translation of the past to popular audiences. The relationship between heritage, history and memory has been the subject of much analysis by geographers, historians and cultural critics (Lowenthal 1996; Tunbridge and Ashworth 1996).

Conventionally a rigid line of demarcation ran between the past as narrated by professional historians on the one hand, and by the heritage industry on the other. While not many historians today would subscribe to the Rankean view of history as a sequence of truths about 'how it [the past] really was' (quoted in Tunbridge and Ashworth, 1996, 6), there remains nevertheless hesitancy about awarding heritage the same status as history. History is widely perceived as the interpretation and documentation of the past based on the careful accumulation of evidence. For some heritage is no more than 'gobbets of unrelated history floating in a stew of time' (Fowler, 1994, 9) or a 're-enacted costume drama rather than critical discourse' (Hewison, 1987, 144). As best it is pap entertainment and as worst it is 'a wall built across our awareness of history, and across the links between past and present' (Ascherson quoted in Samuel, 1994, 262). Samuel (1994, 266) notes however that 'it is not the traditionalism but the modernism and more specifically the postmodernism of heritage that offends. Aesthetes condemn it for being bogus: a travesty of the past, rather than a true likeness, let alone – the preservationist's dream – an original.'

The Heritage Debate

Heritage, as a concept, begins with the highly individualized notion of personal inheritance or bequest (e.g. through family wills and legacies). We are more concerned here, however, with collective notions of heritage, which link a group to a shared inheritance. The basis of this group identification varies in time and in space and can be based on allegiance derived from, for instance, a communal religious tradition, a class formation, geographical propinquity, or a national grouping. Heritage not only provides a sense of time identity, linking the past to the present, but it also provides a sense of place identity. Take for example the recent surge in industrial heritage. Here class identity, manifested through an industrial museum in a specific town or city, can be linked to the broader national drive towards industrial capitalism and its relevance to senses of national identity. As the National Trust has commented, industrial heritage highlights how 'life was no longer dictated by the seasons but by the factory hooter' (National Trust, 1998).

It is with respect to the 'imagined community' of nationhood that heritage is often most frequently connected (Anderson, 1983). While the origins of the nation-state may be relatively recent, ideas of nationhood are often based on the assumption that group identity derives from a collective inheritance that spans centuries and sometimes millennia. National states attempt to maintain this identity by highlighting the historical trajectory of the cultural group through the preservation of elements of the built environment, through spectacle and parade, through art and craft, through museum and monument (Hobsbawm and Ranger, 1983). The heritage industry, then, has often been viewed as a mechanism for scripting and reinforcing nationalist narratives in the popular imagination (Wright, 1985). Lowenthal (1994, 43) claims that 'heritage distils the past into icons of identity, bonding us with precursors and progenitors, with our own earlier selves, and with promised successors.' As such, the historical narratives transmitted through heritage are seen to be selective, partial and distorting. They offer a history which ignores complex historical processes and relationships, and sanitizes the less savory dimensions of the past. Wright (1985, 49) has claimed that 'whatever else it may involve, preservationism has certainly played its part in a nationalisation of history which enables the state to project an idealised image.' This contrasts with the work of professional historians where 'testable truth is [the] chief hallmark [and] ... historians' credibility depends on their sources being open to scrutiny' (Lowenthal, 1996, 120).

The distinction between 'true' history and 'false' heritage, however, may be more illusory than actual when viewed from the perspective of the postmodern turn within the social sciences. Making the claim that all historical narration is interpretative, deconstructivist accounts make problematic the distinction between representation and reality, between fake heritage and genuine history. Postmodernism, it is claimed, involves 'dissolving of boundaries, not only between high and low cultures, but also between different cultural forms, such as tourism, art, music, sport, shopping and architecture' (Urry, 1990, 82). Drawing from the insights of semiotics and literary theory it has been suggested that we should pay more attention to the patterns of signification that appear both in conventional historical texts and other forms of representation (including museums, heritage sites) and consequently understand the mechanisms through which meaning is conveyed.

Heritage Tourism

Increasingly, therefore, there has been a heightened awareness of the links between heritage and history and the fruitful connections that can be made between the two.

This has been particularly evident in the worldwide growth of tourism and the role of the heritage industry in fulfilling tourists' quest for historical understanding. Urry (1990) has linked this new pattern of tourist activity to a number of factors: to the changing circumstances of the work-place where conventional distinctions between work time and leisure time have become blurred; to the rise of a new service class rich in cultural capital; and to rapidly changing technologies which have accelerated a sense of time-space compression. Indeed as heritage consumption has crystallized since the 1960s it has become a type of 'cultural capital on which all are invited to draw' (Samuel 1994, 237). Rather than treating heritage tourism then solely as a commercial transaction MacCannell (1992, 1) notes that 'it is an ideological framing of history, nature and tradition; a framing that has the power to reshape culture and nature to its own needs.'

Moreover while heritage tourism forms a distinct niche market, Ashworth (1994, 21) suggests that it is intrinsically a place-based activity 'whether or not heritage is deliberately designed to achieve pre-set spatio-political goals, place identities at various spatial scales are likely to be shaped or reinforced by heritage planning.' Geographers have long recognized the importance of place-promotion in evoking and disseminating powerful place images (see, for example, Burgess and Gold 1985; Kearns and Philo 1993; Gold 1994; Tunbridge and Ashworth 1996). Analyses of postcards, tourist brochures and advertising literature have all, in various ways, begun to broaden our understanding of the ways in which the influential images of place traded to tourists can be deconstructed (Selwyn 1990; 1996; Cohen 1995; Crang 1996). The significance of these representations does not reside solely in identifying whether they are effective or authentic expositions of place. More particularly it is their role as part of a larger network of circulated ideas about the nature of place and the past that is of import. In this context Britton (1991, 475) suggests that the tourism imaginary is a 'lesson in the political economy of the social construction of "reality" and the social construction of place and people, whether from the point of view of the visitors, the host communities, or the state.' Some commentators have suggested that tourism sites 'are centres of physical and emotional sensation from which temporal and spatial continuities have been abolished' (Selwyn 1990, 24). This may be to overstate the case, however, as there is no necessary connection between the eradication of the historical imagination and the development of heritage sites (Samuel 1994 provides a host of useful examples where popular representations of the past are effectively executed).

In this chapter I will trace the development of a particular heritage site – Ulster Folk Museum (later the Ulster Folk and Transport Museum) – and I will highlight the relationships between the geographical imagination underpinning the idea of a 'folk' museum and the representation of the past offered to a popular audience. In particular the role of Estyn Evans in articulating a vision for the museum and in planning its development will be examined.

The Study of the Folk

In Europe research which centered on folk studies grew significantly through the early twentieth century. While the intellectual origins of investigating the idea of folk culture can be found as early as the eighteenth century, it is not until the formation of the *International Association for European Ethnology and Folklore* in 1935 that such study was put on a firm institutional footing. Emanating from a conference held in Lund, Sweden, in 1935 the aim of this association was the promotion of the study of folk culture and the development and exchange of research methods and findings between Europe's researchers. Parallel to this development was the growth of a number of regional societies including the Ulster Folklife Association.

The folk museum movement, itself an offspring of nineteenth century romantic nationalism, and also having its origins in Scandinavia, gave material expression, in popular form, to the investigations and findings of researchers on Europe's folklife. Under the stewardship of Artur Hazelius and opened in 1873 in Stockholm, Nordiska Museet was the first institution dedicated to the representation of Swedish folk life. Initially a conventional museum, collections were displayed in glass cases but it was soon felt that cabinet constructions were not perhaps the most effective approach to illustrate the vibrancy of a 'living' culture. Consequently an additional open-air folk museum section was added in 1891. It has been contended that this 'transformed the museum from a nineteenth-century curiosity shop into a home of national inspiration' (Peate 1949, 59). Objects of architectural and social significance were re-erected in 'environments suitably planned and planted' (Peate 1947, 201). The new site at Skansen accumulated and represented, for instance, a series of cottages and their interiors typical of the period they represented. This new museum type concentrated on the representation of peasant cultural traditions and customs, and, to some degree, moved the idea of the museum away from the representation of the artifacts and products of elite groups within any given society. Skansen has undoubtedly become the most famed open-air folk museum of its period but within a decade of its opening museums were also founded in Copenhagen, Oslo, Helsinki and Finland. In Britain the idea of the folk museum grew more slowly perhaps reflecting a desire to represent Britain's empire rather than focus on its local folk cultures. However it was in the 'Celtic' areas of the United Kingdom that the folk museum idea initially took hold. In 1939 the Highland Folk Museum, Am Fasgadh was founded, followed by the Isle of Man museum (1953) and the Welsh Folk Museum (1954). The latter emanated from the work of Sir Cyril Fox and his establishment of the Department of Folk Life in 1936.

The open-air museum movement was principally driven by the desire to preserve the ordinary, the mundane and the vernacular. By avoiding a perceived elitist conception of the past, these museums were designed 'not to collect the unusual but to emphasise the typical in an attempt to preserve a picture of the life of a nation or a region in miniature' (Jenkins 1972 509). Indeed Fowler claims '[R]eflecting the contemporary emergence of interest in social history, they [folk museums] cock a snoop at the more traditional "top people's past" projected by cathedrals, castles and country houses' (Fowler 1992, 132). However they have also been described as museums which evoke a 'landscape of nostalgia' (Lumley 1988, 10) by removing visitors from their present day environment into the working and living environments of the past. Whether they do this to any greater extent than other types of heritage museum is open to discussion.

With the open-air folk museum idea taking its lead from the Skansen prototype, these museums have represented and stimulated a continued interest in the study of 'folk' culture. The strong geographical emphasis in the construction of folk study and the important role played by geographers in providing the impetus for some dimensions of folk study and its representation in museums is significant. Indeed in the case of the Celtic 'fringe' geographers have been particularly influential in promoting and preserving the study of local regional cultures. They have done this both in terms of material conservation of cultural artifacts and in the promotion of specific ways of knowing and accessing the past. One geographer particularly concerned with these issues was Emyr Estyn Evans. Influenced by the work of his teacher, H.J. Fleure, Evans' geography found favor with the folk museum movement in general and in particular its development in Ulster. In the following sections I wish address the genesis of the Ulster Folk Museum idea and the influence of Evans in its planning.

Ulster Folk Museum

While there are several important studies highlighting Evans' intellectual development from his early years in Wales to his final years in Belfast (Buchanan 1990, 1991, Campbell 1996, Glasscock 1991, Graham 1994, Crossman and McLoughlin 1994), in this chapter attention will be primarily focused on his vision for an Ulster Folk Museum. The museum was, to some extent, a product of the growing interest in ethnographic collections that emerged in nineteenth century Belfast. The Belfast Municipal Museum, from the 1850s onwards, became a repository of local ethnographic objects some amassed by private collectors and others obtained by the Belfast Natural History Society. By the early twentieth century a reasonable collection of artifacts had been accumulated and the Director of the Belfast Museum Arthur Deane, influenced by his knowledge of the Scandinavian museum, became interested in the idea of establishing one in Belfast. Stendall, Deane's successor, extended both the ethnographic collection and the possibility of developing an open-air museum. His friendship with Estyn Evans encouraged him to promote the idea within the museum world whilst Evans advanced folk study within the university. Stendall inaugurated the Antiquities and Ethnography department within the museum and appointed George Thompson as its Keeper. Thompson was a graduate of Evans' Geography Department at Queen's University and his visit to the International Congress on European and Western Ethnology in Stockholm in 1951 afforded him the opportunity to see a folk museum first hand.

With the appointment of a new museum Director, William Seaby, there was an injection of new enthusiasm for the folk museum idea. He suggested that the museum be a separate national institution (from the Ulster Museum) and that local and central government support it. While Seaby strongly supported the idea, it was Thompson and Evans who actively marketed the proposal to interested parties. This culminated in the passing of the Folk Museum Act in 1955 and the appointment of Thompson in 1959, as Director of the pending Ulster Folk Museum. The Welsh folk museum had already been founded and Evans acknowledged the influence of his mentor Fleur in inspiring him to promote the idea of one in Ulster.

The role of Evans in bringing about a museum in Ulster has been recognized by the commentators on Evans' life. Buchanan claimed that Evans had 'conceived the idea, and carried it through a long period of gestation and labour to a successful birth'

(1990, 3). Similarly Campbell remarks 'above all it is the Ulster Folk and Transport Museum at Cultra, County Down, that stands as the finest monument to Evans' endeavours to rescue for posterity fast disappearing rural traditions' (1996, 242). His ideas formed the cornerstone of early debates about a museum both among academic and political colleagues. Born out of an academic tradition that was wedded to the authenticity of rural folkways and their preservation through artifacts and material culture, Evans set the tone for Thompson's early work at the museum. Buildings at the museum should serve more than aesthetic purposes and each should be an example of regional, vernacular architecture, indicative of local cultural heritage (Robinson 2000). The appointment in 1960 of Alan Gailey, another of Evans' students, as the first senior Curator of the museum helped to consolidate this vision. Of Evans he wrote '[he] opened my eyes to the interest of traditional rural houses in his lectures in human geography' (1984, vi). This is reflected in Gailey's concern that the museum promotes the preservation and conservation of traditional building features and spade types. For both Thompson and Gailey the folk museum became a unique opportunity to promote the cultural geography that Evans had championed at Queen's University and to bring it to a wider, more popular audience.

The evolution of a folk museum in Northern Ireland in the 1960s had, to some degree, to contend with the political and cultural conflicts of a divided society. Although Evans wished to represent the shared common ancestry of Ulster through a celebration of regional identity expressed in material culture, and one which might transcend any other bases for identity or division, the founders of the museum were not unaware of the political realities. Thompson (1978, 3) commented that:

> the Museum is actually bridging the gap by bringing together young people from Catholic and non-Catholic schools in a common educational experience. What is more, we have as our teaching resources the products of a social and cultural history which unlike those of constitutional history, reveal a heritage that they share rather than one which casts them into conflict as it has their forebears (Thompson 1978, 3).

By focusing on a common descent expressed through material culture the didactic role of the museum might strengthen a consciousness of a common social origin and social identity that would surpass political and religious differences. Evans was of the view that 'any aspect of material culture – types of thatch, shapes of spades, styles of field-fences, traditional fashions of food or dress – and you will find endless minor regional variations within what is at first sight a single homogenous culture-province' (1996, 51). Gailey similarly suggested that the 'claim that there is a single Ulster, or even Irish, tradition is false. Instead ... in Ulster and in Ireland generally there is both cultural diversity and cultural uniformity' (1988, 49). Trying to steer the museum project away from what might be considered politics Thompson claimed that 'it is our aim that the Ulster Folk Museum shall become, in theory and in practice, an institution of and for the whole province' (1959, 6). By focusing the museum on the representation of material culture, especially that found in a rural context and one that would illustrate regional variation of that material culture without reference to religious or ideological difference, it was suggested that the museum might overcome any sectarian preoccupations.

The folk museum would both educate the population of Northern Ireland about their past but it would also inform overseas visitors of Ulster's identity. Consequently Evans and Thompson were keen to steer the museum away from a form of representation which would concentrate on a romantic image of bygone days to one which would focus on a living tradition. Evans imagined 'a live museum where buildings and transport were made to function again' (1961, 4), and one where the recording regional variations in building techniques and so on would give the museum a scientific as well as an historical dimension.

Founding the museum meant bringing together the interests of national and local government to fund the project. Evans, amongst others, was an avid campaigner with Northern Ireland's political elite. For politicians the precarious status of Northern Ireland as a relatively new state within the United Kingdom with diverse political and cultural allegiances may have urged them to view the museum as an opportunity to put this new state on a more stable foundation.

The Ulster Folk Museum, established under the Folk Museum Act Northern Ireland 1958, provided the institutional footing for setting up a committee to establish and develop a museum. The government's proposals were very much based on the arguments presented by Evans and Thompson, although there was some wrangling between different regions in Northern Ireland as to whose voice would most influence the museum. Most councilors, outside of Belfast, feared that Belfast Corporation would dominate the decision-making process. Evans and Thompson sought to dissipate fears among provincial councilors. Most of the debate concerned the siting of the museum. It was initially proposed that Belfast Castle be the venue for the folk museum. As Robinson (2000, 177) points out 'Representative of a seat of imperial British power ... and occupied by a series of families affiliated with Belfast City Council, the Castle came to be a prominent landmark over-looking the city from Cave Hill, symbolic of Belfast's political sympathies.' These associations did not conflict with the will of Belfast's councilors and the site was also seen as appropriate because it was close to the main center of population in Northern Ireland. Geopolitical and cultural factors were influencing the location of the museum. Non-Belfast councilors who thought the city was dominating the project however voiced objections. Thompson worried about the small size of the site and given the opposition to Belfast Castle an alternative site was eventually located at Cultra Manor, eight miles east of Belfast city center. Thus proximity to the center of population was maintained while the balance of decision-making power seemed more evenly distributed. The site was approved and bought in March 1961. For Evans the term 'Ulster' would incorporate the three counties of Ulster in the Republic of Ireland, even though *de facto* the museum was administered, funded and conceived by those living and working in Northern Ireland.

Representation: Early Years of the Ulster Folk Museum

The folk museum was a unique institution within Northern Ireland and its inherent concern was with the preservation of the rural landscape. Guided by the foundational tenets of the broader European folk museum tradition, scholars in Ulster envisaged the project as representing past living and working conditions in a bid to preserve 'as much as possible of the material side of Ulster life and culture' (quoted by Robinson,

2000, 211). As Thompson would put it artifacts in themselves were 'not the alpha and omega of the institution; they merely help to illustrate more clearly, in three dimensions, the relevant facts concerning the social development and character of the community' (1959, 12). Indeed in an attempt to distance the museum from a tradition of representing arcane or 'dead' history, Gailey was keen to establish both an archival collection of folklore as well as a contemporary study of folk tradition and folk life. This was achieved in 1964-65 when the archive of the Ulster Folklife Society was amalgamated with the museum. In a first attempt to itemize what the museum should contain the Ministry of Finance produced a booklet *Proposed Open-air Folk Museum* in which they stipulated the inclusion of 'a barn church, a village school and other classes of public or domestic buildings' (quoted in Robinson 2000, 216). The report outlined a five year development plan in which the elements of Ulster's rural material culture would be assembled and displayed, although in practice Evans, Thompson and Gailey made the crucial decisions about which exhibits would be brought to the museum.

Selecting artifacts to exhibit is never a neutral process. As Waitt and McGuirk (1996, 15) point out 'no explicit agenda may shape the interpretation of history as heritage, what is valued as significant reflects the power balance of various interests and their status in both the past and the present.' While the key tenets of the folk preservation movement broadly informed the selection of exhibits and the type of past that the museum sought to represent, individual decisions however could still provoke controversy. The first vernacular building reconstructed and opened at Cultra is a case in point. The building, an eighteenth century two-roomed thatched house, was purchased and removed from an estate on County Londonderry/Derry. One of the museum staff had previously surveyed the building and recorded its traditional features as an exemplar of a cottier's house. The stone building is one of the oldest in the museum and its structural features are characteristic of peasants' dwellings in this region of Ulster. The process of dismantling and re-assembling the house at Cultra acted as a test and the museum was especially vigilant in ensuring that the re-assembly exactly mirrored the original. Although the house was empty when purchased extensive research was carried out in the surrounding area to identify and acquire representative interior furnishings. The attention paid to the internal design reinforced the efforts made by the museum staff to produce a 'living' past. The result was the recreation of a cottage, as it would have appeared at the turn of the twentieth century. This house was further developed as a working exhibit where visitors could watch bread being made on an open turf fire. In all its detail, therefore, this first exhibit at the museum seemed to follow all the principles of a folk museum developed by Evans and his disciples. The Prime Minister of Northern Ireland, Terence O'Neill, opened the cottage to the public in May 1963. The Board of Trustees of the Museum seemed pleased; they recorded in their annual report of 1964:

> The considerable satisfaction of completing our first full-scale removal resulted not only from the visual appearance of the structure, but also from the realisation that the first big hurdle in the work of building an open-air museum … had been accomplished; and although this first house is small by comparison with later removals, future prospects now seem much less formidable (quoted by Robinson 2000, 220).

The house, however, immediately sparked controversy. Because the house was originally occupied by Catholics, critics commented that the museum was being 'furnished with popery' (quoted in Robinson 2000, 223). The Museum's response was that 'Duncrun Cottage was chosen as a result of availability, coupled with the fact that it had constructional features worth preserving. No deliberate symbolism was intended, nor will it remain for long the sole representative of Ulster traditional houses as some people may have thought' (quoted in Robinson 2000, 223).

Whilst there was certainly no intention to cause offence with the choice of vernacular house type, perhaps the folk museum idea itself necessarily tended to ignore the significance that people attach to the symbolic in the constitution of their cultures. Perhaps the focus on re-creating with accuracy the material expression of culture was to lead the museum developers to underestimate the ideological and symbolic baggage attending to material objects on display. In an Northern Irish context, the role of the symbolic as it got played out in the materiality of everyday lives, especially in domestic space, was perhaps just as important as the precise detail of the architecture and design. Rather than highlighting, and perhaps acknowledging, the significance of certain religious iconography in the interior design of houses in Ulster, the museum tended to focus on the argument about accuracy and regional identity at the expense of cultural meaning.

Despite the opening controversy the museum continued in the early years to develop representative aspects of rural material culture. Evans was keen for both the isolated farmstead and the clachan (nucleated settlement) to be represented at the museum. The museum quickly supplemented its original cottier's house by dismantling and transporting another farmhouse from County Fermanagh. This house dated from the early nineteenth century and consisted of a thatched, hipped-roof structure. In its original formation it consisted of two rooms but was later divided into five separate units, and it was chosen to represent the home and domestic arrangements of a fairly prosperous west Ulster farming household. The interior included a dresser and china from the Beleek pottery factory in north Fermanagh. Although the museum, following Evans' vision, was keen to represent Ulster's vernacular architecture, it did include a planter's house whose design was inspired by English vernacular taste. Its steeply pitched roof distinguished it from other buildings in the museum and its inclusion at least acknowledges the role of plantation and English architectural aesthetics on the Ulster landscape. Similarly the museum acquired a hill-farm complex from the Glens of Antrim, where the central dwelling underwent much renovation from its original form in the nineteenth century. Although built as a single storey rope-thatched house, by the 1900s a second storey had been added and the roof was replaced with slate. Local craftwork was represented in this house, with the old red sandstone kitchen floor made locally from fired clay included.

Domestic space was an important feature of Gailey's design policy because he believed that changes in the internal structure of the house and of its interior furnishings represented regional cultural motifs of Ulster. The hill-farm project reinforced in the minds of the museum's curators the need to place the buildings within surroundings that were typical of the site of origin. In other words a contextual or environmental approach to the overall design of the museum was significant. Rather than randomly scattering buildings across the site, as had been the case at the

Welsh Folk Museum, it was preferred at Cultra to situate some of the exhibits in a fashion that bore some relationship to their original location. The *1965-66 Yearbook* made this point:

> Already certain sites have been used to duplicate original sites from which the open-air exhibits have been removed, particularly the Cushendall hill-farm complex from a site in the Glens of Antrim overlooking a beautiful glen ... Where inevitably the open-air Folk Park must differ from the Ulster landscape is in the question of scale, but there is a belief fundamental ... that on a different scale, most of the environments found in Ulster either exist already or can be re-created with relatively little difficulty within the Cultra Manor Estate. An Ulster landscape in miniature is the projected outcome for the open-air museum (Ulster Folk Museum 1967, 5).

To achieve the Ulster landscape in miniature the museum aimed to use the natural slopes of the site to recreate the environment of hill farms, and to use the existing field patterns and trees to reduce the number of buildings visible from the each house. Consequently it was 'the inherent character of the estate, particularly in terms of topography and vegetation, ... that one must inevitably try to mould the Museum into the estate rather than re-model the estate to meet the ends of the Museum' (1967, 6). The Cushendall exhibit lent itself to this philosophy. It was re-erected on a steep slope to resemble the Antrim Glens and scrubs and trees native to the glens were planted alongside the yard. Some other farmhouses added in the opening years of the museum were also positioned in an environmentally sensitive manner. In an effort to authentically recreate a 'living' tradition houses were accurately reconstructed at Cultra, with attention being equally focused on the macro-level architectural superstructure and the micro-level interior design.

The Working Past at Cultra

While the museum would seek to represent Ulster's material culture, it was also keen to be a living museum that would put on display traditional skills and crafts. One of the earliest exhibits of this nature in the museum was the spade mill. Evans (1970, 1) had lamented that 'it is only in recent years that serious attention has been given to the study of the spade, while the plough and the hoe, in various parts of the world, have been extensively studied and endlessly classified.' As the main implement used by Irish farm labor in the nineteenth century particularly, Evans was keen to have the spade represented in the museum. Consequently the spade was the second exhibit brought to the Cultra site. A nineteenth-century water-powered spade mill was removed from Coalisland in 1963 and it was in operation at the museum the following year. To enable demonstrations to be made a water supply was installed at the site. The mill and the accompanying exhibition facilitated the display and interpretation of regional differences in spade sizes and shapes that were common in Ulster. As Robinson (2000, 246) comments 'Bringing the mill to the museum was undoubtedly an expression of Evans's brand of cultural geography, as it calibrated sub-regional variety.' The museum was keen to advertise its acquisition and in the a press release attention was drawn to the fact that:

[it had] the added distinction of being the Museum's first working exhibit and plans are being made to go into spade production in a small way. For this purpose Mr. Arnold Patterson of Glengormley was recently appointed to the Museum staff. Mr. Patterson is one of very few skilled spade platers – probably less than a dozen – left in Ireland. His father and grandfather actually served their apprenticeships to the same craft in the mill now re-erected at the Folk Museum (quoted by Robinson 2000, 248).

The re-erection of a working mill at the site in Cultra served a number of important objectives. It satisfied the museum's desire to create a 'living' environment, which translated Ulster's past. It enabled the visitor to see the spade and its making as an economic and social tool and it also, quite significantly, allowed the museum to display the folklore attached to this artifact in Ulster. For example, the names of specific spades related to regional context. Those manufactured in Coalisland were known as Dungannon spades, whereas in Newry they were referred to as Armagh spades. Indeed the nomenclature of this humble agricultural implement also had wider political import. In the language of sectarianism 'kicking with the wrong foot' can be traced to the type of spade used by different religious groups. Historically Catholics continued to dig with the right foot because of their use of a single-sided spade, whereas Protestant planters, who introduced the two-sided spade, preferred to use the left foot. Hence the contemporary use of this expression as a loaded shorthand for identifying religious background.

The success of the spade mill encouraged the museum to expand its collection of living exhibits. The museum acquired the Lisrace Forge from County Fermanagh and here the skills of the blacksmith were demonstrated. Perhaps one of the most significant early examples of the working life in the countryside to be represented at the museum was the representation of the linen industry. The linen industry formed a crucial part of Ulster's rural, and indeed urban, economic development. Three buildings were chosen to represent the linen industry: a scutching mill, a bleach green tower and a weaver's cottage (the latter was a reproduction based on an exact replica of an existing cottage). These three elements of the industry really represented some of the early phases in commercial linen production and they were grouped together in the museum to replicate how aspects of the industry were localized in the Ulster countryside. The water powered mill and the two adjacent buildings gave a sense of the four processes that flax underwent on its journey to become linen. Weaving demonstrations were given in the cottage.

Conclusion

The first decade of the museum's history indicates how the foundation stones of the museum were laid. The emphasis on the representation of material culture, particularly of rural architecture and lifestyle, was central to the initial exhibits in the museum. The representation of a variety of farmhouse types and their associated outhouses formed one of the principal building blocks of the new museum. Evans, Gailey and Thompson's desire to bring representative aspects of Ulster's agricultural settlement patterns and practices into the museum were quickly realized. It is evident that both exterior and interior design considerations were awarded equal weighting.

Accuracy of detail was important to the museum designers. The environment in which the houses were placed was also of both aesthetic and practical import. The museum sought, as far as was possible, to give something of the flavor of the diversity of environmental conditions under which the different types of cottage and farmhouse were built. While architecture and the domestic interior of houses were important dimensions of the early development of the museum, the desire to create a living museum where the crafts and trade practices of the countryside could be represented were also important. The representation of agricultural equipment and the processes of agricultural manufacture were quickly incorporated into the museum. Demonstrations of technique were put on display, and the regional variation in practices was emphasized. While by the mid-1970s the museum had achieved a considerable amount in terms of representing something of Ulster's material culture, it was not until 1974 that a development plan was put in place. This plan proposed the inclusion of facets of Ulster society that were not part of Evans' original vision and had not, thus far, been represented in the museum. In particular the introduction of elements of urban life was seen to be important. In addition it was proposed that the museum include a representation of the landholding elite in Ulster, an important part of rural life albeit not always considered under the remit of 'folk.' The early development of the museum however demonstrates how the vision of history, embedded in a cultural geography that focused on rural material cultures, found expression in this heritage museum. Although questions of authenticity have animated recent discussions on the role of heritage in contemporary society, the Ulster Folk Museum illustrates how these questions have a long trajectory and have formed part of the general discourse about relating the past to popular audiences.

References

Anderson, B. (1983) *Imagined communities: the origins and spread of nationalism*. Verso, London.
Ashworth, G.J. (1994) 'From history to heritage – from heritage to identity. In search of concepts and models' in Ashworth, G.J. and Larkham, P.J. (eds.) *Building a new heritage: tourism, culture and identity in a new Europe*. Routledge, London, pp. 13-30.
Britton, S. (1991) Tourism, capital and place: towards a critical geography of tourism, *Environment and Planning D: Society and Space* 9, pp. 451-478.
Buchanan, R.H. (1990) Obituary: Emyr Estyn Evans 1905-1989, *Ulster Folklife* 36, pp. 1-3.
Buchanan, R.H. (1991) 'The achievement of Estyn Evans' in Dawe, G. and Foster, R.W. (eds.) *The poet's place: Ulster literature and society*. Institute of Irish Studies, Belfast, pp. 149-156.
Burgess, J. and Gold, J. (eds.) (1985) *Geography, the media and popular culture*. Croom Helm, London.
Campbell, J. (1996) 'Ecology and culture in Ireland: Estyn Evans and the trilogy of the humanities: an appreciation' in Evans, E.E. *Ireland and the Atlantic heritage: selected writings*. The Lilliput Press, Dublin, pp. 225-244.
Cohen, E. (1995) 'Contemporary tourism: trends and challenges: sustainable authenticity or contrived postmodernity' in Butler, R.W. and Pearce, D.G. (eds.) *Change in tourism: people, places and processes*. Routledge, London, pp. 12-29.
Cráng, M. (1996) Envisioning urban histories: Bristol and palimpsest, postcards and snapshots, *Environment and Planning A* 28, pp. 429-52.
Crossman, V. and McLoughlin, D. (1994) A peculiar eclipse: E. Estyn Evans and Irish studies, *Irish Review* 15, pp. 79-96.

Evans, E.E. (1961) Book Review, *Ulster Folklife* 7, p. 80.

Evans, E.E. (1970) The personality of Ulster, *Transactions of the Institute of British Geographers* 51, pp. 1-20.

Evans, E.E. (1996) *Ireland and the Altantic heritage: selected writings.* The Lilliput Press, Dublin.

Fowler, P. (1992) *The past in contemporary society.* Routledge, London.

Fowler, P. (1994) The nature of the times deceas'd, *Journal of Heritage Studies* 1, pp. 1-16.

Gailey, A. (1984) *Rural houses of the north of Ireland.* John Donald Publishers, Edinburgh.

Gailey, A. (1988) 'Tradition and identity' in Gailey, A. (ed.) *Use of tradition.* Holywood, Ulster Folk and Transport Museum, pp.61-67.

Glascock, R.E. (1991) Obituary: E. Estyn Evans, 1905-1989, *Journal of Historical Geography* 17, pp. 87-91.

Gold, J. (1994) 'Locating the message: place promotion as image communication' in Gold, J. and Ward, S. (eds.) *Place promotion: the use of publicity and marketing to sell towns and regions.* Wiley, Chichester, pp. 19-38.

Graham, B. J. (1994) The search for common ground: Estyn Evans's Ireland, *Transactions of the Institute of British Geographers N.S.* 19, pp. 183-210.

Hewison, R. (1987) *The heritage industry: Britain in a climate of decline.* Routledge, London.

Hobsbawm, E. and Ranger, T. (1983) *The invention of tradition.* Cambridge University Press, Cambridge.

Jenkins, G. (1972) 'The use of artefacts and folk art in the folk museum' in Dorson, R.M. (ed.) *Folklore and folklife: an introduction.* University of Chicago Press, Chicago, pp. 497-517.

Kearns, G. and Philo, C. (eds.) (1993) *Selling places.* Pergamon, Oxford.

Lowenthal, D. (1994) 'Identity, heritage and history' in Gillis, R. (ed.) *Commemorations: the politics of national identity.* Princeton University Press, New Jersey.

Lowenthal, D. (1996) *The heritage crusade and the spoils of history.* Viking, London.

Lumley, R. (1988) 'Introduction' in Lumley, R. (ed.) *The museum time machine.* Routledge, London, pp. 1-24.

MacCannell, D. (1992) *Empty meeting grounds: the tourist papers.* Routledge, London.

Peate, I.C. (1947) Swedish folk museums, *Museums Journal* 46, pp. 201-202.

Peate, I.C. (1949) The Folk museum, *Museums in Modern Life* 3, pp. 57-69.

Robinson, E. (2000) 'Making the past: history, identity and the cultural politics of the Ulster Folk Museum.' Unpublished PhD thesis, School of Geography, Queen's University, Belfast.

Samuel, R. (1994) *Theatres of memory: volume 1. Past and present in contemporary culture.* Verso, London.

Selwyn, T. (1990) Tourism brochures as postmodern myths, *Problems of Tourism* 13, pp. 13-26.

Selwyn, T. (ed.) (1996) *The tourist image: myth and myth making in tourism.* Wiley, London.

The National Trust (1998) *Industrial heritage: the people and places of industry – a guide to industrial heritage in England Wales and Northern Ireland.* Pamphlet.

Thompson, G.B. (1959) The Ulster Folk Museum collection: a preliminary report, *Ulster Folklife* 5, pp. 9-13.

Thompson, G.B. (1978) A museum's role in a changing society, *Museums Journal* 78, pp. 2-4.

Tunbridge, J. and Ashworth, G. (1996) *Dissonant heritage: the management of the past as resource in conflict.* Wiley, Chichester.

Ulster Folk Museum (1967) *1965-66 Yearbook.* Ulster Folk Museum, Cultra.

Urry, J. (1990) *The tourist gaze: leisure and travel in contemporary societies.* Sage, London.

Waitt, G. and McGuirk, P.M. (1996) Making time: tourism and heritage representation at Millers' Point, Sydney, *Australian Geographer* 27, pp. 11-27.

Wright, R. (1985) *On living in an old country.* Verso, London.

PART V
REANIMATING GEOGRAPHIES

Chapter 14

Place-Making and Time

Robert David Sack

Often the link between geography and time is sought in the connection of time to space. I will focus here instead on the connection of time to place, or more specifically place-making. I do so for two related reasons. The first is that place and place-making, rather than space, are more central to geographical analysis. Place and place-making underlie the concepts of 'spatiality,' 'the social production of space,' and a host of other geographic concepts that have been employed in discussing time. The second is that we make places for many reasons, and one of the most important is to create a variety of temporal relations or temporalities. That is, we create places in part to create temporal relations or temporalities.

The full argument for the centrality of place and how it undergirds spatiality can be found elsewhere, but in these works, the link to time was not developed.[1] Since there is much to cover in a short paper, I will dispense with a review of other geographic approaches to time, and go directly to the discussion of how place-making creates temporalities. I will do so in two parts. The first examines this connection in the projects of everyday life. The second is more existential. It focuses on the role of place-making in addressing questions of time that are at the core of being human. Before either is addressed, I must explain what I mean by place, and summarize several of the points about its structure and dynamics that will be needed to understand its role in the creation of temporalities.

Place, Place-Making, and Space

Place refers here to something we humans make. A place is made when we take an area of space and intentionally bound it and attempt to control what happens within it through the use of (implicit and/or explicit) rules about what may or may not take place. This bounding and rule-making leads to the creation of place at any scale, from a room to a nation-state.

This is what I will mean by place, but the term has many other meanings. Place, in my sense of the term, is not necessarily the same as an area or region like a forest or a desert. These are often called 'places,' but they would not be unless what happens there were to depend at least in part on humanly created boundaries and rules delimiting and controlling the area. This though is now the case for virtually all such areas on the earth's surface, as we have come to set aside and preserve them by limiting the human activities and affecting many of the natural ones that occur within. For this reason, wilderness areas, and even 'air space' are places: what happens within the wilderness, or what chemicals and particulates are emitted within a column of air, are now influenced by rules. Place is also not the same as location in space, though the places we make are located: a room has location in space, and so does a nation-state. But location alone does not make a place.

Places are intimately related to space. As we noted, a place is carved out of space, it is in space and has location, and as we shall see, it is also involved in generating flows through space and creating surfaces and appearances of varying scales and extent. The last two especially are important for understanding why place and place-making are at the foundation of spatiality and constitute what is meant by the 'construction' of space.

The fundamental reason why place and place-making are foundational to geography and why humans constantly produce and maintain places everywhere (and at all scales) is that places are tools that enable us to undertake virtually any and all projects. Their power as tools comes from their ability to help us weave together in new ways components of reality that cannot be joined any other way – including components of time. Understanding how this tool or instrument works – its structure and dynamics – provides the details of why geography matters, and why it is basic to the construction of temporalities. Even before we confront place's structure and dynamics, the importance of place-making for our daily activities and for our more general existential concerns can be expressed briefly and generally through what has been called 'the geographical problematic.'

The Geographical Problematic

The problematic says that we cannot accept reality as we find it, and so we create places to transform it into what we think it ought to be, and then transform these, and so on. We may want to create something new, or return to a reality we think once was. In any case we are transforming because we must use place to prevent things from happening – removing things, bounding other things – so that the things that take place conform to what we think reality should be. Once we do so, we have then created yet another layer to reality that in turn will be transformed. The process is propelled by our own imaginations and our capacity to think of worlds that once were, or that are not yet, and by our anxiety about what could happen to change what is. Our changes and transformations were modest in the past. Now they are not only vast, but often dangerous.

The problematic points to the central importance of place-making as it addresses our existential concerns of permanence and change, and the development of daily rhythms and pulses. Place-making is central, but why? How does it work as an indispensable tool? The answer is in its structure and dynamics.

The Structure and Dynamics of Place

Place is an instrument or tool that enables us to connect elements of reality without reducing one set to another. The most natural way to think of these connections (and one I have already alluded to) is in terms of weaving, which brings up the idea of a place as something like a loom.

The Loom

The basic point is that in making places, we are making something that possesses the structure and dynamics of a loom, helping us draw together and weave elements or threads of reality in new ways. There is then the loom (or the structure and dynamics of place), the threads (or elements of reality), and there is also us, the weavers (or place-makers.) Geographers have spent a great deal of attention on categorizing the 'us' – who is controlling the place and doing the weaving: is it the rich, the white males, the homophobes? These are important matters, but determining who is doing the weaving does not address why place matters. Simply showing that different groups fight over controlling place and use it differently with different outcomes does not explain how place works and why it is worth controlling. Rather it treats place as though it were a 'black box.' So, let us leave aside who is in control, and focus instead on the structure and dynamics of the loom, and nature of the weave.

To put it as simply as possible, all places have the same structure or loom-like quality consisting of three dynamically interrelated components (that work like the heddles and shuttles of a real loom.) The three are: the *in/out of place rules* (which follow from the role played by the boundary), *spatial flows or interactions*, and the issue of *surface and depth, or appearance and reality*. Each of these is a way of taking the fundamental property of spatial causality and dividing it into three interconnected parts. Let us take up the in/out of place rules first, for they follow directly from the boundaries of place.

The boundaries that we create to define and contain some things and deflect others are based on rules about what we would like in and out. These rules can be explicit or implicit, simple or complex, verbal or non-verbal. They not only help to contain and deflect spatial flows or interactions, but also are responses to them. If we expect a change in spatial interactions, we will change the rules; and the rules in turn change the flows, ... and so on. Suppose, for example, that there is an expected increase in the number of people seeking immigration and asylum to a particular country. This potential increase in the spatial flow of people to a country may result in that country tightening its immigration rules – limiting the possibilities of those outside from entering, and deporting others who have tried. These individuals then would have to move to other places which might initiate changes in their rules, and thus affect further the flows and so on.

The in/out rules and spatial interaction components of the loom are connected to a third one – appearance and reality. Rules and flows result in a weave that affects the appearance of the place's landscape. Immigration laws alter the income, ethnic, and racial mix of a population overall, and its cultural artifacts and practices. These establish an appearance to landscape, one that may be challenged. The place may no longer appear to be what it claims to be, or the place's appearance may belie the reality. A country that claims to be tolerant, and yet does not include members of different races, may not look like what it claims it is. Or, an appearance of a place may in fact disguise the reality. A tourist landscape, established by a complex connection of rules and spatial flows, has an appearance that may claim to be authentically representing a culture, but others may argue it to be staged, superficial, or 'hype' – an appearance only – that disguises or hides a reality found beneath the surface or behind the scene. The same can be said for a place that claims to be preserving nature or wilderness. One might ask about Yellowstone whether it is really authentic wilderness or are its vegetation and wildlife to some degree controlled and cultivated. Do the forest rangers allow forest fires? And where are the grizzlies? Serious challenges to the appearance of the landscape, claiming it to be only a surface and not the reality, will then lead to changes in the rules of in and out and to the spatial flows. Similarly, challenges to the rules affect the appearance and flows, and changes in the flows affects the rules and the appearances.

Creating any place is creating this dynamic interrelationship. We create places, in fact, because we want to create these relationships and cannot do so any other way. Why do we want them and what do they do? The answer is that the relationship among these three components is the mechanism – or loom-like structure – that enables us to weave together elements of reality that are difficult if not impossible to join any other way, and to do so without reducing one set to another.

The Weave

I have said that the interconnections among in/out, spatial interaction, and appearance/reality help us draw together and weave elements of reality. These elements are like threads from three vast spools or realms that coincide with and are reinforced by the structure of the loom. One of these spools or realms is that of nature. A second is the realm of social relations. The third is the realm of meaning. Every place is used by us to weave together threads or elements from the three, but we use each place to weave different proportions to produce different appearances and landscapes. Schools, for example, claim to focus on threads of meaning – conveying and exploring ideas and symbolic systems, with social relations and elements of nature playing supporting roles; though occasionally these become dominant as when a teacher must temporarily put aside the intellectual to address social issue of discipline. Business can be said to focus primarily on elements of social relations – developing wealth, and articulating relationships among workers, managers, owners, and customers – with meaning and nature in the background; while the places we call wilderness areas draw primarily on the threads of nature, with the other two playing supporting roles. (My point is not as much about whether I have necessarily characterized these places correctly, as about the fact that any characterization would be in terms of the weave of these elements.)

It is important to note that we also use these three realms – meaning, nature, and social relations – to describe qualities of the self. We think of our own selves as combinations of them, and often one predominates, as when we say we are biological beings, or that we are social, or intellectual beings. This makes the connection between self and place extremely important.

How though does the weave happen? The key point is that these elements or threads are engaged by the parts of the loom's structure. Each draws elements or threads from the three realms, but in doing so it puts its own spin on them. When the elements are engaged by spatial interaction they tend to become 'naturalized' – making them appear to be part of the realm of nature. This is because it focuses on a spatial quality of causality that operates in the natural world. The causal relations of nature occur via the flow of energy and mass through space. This applies to us too when we think of ourselves as physical and biological creatures moving and interacting in space. The spatial interaction part of the loom makes this the focus of attention, and so it presents us as natural objects in the physical world. (Notice that when geography models these flows it uses natural science approaches. Take for example the case of the 'gravity model' and the 'laws of social physics.') Even when the focus on spatial interaction is applied to the flows of meaning – the circulation of newspapers, the innovation of information and the diffusion of ideas – it naturalizes them by making these meanings appear as objects and relations in space. So whenever the spatial interaction part of the loom engages threads of meaning, or social relations, as well as elements of nature, it naturalizes them.

The rules of in/out can engage elements of all three realms: ideas or meanings can be in or out in that they are appropriate or inappropriate; so too can specific social relations, and so too can elements of nature. When elements are engaged by these rules, they become socially sanctioned (or condemned) – they become 'socialized.' The rules of a church, for example, allow certain meanings to be conveyed and not others. The place's rules also encourage some social relations and not others. And the rules allow only a few elements of the natural world in, but not the rest. Stipulating that some meanings, or social relations, or natural elements are in or out of place is to mark them socially. This is true even with elements of nature, as when water falling from the sky outside the church is rain, a part of the natural world; but this rain becomes a leak when inside the church. Or when soil is soil outside, but becomes dirt when inside.

And the appearance/reality component also engages elements from all three realms, but when it does, it 'problematizes' their meanings. So a landscape that is supposed to be 'natural' can be called into question as artificial, thereby drawing into question what we mean by the natural. The same is true for the social aspects, and for meanings too. A 'cooperative' workplace, advertising itself as a worker-owned and socially harmonious environment, may in reality be riven with social conflicts. Or, the symbols and meanings communicated in a classroom and lending it a 'class-room' appearance can be challenged as an appearance only, for the claim may be that the place is really instilling political propaganda. Thus, when the appearance/reality component engages elements of social relations, meanings, or nature, their meanings become problematized. We look at the landscape as disguising something else – something beneath, or behind – that deserves to be seen.

By engaging these elements, allowing us to combine and weave them together, and by placing a spin on the role of space that is appropriate to a particular realm, place helps weave the three without reducing one to another. Place-making does not require that we translate or explain the natural into the social, or the social into the natural (which is what reduction means.) Rather place allows them to become simply inter-thread. When I am in a room (a place) and hear someone say something, I am also registering the temperature, experiencing the light, weighing the meaning of the words, and so on, without translating temperature or light into meaning or a social relations.)

This capacity to weave without reduction is the reason place is a universal tool essential for virtually any and all undertakings. The fact that it avoids reduction does not mean, as we have shown, that it avoids emphasizing or weighting the weave with elements from one or another realm. Wilderness areas, schools, workplaces and farms all emphasize different combinations. Rather it provides the possibility of a mix of elements that can be joined and made into new things, without reducing one set of elements to another.

Thus we need places not only to clear a place for something to happen, but to weave together components that constitute the project. The place then is part of the project, and the qualities of the place – the weave – also become part of the project. Returning to the geographic problematic, this explains why we use place as the instrument to transform reality into a new one that fits our conception of what it ought to be.

There is much more to the structure and dynamics of place, but here I want to make five quick points before turning our attention to how these characteristics of place help constitute the experiences of time. First, the weave is not about one place only, but a system of places. The flows through space are the strands from places that are woven and re-woven to become elements in yet other places. Second, the issue of weaving leaves unanswered who is weaving and who is determining what the weave should be. Again, I will not address these because they do not help explain why places are important and how they work. Third, as I said in the first page, the structure and dynamics of place clarify what is meant by 'spatiality' and the 'social construction of space.' It is not space that is constructed, but places and their dynamic and changing relations. Place is far more complex, and its structure and dynamics involve all of the spatiality issues without conflating place and space, and making either or both a handmaiden to social relations. Fourth, the places I am discussing can be generic or specific, and can be used to produce things that are alike, or are distinct; some small places can even be movable – cars, trains and planes. Finally, there is no nostalgia to this model. Whether the places are our homes, schools, workplaces, prisons, or nation-states, they can be functional or dysfunctional, places of good or of evil.

How then are these features of place connected to and generate temporal dynamics and rhythms? I will first consider the role of place-making in daily life and then consider the connection of place-making to the broader existential questions.

Time, Rhythm, and Everyday Place-Making

The process of weaving nature, meaning, and social relations through place-making is just that – a process, requiring changes and adaptations; still its geographic role is to have things come together in discernable arrangements. This means that no matter how dynamic (or contested) this process becomes, place-making has a built-in tendency to provide some degree of stability and coherence. This is reinforced by the fact that the weave is partly material, leaving at least some physical traces on the landscape. The weave lends stability to the ebb and flow of life – an ensemble or context which we can leave and reenter.

But against this are the multiple types of dynamism that result from two general features of weaving: the temporal qualities of the strands or elements that are woven, and the dynamics of the loom. Consider first the temporal qualities of the strands from nature, meaning, and social relations. Even though they are difficult to separate precisely because they are everywhere inter-threaded and woven together through our place-making, still the strands can be isolated to some degree so that the sense of time they bring to the mix can be discussed.

The Rhythms from Nature, Meaning, and Social Relations

Nature is process, from the big bang on. Any causal sequence is a temporal one, and nature is about causes. We can alter and control some of them, but many we cannot. They are simply part of the givenness of the natural world. Within these causal systems, we often notice a qualitative temporal difference between the inanimate and animate. With important exceptions, such as the seasons and the phases of the moon, the physical seems overall to be more linear than the biological which appears almost entirely governed by cycles. This is reinforced within our own selves by the cyclical beats of our hearts, the cycles of respiration, of growth and decay, of birth and death. So, when in our place-making we weave strands from the natural world – as we must in all places – along with them come parts of their rhythms and cycles. Nature pulsates with change, but from our own experiences, many of the large-scale physical and ecological process appear to us to be so slow, or so cyclically repetitive, as to be 'timeless,' and the exceptions are often experienced as catastrophic – a hurricane, earthquake, volcanic eruption, or a forest fire or an devastatingly early frost.

Meanings are Janus-faced. On one side they are the mental representations that help us to think our own thoughts. These are experienced mostly as temporal events – the ordering of ideas and reasons; or in a looser sense, the temporal drifting of a daydream; or the unbridled and chaotic thoughts of a fevered imagination. On the other side is the public communication of these meanings through symbolic systems such as ordinary language or music. These possess their own temporal structures which are combined with those of our thoughts. So, the internal temporal structure of our ideas must then be connected to the grammatical structure of an ordinary language such as English, which provides its own temporal orderings and sequences, or to another form of communication, such as musical. Language and music emphasize the linear, while symbolic systems that seem to be more cyclical and static include painting, sculpture and architecture. Their representations can be taken in by us almost all at once. The strands of meaning we weave, and the symbolic systems used to convey them, coupled with those of nature, enrich further the temporal qualities of the weave.

The strands of social relations, whose primary temporal function is to time or to schedule the sequence of the weave through devices such as calendars and clocks in conjunction with the rules of place, can help bend the other strands to organize a particular temporal arrangement. For example, the orchestra must come together to make music and perform for an audience in a place that defines and schedules these social relationships; and in a wilderness area, the visitors, park rangers, fire fighters, and even some of the wildlife must be scheduled. When a place's focus is primarily on social relations, the scheduling role becomes built into the very fabric of the relation: a trial in a law court defines itself in terms of stages of events, with the opening arguments, the presentation of the prosecution and then the defense, the closing arguments, the deliberations of the jury, and then the verdict; and, in a prison, the prisoner behind bars describes being in the place as 'serving time.'

Even though every place contains strands from all of the realms, most emphasize the strands from one – and some do so for the purpose of emphasizing a quality of time. But then, that quality is still related to and altered by the others with which it is entwined. We may establish places such as natural parks in order to provide a feeling of entering a timeless and eternal world. Yet we enter this for only a socially defined period, such as a vacation, and experience these 'eternal' rhythms through attending to our own biological ones – we wake, eat, rest, and sleep. Or, an art museum can hang paintings chronologically in order to use art as a means of conveying an idea of historical progression if not progress itself. Interconnections such as these in the strands and weave of temporal components can go on and on within places, and thus enable us to use them to create new twists to temporality.

Another component affecting the mix of times is the relationship between place and self. Although we are always in one place or another, when the place and one's personality mesh, a person might then lose himself in the projects supported by the place, and lose a sense of time. That person can be immersed in the flow of activity. Often there is a mismatch from the start. A person who is not interested in intellectual matters will not find a university to be the right place; and someone not interested in the 'timeless' rhythms of nature will not be at home in wilderness. In such cases, the mismatch may come to feel as though we are marking time. Even when our interests and the structure of place are complementary, they are so only for a while, for not only do places change, but so do we. To refer again to the problematic, we, singularly and socially, cannot accept things as they are; there are many weavers with different views about what ought to be.

Time and place's structure can become even more intimately connected. The place can project a sense of time by not only focusing on the mix of elements, but on the particular component of the loom. Some places focus on the spatial interaction component. These places are designed to encourage flows of ideas or information, or people or things. Roads are such types of places: their in/out rules pertain to thin, long, stretches of space that stipulate direction (and often the speed) of flows and provide sanctions against those things that do not move. Bus depots, train terminals, and airports are other types of place that focus on flows. They allow people to remain there only as long as it takes them to find and board their rides. The fact that there are more such places than ever before, suggests to some that our geography can now be described as a 'space of flows.' Such a description, though, in no way diminishes the fundamental role of place, for it is essential in helping to create and sustain these flows. Indeed, whether the place specializes in flows or not, place is linked to other places, and virtually all flows and interactions through space (with the exception of some of the components of nature) involve strands of elements 'woven' in places. So the role of spatial diffusion in particular and the more general idea of spatial interactions and spaces of flows focus only on the movements generated by place, and in so doing run the risk of obscuring the fact that place is generative of these processes.

The relationship of place to time goes even deeper when each component of the loom may itself impart a concept of change and impermanence that then is used to categorize selves. For instance, spatial interactions can be used to label people. This happens when we call them 'commuters,' 'travelers,' 'migrants,' 'hoboes,' and so on. The in/out of place loop can be used to label people when they are called 'homeless,' 'displaced,' or 'refugees.' People can also be labeled by the appearance reality loop when they are called 'usurpers' or 'interlopers' – those whose presence in a place is disputed, and whose position ought to be temporary.

The entire set of relationships between time and place becomes even more pronounced when the loops are used not only to label, but also to destabilize, contest, and disrupt what is taking place. For example, boycotts focus on the disruption of spatial flows and interactions; 'sit-ins' challenge the in/out of place rules; and critiques of the meaning and appearance of landscape draw its reality into question. Using one activates the others, and then changes the threads and resulting weaves. This is how we unweave, change the weave, or generally disrupt, destabilize and transform place. We do this for all sorts of places – those that are predominantly about meaning, or social relations, or nature. But if this were to occur ceaselessly for virtually all places and all scales we would then have chaos: nothing could take place – no project, including the projects of living and creating self.

Knowing how places work – understanding their structure and dynamics -- provides an empirical device for studying the historical geography of a place. We know what to look for in order to untangle the elements of the fabric, to see the order in which they were inter-threaded, to understand the dynamics of engagement of the loops, the history of the flows, the transgressions and contestations.

So, while place may have an underlying stabilizing tendency, it can be used not only to create mixes of temporal pulses and rhythms, but its loom-like structure produces its own dynamics, rhythms and pulses, and this same structure is also often used to create dramatic temporal disruptions because of social and cultural conflicts, or because of changes in our own character and interests. In all of these ways, we intentionally (and unintentionally) create temporalities through the creation of places.

Existential Issues of Place and Time

But what is it that makes us continuously change places and create new ones? Here the following implication of the problematic can help us frame this issue, for the problematic itself creates a continuous dynamic: if we cannot accept reality as it is and create places to make it into what we think it ought to be, then we will soon change these and so on. No matter what our goals may be, we seem to be involved in a continuous process of transformation, where place-making is involved at all of the stages. This is what I mean by an existential relationship between place and time. This fundamental, and potentially unsettling, dynamic itself has lead to several important reactions. Here I will consider four that have arisen in the Western world (though we can find variants of these in other cultures). The first two advocate escaping the dynamic, the second two are attempts to use it. Of the four, only the last one is a realistic and morally responsible one.

The first seeks to escape the continuous and imperfect process of place-making by leaving place and entering the apparently pure, immutable, and eternal rhythms and cycles of the world of nature. This though is unrealistic on at least three counts. First we cannot live without making places, and even the simplest ones would introduce their own dynamics. Second, escaping to nature, as in wilderness, is now simultaneously an escape to place – one that we set aside to conform to our conception of nature. Third, the natural is itself dynamic and punctuated by discontinuities and uncertainties. Those who seek this escapism as a way of life are entering a romantic fantasy.

The second recognizes our place-making, but tries to freeze its dynamics in order to create places that are permanent and perfect – it offers a fixed and unchanging 'heaven on earth.' This is the motivation behind many of the attempts in ancient civilizations to create perfect and lasting monuments and cities, and it is found in many utopian arguments from the Greeks onward. Even though these may begin with good intentions and even an intimation of what is good, the impulse to fix the good permanently and immutably in place produces a parody of the good by creating a closed and static world that becomes oppressive and absolutistic. Instead of utopia, it produces dystopia. Its worst examples are the totalitarian states of the twentieth century which attempted to create 'perfect' and permanent regimes through imposition of inflexible rules, rigid landscapes, and colossal monumental architecture.

The third sees virtue neither in escaping place-making, nor in ossifying it, but rather in continuously contesting and destabilizing it. This position suggests that accelerating and exacerbating the dynamic tendencies of place will sufficiently destabilize the power behind places and thus make them more open and transparent, and that this will liberate us from oppressive power relations. Many forms of post-modernism seem to advocate such a position. I say 'seem' because there is a strong streak of relativism and anti-foundationalism in their positions that prevents them from stating things positively, for fear then of closing and bounding the conversations. Still this idea of a moral emancipation through destabilization and transgression is implied in many postmodern tactics and methods. While open, transparent, and virtually boundless places may be virtues, it is important to articulate why this might be the case. And then would this be so for all places? And once places are sufficiently destabilized, are they all supposed to remain that way? Instead of confronting these issues, advocates of this position hold on to a naive hope that something better will emerge once things are sufficiently shaken up.

The fourth is a more realistic alternative, grounded in the geographic conditions of the problematic. It does not advocate, as does the first, an escape from place to nature; nor does it advocate as does the second, freezing place to create perfection on earth; and it does not advocate, as does the third, a blind faith in the positive outcome of destabilizing and 'deconstructing' place. Rather it tries to navigate among the three, and particularly between the absolutism of the second and the relativism of the third by advocating that our inescapable place-making be guided by the joint application of two criteria: that *we should create places that at the same time increase our awareness of reality, and increase the variety and complexity of reality.* Their joint application can guide our place-making for the better. It recognizes that places can be designed to hinder or help us become more aware, and that they also can be designed to increase or decrease the variety and complexity of reality; and it recognizes that it is better to be more than less aware of reality, and to live in a world that is varied and complex. The need to apply the two jointly arises from the fact that they are mutually reinforcing. We cannot increase our awareness of reality unless we have different places with different projects from which to view it, and we cannot know if our knowledge is real unless it is open, public, and subject to other views that arise from variety and complexity. If a place that we create contributes to the variety of reality, but diminishes awareness of it, then it can do more harm than good. The joint application promotes a public and free access to knowledge. Increasing awareness and variety and complexity is not something that can be done from only one viewpoint and for only one interest.

These two jointly can guide our place-making for the better. They provide a direction, but not an end point. They encourage us to make places that are more democratic, transparent, and open. They have us see that place-making itself is a contribution – a gift – that is part of the circulation of awareness. In this way they encourage us to replace a model of ourselves that is fixated on self-interest, with one that promotes altruism. By taking them up, the rhythms and dynamics of everyday place-making become informed and animated by them and this may help us move toward a better world.

Note

I have benefited from discussions with my student, Erin Olsen, on how place-making creates new experiences of time.

1   The theory of place that I am drawing upon is developed in my books: *Homo Geographicus: A Framework for Action, Awareness and Moral Concern* (1997, Johns Hopkins University Press, Baltimore) and *A Geographical Guide to the Real and the Good* (2003, Routledge, New York). Place and place-making in these works and in this paper subsume the idea of territory and territoriality. That is, every instance of a territory is also an instance of a place whose properties I am describing in the paper.

Index

abyss 202, 205-6, 208-11
Actor-Network Theory 36-7
Adorno, T.W. xiii, 9-10, 25
aesthetics 26, 46, 49, 51, 204, 233, 235, 238
Alexander, C. 30, 123-5, 127-33, 139-41, 141n, 142n
Algonquin 183-6
Almighty 209
Althusser, L. 38
Americanization 22
Anne of Denmark 215
annihilation of space by time 19, 23
Antrim, Glens of Antrim 235-6
Apsu 201-2, 205-6, 208, 210
architecture 29, 129, 134, 142n, 228, 232, 235, 237-8, 249, 252
Arizona 49, 54
Augustine 55
authentic care 119
authenticity 25, 33, 229, 232, 236, 238
avant-garde 25
Babylon, Babylonian 201, 204-5, 207, 209, 211-12
Bachelard, G. 12, 24, 36n; see also topoanalysis
  epistemological break 24, 38n
Bacon, F. 218
Baltic Sea 71, 73, 77-80, 82
battle of the Boyne 87, 93, 101-2, 104
beady-ring structure 135-36
Being and Becoming 19-20, 37n
Belfast, Belfast Castle 101-2, 231, 233
Bergson, H. 13, 37n, 38n, 211n
Berman, M. 21
Beston, H. 72
birdsong 74, 81, 82
body 3, 5-6, 9-13, 15, 17-18, 24-5, 30, 34, 51, 72, 76, 80, 127, 141, 176-7, 179-80, 183, 194, 196, 203-4, 208
  and will 181-3, 198
  as place 5-6, 179, 183
  as prison 179, 198
  gendered 168, 179
  geographical 32-3

rejection and asceticism 176, 179-80, 195-6, 198
  social 24, 180
body politic 31-2, 215, 217-21, 223, 224n
body subject 11-13, 15, 17, 34, 37n, 57
Boyne Valley 29, 87, 93, 102-4
Braveheart (film) 87, 98-9
Britain 32-3, 97, 101, 215-21, 223, 230
Britannia 220-221
British Columbia 28, 60, 62-4, 66, 69
Brunner, O. 222-3
Burnham, S. 53
Buttimer, A. xiii, xiv, 4-18, 23-6, 28, 32, 34-6, 37n, 38n, 57-8, 67-8, 123, 126, 158n, 175, 216
  dwelling 5-7, 25
  experience 4-6, 9-12, 15-16, 18, 24-5, 124, 175
  genre de vie 12, 16-18
  Geography and the Human Spirit 36n, 87
  Grasping the Dynamism of Lifeworld 4, 28, 57, 167
  lifeworld 4-5, 10, 15, 17, 18, 123, 141
  metaphor 6, 36, 215
  rhythms 4-11, 15-19, 25-6, 34, 167
  social space 5-6, 8, 10-13
  story telling 87
  sustainability 111, 91
  Sustainable Landscapes and Lifeways 111
  time-geography 8, 15-18, 37n
Bythos 211
California 48, 78-9, 82-3, 129
Camden, W. 220-221, 224
Canada 28, 57, 59-69, 69n, 70n, 114, 175, 181, 185-7, 189-94, 198
capital, capitalism 9, 18-20, 58, 117, 147-56, 228-9
  late 22, 25
  monopoly 22
  rhythms of, see rhythms
cartographical reason 202, 210
Casey, E. 12, 202
Catholicism 32, 97, 99, 105n, 176-7; see also Christianity; Jesuit Order; Ursuline Order; mission

chaos 32, 204-5
Cheyenne 75
Chief Arapooish 84
child, childhood 27-8, 48, 72, 76-82, 84,
    136, 164-5, 168-71, 172n, 177-8, 180,
    188, 197, 203, 211
China 97
Chinese diaspora 28, 59
choreography, chorography, choros 22,
    31, 220-221
Christ, *see* Jesus Christ
Christian IV (king of Denmark) 220, 222
Christianity 10, 93, 96-7, 176, 179, 183; *see*
    also Catholicism; Jesuit Order
    Ursuline Order; mission
city, *see* urban
Clark University 123, 127, 141
cognitive map 20, 35
Collon 101-2
colony and colonialism 19, 20, 22, 96, 177,
    184-85, 191, 193, 198
colonization of lifeworld; *see* lifeworld
commodification, commodity 22, 24, 29,
    111, 115, 117, 119
Cornell 116-19
corporeality 5, 31
cosmos 32, 204
country 27, 28, 33, 72, 76-7, 81, 84, 218,
    221, 224, 225n
County Down 232
County Fermanagh 235, 237
Court 33, 215, 219-20, 222-5
Crang, M. 37n
creative class 31, 162-6, 171-2
creativity 22, 31, 162-4, 171
cri de coeur 7, 22
Culbertson, F. 47
Cultra 232-4, 236-7
cultural geography 33, 232, 236, 238
culture 5, 21, 25, 32, 37n, 38n, 48, 53, 66,
    72-6, 88-9, 91-2, 104, 113, 125, 167-8,
    179, 202, 212n, 220, 247, 251
    business 62-3, 68
    living 33
    material and symbolic 33
custom 10, 33, 75, 215, 222-3, 230
cycles 3-5, 21, 52, 55, 73-6, 147, 153, 155,
    157n, 165, 249, 251
daily life, *see* everyday life
dance macabre 8, 10, 22, 34
deconstructing, deconstruction, 15, 32,
    208, 228-9, 252

deformed grid 134, 137-40
Deleuze, G. 175, 210
Derry, Londonderry 101, 234
desert 48-9, 54
design, designs 124-6, 128-9, 133-4, 138-
    41, 142n, 143n
Devil 193-4
dialogue 4-5, 7-8, 15-16, 19, 30, 140-141
Dillard, A. 72, 84
displacement 27-8, 31, 58, 71-3, 251
Dom Claude Martin 188, 195
Dom François de Saint-Bernard 194
Dom Raymond de Saint-Bernard 182, 194
Dowth 90, 93-4, 102
Drogheda 90, 92, 101
dualism and dichotomy 9, 20-21, 33
Duayne, A. 143
Dublin xiii, 87, 101, 105n
Dun Laoghaire 3-4, 35
Durkheim, E. 12-13, 17
dwelling 4-7, 9-10, 25, 29-30, 33-4, 36, 50,
    126, 137, 177, 234-5; *see also* home;
    earth; sustainability
dynamism 4, 7-8, 10, 16, 20-22, 26, 124; *see*
    also rhythm
    of place 34-5, 248
Ea, *see* Nudummud-Ea
earth, earthly 7, 27, 36, 46-8, 51, 188, 196,
    201-2, 205, 209, 211n, 219, 221
Eco, U. 98
Ehrenreich, B. 166
embodiment 27, 31; *see also* body
England 98, 105n, 185, 215-16, 218-22,
    224n; *see also* Britain; Britannia
Enuma Elish 32, 201-2, 204-11, 212n
environment 3, 4, 11, 12, 18, 26, 29, 34,
    36, 37n, 58, 62, 68, 81, 84, 90-91, 93,
    102, 104, 112, 119-20, 123-5, 134,
    230, 236, 238
    built 29, 53, 147, 152-3
epistemological break, *see* Bachelard, G.
epistemological journey 203
epistemological separation 19
epistemological transparency 22
Europe 20, 48, 58-9, 66, 78, 81, 101, 198,
    221, 224n, 229-30
European Union 90
Evans, E.E. 33, 229, 231-38
everyday 3-4, 15, 22-3, 25-6, 29, 37n, 119,
    123, 135, 248, 253
everyday life, daily life 4, 14, 16, 18, 20-21,
    23, 25, 29, 31, 35-6, 38, 40-41, 58-9,

72, 82, 119, 161-8, 170-171, 188, 197, 243, 248
existentialism 15, 18, 23, 57, 123
experience 3-5, 10, 13, 15, 18-20, 22-3, 26, 28-9, 34-5, 48-50, 53, 57-9, 64-6, 68-9, 72-7, 80, 83, 87, 96, 103-4, 111, 113-14, 117-19, 123-8, 136, 149, 153, 161, 163, 167-8, 170-171, 175, 177, 183-4, 188, 191, 193, 196-7, 215, 248-50, 253
  lifeworld 3-4, 10, 22, 34, 35
  time-space 19
explanatory-diagnostic moment 17
Faust 36
feminism and feminist geography 161, 166-7, 170
fieldtrip 87, 90, 104
film, *see* movies and films
Fleure, H.J. 231
Florida, R. 31, 162-6, 171
flow 3-5, 7, 15-16, 21, 24, 27, 30, 32, 37-8, 147-8, 151, 155-6, 157n, 158n, 250-251
Folk Museum Act 231, 233
Foucault, M. 25, 36n, 175-6
fragmentation 3-13, 19, 29, 36n, 55, 84
France, 98, 177, 181-2, 184-5, 188, 190, 193-4, 198, 221
Frederik II (king of Denmark) 222
Freud, S. 13
future 5-7, 9, 22, 24, 29, 111-14, 116-20, 129, 152, 156-7, 158n
Gailey, A. 232, 234-5, 237
Gassin 134-8
gender 5, 6, 29, 31, 36, 64, 161-2, 166-7, 176-9, 189, 192, 197, 215, 219; *see also* feminism
  division of labor 161, 168, 170
Genesis 201
genre de vie 12, 16-18
geographical imagination 6, 32-3, 227, 229
geographic body 219, 223
Glasgow 123
globalization 21-2
God 55, 176, 178-82, 187-90, 192, 194-8, 207, 211n, 219; *see also* Holy Spirit; JHWH
gods 93-4, 97, 103-4, 201-10, 211n, 212n
Goethe, J.W. von 147
Gone with the Wind (film) 51, 97
Gregory, D. 17-18
Gren, M. 16

Gullberg, A. 147-8, 157n
Gurevich, A. 12
Guyart, M. 176, 178-9, 195
Habermas, J. 17
habitat 48
habitus 223
Hägerstrand, T. xiv, 15, 16, 37n
  time consciousness 5
  time-geography 15, 17, 37n
Haraway, D. 161
Harvey, D. 19-21, 23, 37n, 153, 158n
Hayden, D. 166
Hazelius, A. 230; *see also* Skansen
head of state 215, 217
healing of the city 125-6
Heaven 201-2, 209-11, 221
Hebrew Bible 212n
Hegel, G.W.F. 211n
Heidegger, M. 13, 119-20, 126
Hell 194, 210
Hellman, L. 89
Heritage 11, 20, 32, 73, 75, 89, 91-3, 96, 100, 102-5, 111, 113-14, 117-19, 178, 227-30, 232, 234, 238
  and authenticity 232, 238
  and history 33, 99, 227-9, 234, 238
  and postmodernity 227-8
  and tourism 96-7, 102, 228-9
  industry 22, 227-9
Heritage Estates 114-17, 119, 121
Heritage Town 115-17, 119
highest and best use 154-5
Hillier, B. 30, 123-4, 133-43
Hobsbawm, E. 112, 119-20
Holy Spirit 194; *see also* God
home 3, 7, 14, 18, 25, 35, 37n, 45-8, 53-5, 60, 64-5, 67-9, 72, 80-82, 84, 178-80, 184
homeland 68, 72, 76, 79, 82, 84, 183
Homer-Dixon, T. 112
Hong Kong 28, 59-62, 64-5, 67-9
horizon 73, 79, 124, 217
Horkheimer, M. 9
house 46-7, 50-1, 54-5, 234-8
housework 161, 165-8, 170-171, 171n
Howard, E. 166
humanism, humanistic geography 7-8, 10-11, 14-18, 30, 34-5, 57-8, 72, 90, 216
humanistic Marxism 23; *see also* Marxist geography
Huron 183, 186, 190
Husserl, E. 5, 10, 12-13, 15, 36n

identity 20, 27-8, 30-31, 33, 58, 64, 87, 91,
    102, 104, 175-6, 197, 228, 231-3, 235
identity thinking 109; *see also* topological
    thinking
ideology 14, 34, 38n, 58, 209
imagination 33, 46, 48, 51, 73, 111, 193-4,
    203-4, 211n, 219, 244, 249; *see also*
    geographical imaginations
imagined community 228
immigrants in Canada 28, 59-70, 70n
IGU, *see* International Geographical Union
indigenous people 177, 180, 183-4, 186,
    190, 194, 197-8
ingenuity gap 112
instrumental rationality 10-11, 18, 34
International Geographical Union 57
intersubjectivity 11, 17, 57-8
Ireland 87, 91-2, 94, 96-8, 100-104, 232-3,
    237
Irish Sea 3, 92
isolated state, *see* state
Jackson, J.B. 74
James I (king of England) 215-16, 218-19,
    221-3, 224n
James VI (king of Scotland) 215-16, 218,
    220, 222
James, W. 118
Jameson, F. 21-2, 36n
Jesuit Relations 189, 191-2, 199
Jesuits 177, 179, 181, 184-92, 194-5, 198
Jesus Christ, Christ, Jesus 103, 182, 196
JHWH 210; *see also* God; Holy Spirit
Jones, I. 217-20
Jonson, B. 217, 219-21, 224n
justice 27, 37n, 221
Kant, I. 202-3
Kemmis, D. 30, 123-7, 133, 140-141, 142n
Kierkegaard, S. 202, 204
Kingu 206-7, 209; *see also Enuma Elish*
Knowth 93-4, 102-3
labor, work 6, 13, 19, 25, 62-3, 161, 164,
    236; *see also* housework
Lalemant, Father Jérôme 180-85, 187, 189,
    192, 195-6
land 72-3, 78, 114, 118, 189, 217-18, 220,
    222-3, 224n
landlord 87, 96-7, 101-2
land rent 147-8, 150-157, 158n
landscape 8-9, 11, 13-15, 23, 26, 30-31, 34,
    48-50, 54, 72-79, 81-4, 87-97, 100-
    105, 113, 117-18, 215-23, 227, 233,
    235-6, 245-8, 251-2

and make-believe 216
and sustainability 29-30
etymology 224n
history 13, 29, 89
political 32-3, 35, 221, 223
reification of 9, 23
representation 23, 32-3, 49, 55, 87
Landschaft 222, 224n; *see also* country; land;
    landscape
language 4, 18, 22, 63, 65, 69, 97, 126, 179,
    186, 189, 192, 202-3, 217, 219-24,
    237, 249
Latvia 71-3, 76-7, 79, 81-3
law 9, 32-3, 162, 215-16, 219, 222-3, 245,
    247, 249
Lefebvre, H. 8, 22, 25, 38n, 147, 156
    abstract space 13, 25
    differential space 35n
    everyday life 23, 37n
    reification 13, 23
    rhythmanalysis 8, 22, 24-5, 37n
Leibniz, G.W. 12, 20
lifestyle 91, 98, 102, 163, 166, 237
lifeworld 3-6, 8-11, 14-15, 17-19, 22, 28,
    34-5, 57-8, 123, 126, 141, 142n; *see
    also* Buttimer, A.
    colonization of 14, 18-19, 22, 25, 35
    eclipse of 9
l'Incarnation, M. de 175-7, 180-98
Lopez, B. 72
lord 204, 206-10, 212n, 222
love 48, 76, 181-2, 188, 196-7
Lukács, G. 13, 36n
Lund University 16
Lynch, K. 113
Maine 73, 76, 78-80, 82, 84
Malmö 30, 148-9, 151, 154, 156, 158n
Malpas, J. 12-13, 175, 197
maps, mapping 6, 15, 35, 184, 201, 220
Marduk 32, 204-10, 212n; *see also Enuma
    Elish*; lord
    50 names of 2, 10, 212n
Marie de l'Incarnation, *see* l'Incarnation, M.
    de
Marie de Saint Joseph 187
Market Village 115, 117
Markham, township of 113-15, 117-18
Martin, C.L. 75
Marx, K. 13, 19, 30, 36n, 38n, 148, 150,
    158
Marxism 18-19, 23, 57, 170
Marxist geography 12, 34, 35

and humanism 8, 12, 17-19
Masque 32, 215, 217, 219-23, 224n
meaning 4, 6-8, 12-13, 15-16, 19-20, 23,
    34, 36n, 57, 59, 91, 93, 100, 103, 119,
    125, 162, 176, 178, 180, 189, 192,
    197, 201-2, 210, 216, 220-223, 224n,
    228, 235, 243, 246-9, 251
Mellifont, Mellifont Abbey 89-90, 92-3,
    98-9, 102
memory 3, 9, 27, 53-4, 75, 81, 84, 113,
    101, 186, 217
Mère de Saint Augustin 193
Merleau-Ponty, M. 12-13, 15, 57, 175
metaphor 6, 7, 30, 36n, 67, 89, 91, 104,
    215, 217, 224n
    root 215-17
migrants 59, 68, 251
Minneapolis 47
Minnesota 47
mission, missionaries 97, 177, 179-86, 188-
    91, 194, 198-9
Missoula 125-7
mobility 23, 27-8, 53-4, 58, 64, 72
modernism 19-20, 114, 227, 252
Molinos, Miguel de 192, 195
Monasterboice 90, 97-8, 102
movies, films 50, 52, 88, 98, 100; *see also*
    Braveheart; Gone with the Wind; Name of
    the Rose
Mummu 201-2, 205; *see also Enuma Elish*
music 4, 30, 32, 47, 50, 52-55, 72, 74, 87,
    113, 147-8, 153, 178, 215, 249
mysticism 177-8, 195-7
Name of the Rose (film) 98
narrative 33, 178, 189, 191, 197-8, 228; *see*
    also stories, story-telling
nationalism 33, 224n, 230
National Trust 228
naturalization 23, 31
nature 4-5, 7-15, 20, 23-4, 26-9, 34-7, 47,
    49, 52, 55, 71-5, 77-9, 81, 84, 88, 90-
    91, 95, 102-3, 111, 119-20, 168, 179,
    217, 219, 229; *see also* environment;
    rhythm
necropolis 90-91, 93, 95
New France 177, 180-183, 186, 193, 197
Newgrange 88-91, 93-6, 100, 102, 104, 106
Newtonian time-space 12, 15
North America 32, 59, 75, 80, 176-7, 180,
    190, 193, 197
Northern Ireland, Northern Irish 101-2,
    232-5

nostalgia 92, 100, 230, 248
Nudummud-Ea 205-6; *see also Enuma Elish*
O'Connell, D. 97
Old Testament 211
Olsson, G. xiv
Olwig, K.R. 15
ontological transformation 32, 204, 207,
    209-10
Pacific Mall 115, 118
Paris 182, 185-6, 188-91, 198
Park, R. 59, 72
past 3, 6, 7, 9, 22, 24, 28, 32, 35, 79, 84-5,
    111, 113-20, 124, 197, 209, 211, 227-
    31, 233-4, 236-8, 244
    working 236
patriotism 47
pattern language 124, 129, 142n
paysage 222, 224n
Peltrie, Mme de la 181, 185, 189-90, 192
pentimento 89-90, 93, 95-6, 98, 101, 104
permanence 20-21, 23, 30, 35, 37n
perspective 33, 217, 220, 223; *see also*
    landscape; scenery; theater
phenomenology 15, 57-8, 61, 111, 118-20,
    123-4, 126, 133, 141, 143n; *see also*
    Humanism; Husserl, E.
Phoenix 8, 36n, 207
photographs 27, 49-50
place, platiality 3-37, 37n, 38n, 45-55, 58-9,
    67-9, 72-9, 81-4, 113, 118, 120, 123,
    156-7, 96, 103, 163-4, 168, 175-7,
    179, 181, 183-9, 191, 193, 197-8, 202,
    205, 209, 211n, 217-18, 220, 222-3,
    229, 238, 243-53
    and body, *see* body
    and space 10-13, 18, 20-21, 24, 36n,
        123, 177, 183, 185, 189, 191, 198, 243
    and time 4, 11, 20, 89
    ballet 6, 147, 157n
    enclosed 176, 198
    erasure of 13, 38n
    nested 32, 175-6
    of writing 185, 191
    open, openness of 183, 252
    pause 45, 50
    recalcitrance of 12
    sense of 27-9, 45-8, 50-51, 53-4, 69, 74,
        76, 89, 167, 228
    shared 191
    spirituality 55
    weaving 245, 248
placelessness 141

place-making 27, 30, 34, 124, 243-4, 247-9, 251-3

planning 91, 99, 134, 138, 141, 229, 231; *see also* urban

poems 27, 51, 211n, 224n

poetry 71, 87, 92, 119-20, 204, 211n

poiesis 119

postmodernism 21, 33-4, 84, 102, 227-8, 252

poesis 7

power 6, 10, 18, 21, 23, 25, 29, 32n, 87-90, 92-4, 98-100, 102-4, 126, 157, 175-8, 181-4, 192, 195, 197-8, 204, 206-8, 210, 211n, 215, 217, 224n, 233-4, 244, 252

power-geometries 21

Pred, A. 18

present 84-5, 111-14, 118-20, 163, 197, 227-38

Protestantism 192, 197; *see also* Catholicism; Christianity; religion

Proust, M. 197

quantative revolution 11, 37n

Quebec 32, 69n, 176-7, 180, 182-6, 189, 191, 193, 195, 198

Queen's University, Belfast 231-2

quietism 192

rationalism and rationalization 36n

reanimation 8, 10, 13, 23, 26-7, 36

redevelopment 30, 127, 147-9, 150, 152, 155-6, 158n

rehumanization 30

reification 9-11, 13-14, 16-17, 23-4, 34, 36n, 37n; *see also* topological thinking
   as spatialization of time 13, 37n
   rhythms 14

religion and spirituality 6, 32, 55, 187, 189, 191, 205, 210, 219, 221; *see also* Christianity

rent 30, 147-8, 150-156, 157n, 158n
   differential 148, 150-151, 155-7

rent gaps 30-31

representation 4, 6, 17, 19, 21-3, 35, 37n, 87, 89, 90, 162, 165, 171, 176, 203, 210, 211n, 216, 223, 228-33, 237-8, 249
   pictorial 49-50, 249
   political 33, 220, 223
   theatrical 32, 215
   through material culture 33, 232, 234-8

rhythm 3-11, 13-36, 36n, 37n, 38n, 52, 58, 67, 69, 88-91, 93, 99, 103, 113-14,

118, 205, 215, 244, 248, 251; *see also* dynamism

capitalism 21-2

ethymology of 4, 157n

generational 90

geographical 7, 26

geography of 3, 7, 9, 16, 18, 26, 31, 34

meter and rhyme 205

natural, nature 7, 24, 26, 58, 71, 72-5, 79-84, 87-8, 91, 111, 168, 171, 249-50

of geography 7, 8, 14, 17, 34

physiographic 80

place 27, 30-31, 72, 78, 223, 253

politics 36

positivism 24, 57

purified 7

rent 147, 155

repetition, repetitive 5, 6, 24-5

timeless 250

time-space 4-6, 8-9, 14-16, 22-3, 26-7, 29, 31, 35, 123, 161-2, 165, 167-8, 170-171

rhythmanalysis 8, 22-6, 35, 37n, 38n; *see also* Lefebvre, H.

Ricoeur, P. 118

Rome 185, 190, 193

Rostock, Malmö 149

routine 5-6, 17-18, 28-9, 66, 69, 123, 141, 168; *see also* lifeworld

Russell, B. 202, 211n

Saint-Bernard, Dom François de, *see* Dom François

Saint-Bernard, Dom Raymond de, *see* Dom Raymond

San Francisco 129-32, 139

Sassen, S. 196

scenery 215-23; *see also* perspective

Schama, S. 101

Schopenhauer, A. 211n

Schuetz, A. 58-9, 63-4, 69

Scotland 96, 215-16, 218-22

Seamon, D. 6, 157n

seasons 27-8, 73-9, 81, 85, 118, 249
   and mass consumption 75

sectarianism 232, 237

self 45-8, 51-5, 76, 179, 181, 197-8

self-knowledge 53, 203

Sen, A. 112, 120

Shakespeare, W. 47, 224n

Shields, R. 37n

Skansen 230; see *also* Hazelius, A.

Slane, Slane Castle 87, 96, 101-2

Sloan, H. 113-15, 121
Smith, D. 170-171
Smith, N. 157
social relations 34, 209-10, 246-9, 251
Sorre, M. 12; *see also* social space
space 3-36, 36n, 37n, 38n, 46-7, 84, 89, 95-
    8, 101-4, 117, 123, 128-9, 133-7, 139,
    141, 142n, 153, 155-6, 168, 176-7,
    183, 187-9, 191, 197, 198, 202, 205,
    210, 218, 221, 227, 243-4, 247-8, 250
    abstract 8, 10-13, 25, 33, 36n, 38n
    axial and convex 136-9
    bounded 50
    dead 117
    domestic 235
    enclosed 183
    Euclidean 9
    geometrical 15
    infinite 11
    lived 4, 17, 19, 22, 25
    measuring 11
    Newtonian 12
    of flows 250
    of writing 185
    open 128, 135
    relational 12, 15, 20, 37n, 151, 155
    social 6, 12-13, 17, 24, 65, 123, 167
space syntax theory 134, 139
spatial 'fix' 152, 154
spectacle society 22
stability 45, 47-8; *see also* flow; permanence
state 93-4, 97, 99, 101-2, 215-23, 224n
    isolated 148
    throne 217, 219, 223n
Sterba Farm Market 118
stories, story-telling 22, 27, 32, 51-2, 77,
    84, 203-4, 227
St Patrick 87, 96-7
Stralsund, Malmö 148-9, 151-2, 154-6
stranger 57-9, 63, 64, 66, 69, 70
structuration theory 18
structure 23, 34, 126-8, 133-9, 141, 142n,
    152, 164, 171, 178, 197-8, 205, 210,
    234-5, 243-6, 248-51; *see also* agency;
    system; structuration theory
    politico-economic 18
structure-agency dialectic 19, 58
Stuart court 32, 215, 219-20, 222-3
students and experience of place 87-105
subject, subjectivity 3, 8, 10-11, 14, 16-17,
    19, 21-2, 25, 57-9, 175-6, 187-8, 197,
    203, 210

surface 10, 25, 28, 218, 222-23
sustainability, sustainable development 7,
    29-30, 87, 90-1, 102-4, 111-12, 118-
    20
Sweden 30, 157, 158n, 165-6, 230
symbols, symbolism 5-7, 9, 12, 26, 29, 33,
    38n, 91, 93, 96-8, 101-3, 182, 195-6,
    218, 220, 235, 246-7, 249
system 13, 17-18, 21-2, 25, 35, 164, 209,
    246, 248-9
Taiwan 59-61, 64, 66-9
taken-for-granted 11, 14, 17, 22, 28, 36,
    38n, 58-9, 123, 202; *see also* lifeworld
Tara 87, 90, 92-3, 96-7, 100, 102-4
Techne 119
temporalities, creation of 243-4, 251
temporality 22, 29, 36n, 111-14, 117-20,
    197, 250
Tennyson, A.L. 120
Teresa of Avila 177, 195
Thompson, G. 231-4, 237
Thoreau, H. 72, 76, 84
Thünen 148, 157n, 158n
Tiamat 32, 201-2, 205-9
time, temporal, temporalization 3-6, 8-27,
    29-31, 33-6, 37n, 38n, 45-55, 72-84,
    89, 96, 103, 111, 115, 117-20, 124,
    128, 152, 155, 158n, 164-6, 170-171,
    175, 201-2, 209-10, 219, 223, 227-9,
    243-4, 248-53
    abstract 17, 24
    clock 9, 23
    commodity 22
    cyclical 22
    lived 4, 19, 22, 24, 113
    Newtonian 16
    pseudo-cyclical 22
    social 14, 24, 111
    spectacular 22
time-diaries 168-9
time edges 29, 114, 117-18, 120
time-geography 8, 15-18; *see also* Buttimer,
    B.; Hägerstrand, T.
timelessness 117
time scales 147, 157n
time-space xiii, 4-6, 8, 14-20, 22-7, 29, 31,
    37n, 38n, 90, 123, 147, 161-2, 165,
    167-8, 170-171; *see also* rhythm
    compression 19-20, 229
    horizon 16
topoanalysis 28
topological thinking 9-11, 14, 34-5

Topos 32, 205, 210
Toronto 29, 64, 112-14, 116-17
tourism and place 29, 33, 229
Tours 180-181, 185-7, 192, 195-6
townscape 113, 115, 120
tradition 97, 102-3, 105, 218, 224n, 228-30, 232-6; *see also* custom
trees and sound 71, 73-4, 82-5
Trevelyan, G.M. 216
Trim 87, 90, 98-100, 102
Tuan, Y-F. 5, 57, 216
Ulster 32, 93, 105n, 227, 229-38
Ulster Folk and Transport Museum, Ulster Folk Museum 33, 229, 231-3, 236, 238
Ulster Folklife Association 230
uneven development 21
urban 6, 20, 26-31, 37, 46-8, 51, 72-3, 75-6, 78-9, 90, 115-16, 120, 123-5, 127-9, 133-4, 138-43, 147-9, 151-7, 158n, 161, 163-4, 168, 209, 228, 233, 237-8
    development 114, 127-9, 131, 142
    flamenco 155-6
    liveliness 125, 141
    planning 138, 141
    redevelopment 30, 127
Ursuline Order, Ursalines 31-2, 176-80, 182-7, 189-93

contacts with Iroquois tribes 183-4, 186
contacts with Jesuits 184, 189-92, 196-7
values 6, 18, 21, 26, 29, 31, 48-9, 53, 57, 61, 90, 92, 95, 97, 100, 103, 112, 137, 142, 152, 162, 171
    and geography 11, 14, 16, 156
    exchange 10, 23, 25, 27, 36n, 158n
    surplus 150
    use 25, 27, 158n
Vancouver 59-68
Virgil 72
Virilio, P. 118
void, void-in-itself 201, 206, 209
Weber, M. 18, 36n
Wellesley, A. *see* Wellington, Duke of
Wellington, Duke of 99
Welsh Folk Museum 230-231, 236
Wesley, J. 117
Whitehead, A.N. 20, 36n
wholeness 13, 30, 36n, 124-9, 133, 139-41, 142n
Williams, R. 19-21
woods 79, 82; *see also* trees and sound
work, *see* labor
World Heritage Site 93
Ziedonis, I. 71-2
Zweckrationalität 18